Introduction to Stochastic Level Crossing Techniques

Introduction to Stochastic Level Crossing Techniques describes stochastic models and their analysis using the System Point Level Crossing method (abbreviated SPLC or LC). This involves deriving probability density functions (pdfs) or cumulative probability distribution functions (cdfs) of key random variables, applying simple level-crossing limit theorems developed by the author. The pdfs and/or cdfs are used to specify operational characteristics about the stochastic model of interest. The chapters describe distinct stochastic models and associated key random variables in the models. For each model, a figure of a typical sample path (realization, i.e., tracing over time) of the key random variable is displayed. For each model, an analytic (Volterra) integral equation for the stationary pdf of the key random variable is created–by inspection of the sample path, using the simple LC limit theorems. This LC method bypasses a great deal of algebra, usually required by other methods of analysis. The integral equations will be solved directly, or computationally. This book is meant for students of mathematics, management science, engineering, natural sciences, and researchers who use applied probability. It will also be useful to technical workers in a range of professions.

Key Features:

- A description of one representative stochastic model (e.g., a single-server M/G/1 queue; a multiple server M/M/c queue; an <s, S> inventory system; etc.)
- Construction of a typical sample path of the key random variable of interest (e.g., the virtual waiting time or workload in queues; the net on-hand inventory in inventory systems; etc.)
- Statements of the simple LC limit theorems, which connect the sample-path upcrossing and downcrossing rates across state-space levels, to simple mathematical functions of the stationary pdf of the key random variable, at those state-space levels
- Creation of (usually Volterra) integral equations for the stationary pdf of the key random variable, by inspection of the sample path
- Direct analytic solution of the integral equations, where feasible; or, computational solutions of the integral equations
- Use of the derived stationary pdfs for obtaining operational characteristics of the model with respect to the key random variable of interest.

Introduction to Stochastic Level Crossing Techniques

Percy H. Brill

CRC Press
Taylor & Francis Group
Boca Raton London New York

CRC Press is an imprint of the
Taylor & Francis Group, an **informa** business
A CHAPMAN & HALL BOOK

Designed cover image: © Percy Brill

First edition published 2023
by CRC Press
6000 Broken Sound Parkway NW, Suite 300, Boca Raton, FL 33487-2742

and by CRC Press
4 Park Square, Milton Park, Abingdon, Oxon, OX14 4RN

CRC Press is an imprint of Taylor & Francis Group, LLC

Library of Congress Cataloging-in-Publication Data

Names: Brill, Percy H., author.
Title: Introduction to stochastic level crossing techniques / Percy H. Brill.
Description: First edition. | Boca Raton : CRC Press, 2023. | Includes bibliographical references and index. | Identifiers: LCCN 2023000351 (print) | LCCN 2023000352 (ebook) | ISBN 9780367277352 (hardback) | ISBN 9780367343941 (paperback) | ISBN 9780429297601 (ebook)
Subjects: LCSH: Stochastic models. | Distribution (Probability theory) | Operations research.
Classification: LCC TA342 .B74 2023 (print) | LCC TA342 (ebook) | DDC 003/.76--dc23/eng/20230329
LC record available at https://lccn.loc.gov/2023000351
LC ebook record available at https://lccn.loc.gov/2023000352

ISBN: 978-0-367-27735-2 (hbk)
ISBN: 978-0-367-34394-1 (pbk)
ISBN: 978-0-429-29760-1 (ebk)

DOI: 10.1201/9780429297601

Typeset in CMR10
by KnowledgeWorks Global Ltd.

Publisher's note: This book has been prepared from camera-ready copy provided by the authors.

To the memory of my parents

Contents

Preface

This book considers stochastic models such as queues, inventories, dams, and others. It analyzes the models mainly using the "stochastic level crossing method" (aka: "system point method", "system point level crossing method"; "level crossing method"). The book states and applies basic level crossing limit theorems and concepts, and reviews useful companion methods of analysis. It provides references for mathematical proofs of stated theorems. The topics covered are meant for; anyone interested in stochastic modelling, university students, postdoctoral fellows, and career researchers. The "stochastic level crossing method" (abbreviated "LC" or "SPLC") is applicable in many fields including: Operations Research, Management Science, Pharmacokinetics, Insurance Risk Models, Natural Sciences, Engineering, Applied Probability, Applied Mathematics, and related areas.

Acknowledgments

I thank all my teachers, coauthors, postdoctoral fellows, graduate students, and colleagues, for learned research discussions. I thank David Grubbs, Senior Editor, Statistics, Chapman & Hall/CRC, Taylor & Francis Group, for his encouragement. NSERC (Natural Sciences and Engineering Research Council of Canada) provided long-term practical support of my research.

Percy H. Brill,
Odette School of Business. Mnagement Science area (and)
Department of Mathematics and Statistics
University of Windsor, Canada

1

Preliminaries

This chapter outlines the properties of a generic stochastic model. It introduces the nature and goals of stochastic level crossing method (LC). It includes for easy access by the reader: statements of simple basic-level crossing limit theorems with references to proofs; brief reviews of "companion analytic methods" which are useful when applying LC.

1.1 What Is a Stochastic Model?

A stochastic model is a mathematical model of a real-world process or potentially new real-world process, where at least one variable is uncertain or random. Random variables are defined and discussed on p. 21ff in Chapter 2 of [52] (2010) ("random Variable" is abbreviated "r.v."). The r.v.s are either "input" or "output" variables of the model. The input r.v.s are assumed to have a known probability cumulative distribution function (abbreviated "cdf") and/or probability density function abbreviated "pdf"), with known parameter values. One goal of the analyst is to derive the pdf and/or cdf of the output r.v.s in terms of the given parameters of the input r.v.s.

1.1.1 Brief Description of the Stochastic Level Crossing Method (LC)

In LC, an analytic integral equation for the pdf of a key output r.v. is obtained by a "thought process" about the "meaning" of the derived integral equation. In previous classical methods the integral equation for the same pdf is obtained using a step-by-step succession of algebraic operations, which ultimately reach the "same integral equation" as when using LC. For complex models the sequence of algebraic operations may be lengthy and time consuming. The "thought process" in LC "pictures the SP motions" in the state space, which are in "one-to-one correspondence" with each analytic term, on both sides of the equal sign of the derived integral equation. LC utilizes "SP rate balance" across state-space levels or boundaries, to build the integral equation for the pdf, term by term. This results in a much more efficient way for deriving the integral equation. In LC, the integral equation is completely built "by

DOI: 10.1201/9780429297601-1

inspection" of sample paths. The thought process in LC frees time for the analysts to envisage new viewpoints and new potential applications of the currently studied stochastic model.

Using LC, an analysis of a stochastic model usually begins by sketching a "typical sample path", which is a single realization (tracing) over time of a key output r.v. of the model (see Section 1.3). Simple "LC limit theorems," originally developed by the author in [10] (1975), greatly simplify transforming the SP level-crossing rates in the sample path (thought of dynamically over time) into integral equations for the pdf of the output r.v. (see stages (a)–(c) in Figure 1.1). (See also [11] (1975).) The solution of the integral equation is the stationary pdf of the key output r.v. This stationary pdf can be used to predict the performance of the model in relevant environments. (See Section 1.2.2, for further discussion on the LC method.)

Some proofs of LC limit theorems utilize fundamental properties of sample paths (aka "sample functions") in stochastic processes developed in reference [30] (1967). LC applies those fundamental properties in a completely novel way. That is, LC connects SP System Point motion in a state space with each term of the analytic abstract integral equation for the pdf of the key random variable of a stochastic model. Also, LC uses the important universal concept of "rate balance across a boundary line" to build the analytic integral equation.

Aside from LC, there are other relatively new and classical methods for analyzing stochastic models. For example, a useful technique developed soon after LC is the method of "Matrix-Geometric Solutions, an algorithmic approach" (see [47] (1981)).

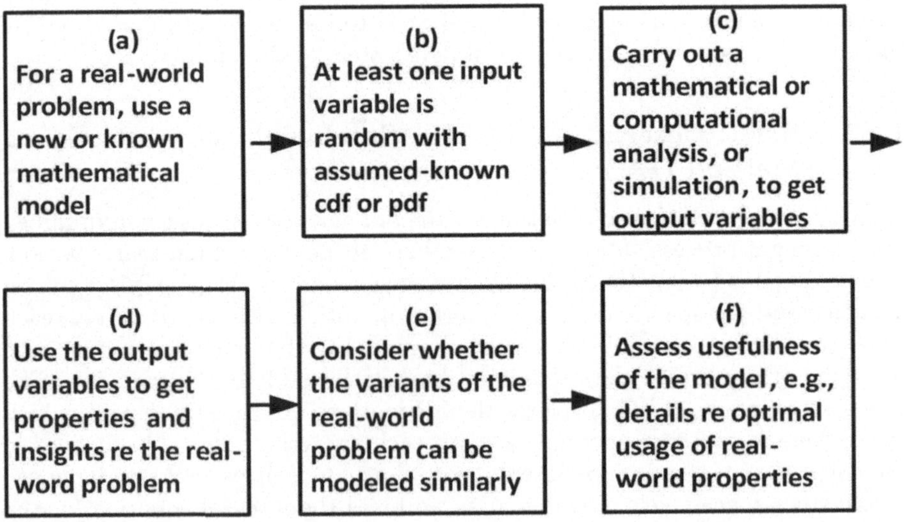

FIGURE 1.1
General sequential stages (a)–(f) of a stochastic model.

1.1.2 Examples of Input and Output Random Variables in a Stochastic Model

There is a vast number of potential stochastic models and their evolutions over time (i.e., sample paths). Some examples of stochastic models, with their input and output r.v.s, are:

1. Queues: Input r.v.s. – customer interarrival times and service times; output r.v.s. – pdf of customer actual waiting time; pdf of virtual waiting time (how long an arriving customer must wait in line until its service begins): pdf of workload (how much work to be done by the server–same as virtual wait in single server queues); pdf of system time (customer's "wait + service", the total time that a customer spends in the system; probability distribution of the number of customers in the system at an arbitrary timet, etc. (See Section 2.1.4, and pp. 9–10 in [22] (2017) for discussions on actual wait, virtual wait, workload, and sample path)
2. Inventories: input r.v.s.– inter-demand times for product; demand sizes; output r.v.s.– pdf of net inventory (amount of stock on hand), expected time between reordering a product (expected ordering cycle), pdf of deficit of net inventory just before an order instant, etc.
3. Dams: input r.v.s.– inter-input times of fluid; influx into and efflux out of the dam; output r.v.s.– pdf of amount of fluid in the dam; expected time until content is empty or overfull; etc.
4. Ruin models in Insurance risk model: input r.v.s.– inter-claim times; claim amounts; output r.v.s.– amount of Insurance company funds available (aka "buffer size"), time until ruin (when remaining funds of Insurance company reach zero from above); financial deficit at ruin time; effect on investors of a dividend barrier, etc.
5. Pharmacokinetics: input r.v.s.– drug inter-dose times; multiple dosing; dose size; absorption rate; elimination rate; output r.v.s.– pdf of concentration of drug in the bloodstream; probability that the drug concentration is within therapeutic limits, etc.

1.1.3 Diagram of General Stages in Analysis of a Stochastic Model

Figure 1.1 depicts general sequential stages for analyzing and assessing the usefulness of a stochastic model.

Insightful discussions on the usefulness and essence of a stochastic model, are given on pp. 1–4 in [48](2011). The present monograph will focus mainly on stages (a)–(c) in Figure 1.1.

1.2 Sample Path of Key Random Variables of a Stochastic Model

LC utilizes sample paths as "concrete templates" for building "analytic integral equations" for the pdfs of the key random variables in a stochastic model, by inspection. The sample paths lead to straightforward intuitive, efficient techniques for obtaining analytic pdfs of the key random variables. The intuitive logical arguments made throughout this book are applications of simple mathematically proved limit theorems about the motion of sample paths (thought of dynamically over time) with respect to state-space levels, or boundaries of state-apace sets.

Definition 1.1 *A sample path is a bounded real-valued, right-continuous function* $X(t)$, $t \in [0, \infty)$. *(The time-t domain is* $[0, \infty)$ *and the range in the state space is the set of values that* $X(t)$ *can take.) Left limits of* $X(t)$ *exist for all* $t > 0$. *During arbitrary finite time intervals, the number of vertical jumps is finite with finite expected values.*

In this book, most sample-path discontinuities occur at vertical jumps, upward or downward. (Other types of sample-path jumps, e.g., "parallel jumps" may occur in stochastic models with composite state spaces, see "Method of Lines and Pages" in Section 4.1.2 in Chapter 4).

A typical sample path represents a non-rare outcome of a random variable over time associated with a stochastic model (non-rare means having a positive-valued probability). Sample paths are also called: sample functions; realizations; trajectories; tracings; orbits. Reference [30] (1967) contains a comprehensive treatment of sample functions. (Note : In LC, the "leading point" of a sample path thought of dynamically over time, is called the "System Point", or "SP". The motion of the SP includes vertical jumps, and more general jumps, in the state space (see Chapter 2 in [22](2017).)

Remark 1.2 *In this book, the word "crossing" can be a noun ("the SP makes a crossing of level* x*"), an adjective ("the crossing rate is* $f(x)$*"), or a verb ("when crossing level* x*") depending on the context. Similar remarks hold for the words "upcrossing", "downcrossing", "upcross", and "downcross". These "crossing" words occur often in the book. Each use of a "crossing" word describes some physical motion of the SP in the state space. Each SP motion relative to a fixed level set in the state space, corresponds to a term in an analytical integral equation for the probability density function (pdf) of the associated output variable (r.v.). The equation is derived using rate balance across a state-space level, e.g., "level* x*". This connection between SP motion and the analytical terms permits the analyst to write down the integral equation by inspection of the sample path.*

Example 1.2.1 *Consider a sample path of the virtual wait process in a standard single server queue with queue discipline: "first come first served" (same as "service in order of arrival"). See Figure 2.1 in Section 2.1.5 in Chapter 2. The virtual wait := time that a would-be time t arrival to the system must wait until its service begins ("virtual wait" is the same as "workload" in single server queues, e.g., the M/M/1 standard queue considered in Section 2.1 in Chapter 2). In multiple-sever queues like the standard M/M/c queue (c parallel servers), the definitions of virtual wait and workload require more explanation than in M/M/1 (see Chapter 4).*

The virtual wait is a continuous-time continuous-state stochastic process denoted here by $\{W(t)\}_{t\geq 0}$ (see the sample path in Figure 2.2 in Chapter 2). The "Time set" is $\{t| \ 0 \leq t < \infty\}$. In the standard M/M/1 queue (and many other stochastic models) the state space of $\{W(t)\}_{t\geq 0}$ is a subset of $[0, \infty)$. (See p. 10 and pp. 20–24 in [22] (2017) for examples of sample paths, state-space levels, and crossings of levels by the SP.)

1.2.1 Stationary and Transient (Time-Dependent) Probability Distributions Using LC

This book deals with "stationary" probability distributions, e.g., "limiting" pdfs of r.v.s as time $t \to \infty$, unless stated otherwise in a particular section. The author developed and described the LC method to derive "time-dependent" (aka "transient") probability destitutions of a stochasticmodel denoted by $\{Vrbl(t)\}_{,t\geq 0}$, for any "fixed finite time" $t \geq 0$, in 1983–1988 (see [12] (1988); [13] (1988); [14] (1988), or the Indexical entries for "Theorem B" and "transient cdf/pdf" in the Indexes of [21] (2008) or [22] (2017)).

1.2.1.1 Quantities Based on Knowledge of Pdfs of Output Variables

In Example 1.2.1, the derived stationary pdf of the virtual wait can be used to obtain various quantities of interest, e.g., expected waiting time before service; variance and higher moments of waiting time before service; the probability (long-run proportion of time) that the virtual wait is in any given state-space interval; the pdf of an arbitrary customer's system time := "wait before service" + "service time", etc.

1.2.2 The Stochastic Level Crossing Method (LC)

LC provides an intuitive, efficient method of analysis, which mitigates a significant amount of time-consuming algebra. Its power depends on seven basic properties, and others which come to light when analyzing stochastic models. The basic properties are: (1) Sketching a "logiccally correct" sample path of the random variable of interest over time (e.g., the "generic process denoted by $\{Vrbl(t)\}_{t\geq 0}$", discussed above); (2) Knowledge of the rates, slopes, and types, of the sample-path "downcrossings" of an arbitrary state-space level, e.g., level x, and their connection to terms of an integral equation for the stationary pdf of "$\{Vrbl(t)\}_{,t\geq 0}$"; (3) Knowledge of rates, slopes and types of the sample path "upcrossings" of the same arbitrary state-space level, and their connection to additional terms of the integral equation for the stationary pdf of $\{Vrbl(t)\}_{,t\geq 0}$; (4) Application of sample-path "rate balance" across the arbitrary state-space level x, which equates downcrossing and upcrossing rates across level x; (5) The one-to-one correspondence between the "physical" downcrossing and upcrossing rates of the sample path across level x, and the analytic terms as a function of x in the integral equation for the stationary pdf of $\{Vrbl(t)\}_{,t\geq 0}$; (6) Construction and solution of the integral equation for the stationary pdf; (7) Sample-path rate balance into and out of state-space sets.

Note: we distinguish between the two measures: "rate at which an event occurs" and "speed or velocity of a point or object". (See Section 1.8.6 on dimensional analysis.)

1.3 Sample Path and Level Crossings: Details

1.3.1 Sample Path

An LC analysis begins with a drawing (or sketch) in a Cartesian coordinate system, of a "typical" sample path (realization over time) of an output r.v. For a two-dimensional coordinate system, the horizontal axis represents usually time $t \geq 0$. The vertical axis represents usually the state of the system, $x \geq 0$, corresponding to a time point $t \geq 0$. Point (t, x) is the "system point" at time t. A sample path is a function: $t \to x$ in an t-x coordinate system. A particular sample path represents just one possible outcome of the underlying stochastic process. There is often an infinite number of possible sample paths when the stochastic process has components which are continuous r.v.s., e.g., in a queue with continuous customer interarrival times, service times, and waiting times, etc. A typical sample path is one that is "reasonable" or "not rare" (see p. 20 in [22] (2017)). A sample path depicts the evolution over time the point (t, x) in the state space. For the analyst using LC, it is immaterial whether the physical sample path is exact, as long as it is a logic-preserving sketch of the system. The analyst should be mindful of the type of r.v.s behind the sample path's construction, i.e., continuous or discrete. The advantage of portraying a "physical" sample path as a motion with a leading point (SP) moving from left to right in the time domain, provides intuitive information about the model, and a concrete picture for analyzing a stochastic model.

1.3.2 Focus on Level Crossings Made by Sample Paths

To describe the nature of sample-path crossings of a state-space level x by a sample path over time, it is useful to overlay a horizontal line at the ordinate level x of the t-x coordinate system. We then observe the number of times the sample path (or SP) downcrosses or upcrosses level x during a fixed time period (with the precise slope for "smooth" crossings, or with either vertical "upward" or vertical "downward" jumps).

The virtual wait process in Figure 2.2 in Section 2.1.5 of Chapter 2 illustrates these ideas. In Figure 2.2 the sample path downcrosses level x smoothly four times: once during the first busy period and three times during the second busy period, all with slope -1. The sample path upcrosses level x by upward jumps five times: once at the end of each of three idle periods, and twice during the second busy period. The sample path hits level 0 from above with slope -1 four times, at the ends of four busy periods, and remains at level 0 for a positive random time (idle period) because the state of the system does not change from level $x = 0$ during idle periods. The SP jumps upward from level 0 three times, once at the end of each idle period. For a fixed state-space level, the number of downcrossings and number of upcrossings (smooth or jump) differ by 0 or by ± 1 during any fixed time interval $(0, t)$.

The next three sections continue with Example 1.2.1 in Section 1.2.

1.3.3 Focus on Sample-Path Crossing Rates of State-Space Levels

Note: We often use "Sample-path Crossing" to mean "Sample-path or SP Crossing" depending on the context. Consider a sample path of the virtual wait process $\{W(t)\}_{t\geq 0}$ in a single-server queue (see Figure 2.2 in Section 2.1.5 of Chapter 2). Let $\mathcal{D}_t(x)$ and $\mathcal{U}_t(x)$ denote the number of sample-path downcrossings and sample-path upcrossings of level x, respectively, during the time period $(0, t), t > 0$. The long-run sample-path downcrossing and upcrossing rates of level x are expressed by

$$\lim_{t\to\infty} \frac{\mathcal{D}_t(x)}{t} \text{ and } \frac{\lim_{t\to\infty} \mathcal{U}_t(x)}{t}$$

respectively, for all $x \geq 0$. For the lower state-space boundary level $x = 0$, we consider, without loss of generality, that a downcrossing of level 0 is the same as a hit of level 0 from above; an upcrossing of level 0 is the same as an upward exit out of level 0. (See Section 2.5 pp. 45–47 in Chapter 2 of [22](2017) for different transitions by a sample path or SP with respect to a state-space level)

1.4 Some Basic Level Crossing Limit Theorems and Concepts

Basic LC theorems connect level-crossing rates of the "physical" sample path of the r.v. of interest, to corresponding "analytic" terms of an integral equation, whose solution is the analytic pdf of the r.v. This connection bypasses time-consuming algebra for deriving the integral equation. Thus the LC method is more efficient than classical methods while yielding analytic solutions. A logically sketched sample path acts as a template for writing down analytic integral equations The integral equation gives the analyst quantitative results for the stochastic model under consideration.

1.4.1 A Basic LC Limit Theorem for the Virtual Wait Process $\{W(t)\}_{t\geq 0}$ in M/M/1 Queues

In "M/M/1" queues the first "M" is often an exponential r.v. with rate λ (arrival rate); the second "M" is often an exponential r.v. with rate μ. Let $P_0 := \lim_{t\to\infty} P(W(t) = 0)$ and let $f(x), x > 0$, be the continuous part of the mixed pdf $\{P_0, .f(x)\}_{0<x<\infty}$, the limit as $t \to \infty$. A typical sample path of $\{W(t)\}_{t\geq 0}$ is shown in Figure 2.2 in Section 2.1.5 in Chapter 2. In the standard M/M/1 queue the mixed pdf $\{P_0, f(x)\}_{0<x<\infty}$ exists if the "traffic intensity" $:= \lambda/\mu < 1$ for stability. Alternatively stated, "if the expected number of arrivals during a service time is less than 1" (see Section 2.1 in Chapter 2).

The basic LC theorem stated in Theorem 1.3 connects the analytic continuous part of the stationary pdf $f(x)$, with the physical sample-path. That is, the long-run left-continuous sample-path crossing rate of level x is a very simple function of $f(x)$ (see Figures 1.2 and 1.3). (Note: The basic LC theorem for M/M/1 queues holds also for M/G/1 queues where "G" is not an exponential r.v.)

Let $\mathcal{D}_t^c(x)$ denote the number of physical sample-path continuous (left-limit) downcrossings of level x during time period $(0,t)$. Let $E\left[\mathcal{D}_t^c(x)\right] :=$ expected value of $\mathcal{D}_t^c(x)$. Assume that the M/M/1 queue of interest is stable.

Theorem 1.3 *In a typical sample path of the virtual wait $\{W(t)\}_{t\geq 0}$, the left-limit continuous downcrossing rate is*

$$\lim_{t\to\infty} \frac{\mathcal{D}_t^c(x)}{t} = f(x), 0 < x < \infty, \text{ with probability 1;} \tag{1.1}$$

$$\lim_{t\to\infty} \frac{E\left[\mathcal{D}_t^c(x)\right]}{t} = f(x), 0 < x < \infty. \tag{1.2}$$

Proof *See pp. 13–16 in [22](2017) for a proof of (1.1).and (1.2). Note: the slope of the sample path at level x is $-1, 0 < x < \infty$, in the virtual wait process.*

Corollary 1.4 *For left-limit continuous downcrossings of level x*

$$\lim_{t\to\infty} \frac{\mathcal{D}_t^c(x)}{t} = -(\text{sample-path slope at level } x)\times f(x), x > 0, \text{ with probability 1,}$$

where the sample-path slope at level $x > 0$ is $\frac{dV(t)}{dt}$ at time t.

In (1.1) "With probability 1" means "for all sample paths except for ihose sample paths which have a total probability (i.e., measure) = 0").

1.4.2 A Basic LC Limit Theorem for the Extended Age Process $\{V(t)\}_{t\geq 0}$ in G/M/1 Queues

The "extended age" process, denoted by $\{V(t)\}_{t\geq 0}$, expresses how long a customer has been in the system, given that the customer is in service. Process $\{V(t)\}_{t\geq 0}$ is useful for deriving the stationary pdf of the virtual wait in G/M/1 queues, where "G" represents an arbitrary r.v. for customer interarrivals, which is usually not an exponential r.v. This is because a duality exists between $\{V(t)\}_{t\geq 0}$ in G/M/1 queues and the virtual wait in M/G/1 queues (See Section 5.1.1 in Chapter 5.)

Theorem 1.5 states the connection between the physical sample-path upcrossing rate of $\{V(t)\}_{t\geq 0}$ and its stationary pdf $f(x)$. The pdf $f(x)$ is also equal to the stationary pdf of he virtual wait in the G/M/1 queue by duality.

Theorem 1.5 *Consider a typical sample path of the extended age* $\{V(t)\}_{t\geq 0}$. *For continuous (left-limit) upcrossings of level* x

$$\lim_{t\to\infty}\frac{\mathcal{U}_t^c(x)}{t} = \textit{(sample-path slope at level } x\textit{)} \times f(x), \textit{ all } x, \textit{ with probability 1;}$$

$$\lim_{t\to\infty}\frac{E\left[\mathcal{U}_t^c(x)\right]}{t} = \textit{(sample-path slope at level } x\textit{)} \times f(x), \textit{ all } x,$$

where the sample-path slope at level x *is* $\frac{dV(t)}{dt}$.

1.4.3 A Basic LC Limit Theorem for Dams, Inventories, and Related Models with Poisson Inputs

Note: "Poisson inputs" is the same as inputs where the inter-input times are iid (independent and identically distributed) exponential r.v.s (see Section 1.5).

Let $\{X(t)\}_{t\geq 0}$ be a generic process in the dam, inventory or related model. Assume that each sample-path downcrossing of level x in the state space is smooth (left-limit) with slope $r(x)$. The slope is a function of level x (Figure 1.2 in this section). Assume that $r(x)$, $x \geq 0$ is a piecewise continuous real function. Formulas (1.1) and (1.2) in Section 1.4 above are generalized by multiplying slope $r(x)$ by $f(x)$ as follows.

Theorem 1.6

$$\lim_{t\to\infty}\frac{\mathcal{D}_t^c(x)}{t} = r(x)f(x), x > 0, \textit{ with probability } 1, \tag{1.3}$$

$$\lim_{t\to\infty}\frac{E\left[\mathcal{D}_t^c(x)\right]}{t} = r(x)f(x), x > 0, \tag{1.4}$$

Proof *For any fixed finite level $x > 0$, $r(x)$ = "– (slope of the downward-moving SP)" when it downcrosses the fixed finite level x in the state space. The SP moves from left to right in the direction of Time. Then $r(x) = -\frac{dX(t)}{dt} < 0$ at level x, which is.given by:*

$$r(x) = -\ \textit{(slope of tangent to smooth sample path at the point } (t,x)) > 0. \tag{1.5}$$

Note that $r(x) > 0$ because the slope of a smooth downcrossing is negative. If, for all $x > 0$, $\frac{dX(t)}{dt} = -k$ where k is a positive constant then the sample path of $\{X(t)\}_{t\geq 0}$ declines at rate k until it hits level 0 (after which $X(t)$ may be 0 in dams). The point (t, x) is called the "system point" in the original LC theory (see [10] (1975)).

1.4.4 Dimension of $r(x)f(x)$

An important property of $r(x)f(x)$.is its "dimension" (see Section 1.8.6 on "dimensional analysis" in this chapter). Briefly, measurements in Physics are given in terms of three basic quantities: Mass, Length, and Time. Let [physical quantity] denote the dimension of "physical quantity." The dimension of $r(x)f(x)$ is:

$$[r(x)f(x)] = [r(x)] \times [f(x)] =$$

$$\frac{\left[\text{state variable of } \{X(t)\}_{t\geq 0}\right]}{[\text{Time}]} \times \frac{1}{\left[\text{state variable of } \{X(t)\}_{t\geq 0}\right]},$$

$$[r(x)f(x)] = \frac{1}{[\text{Time}]}. \tag{1.6}$$

In explanation of (1.6), note that

$$[f(x)] = \frac{1}{\left[-\text{state variable of } \{X(t)\}_{t\geq 0}\right]} = \frac{1}{\left[\text{state variable of } \{X(t)\}_{t\geq 0}\right]} \tag{1.7}$$

because "–1" is a number without a unit, which is dimensionless. Also, the pdf $f(x) = \frac{dF(x)}{dx}$, where $F(x)$ is the associated stationary cdf of the state variable of $\{X(t)\}_{t\geq 0}$ (see item 2 in Section 1.4, p. 7 in Chapter 1 in [22](2017)). Importantly, from formula (1.6) $r(x)f(x)$ is a "rate" in the sense of Physics. The rate is given by the following corollary.

Corollary 1.7

$$r(x)f(x) = \lim_{t\to\infty} \frac{\mathcal{D}_t^c(x)}{t} \tag{1.8}$$

Proof *Assume that* $\{X(t)\}_{t\geq 0}$ *is stable (see Section 1.7). Let* d_x:= *time period between two successive sample-path downcrossings of level* x. *By stability,* $E[d_x]$ *is finite. Just after each downcrossing instant of level* x, *the SP restarts moving in the same manner, independent of the past, for a time* $\stackrel{d}{=}$ d_x *until the next level-*x *downcrossing, due to the memoryless property of the exponential distribution (Poisson inputs) following each level-*x *downcrossing. The successive downcrossing instants of level* x *form a renewal process* $\{d_{x,n}\}_{n=1,2,...}$ *where* $d_{x,n} \stackrel{d}{=} d_x$, $n = 1, 2, ...$ *(see Sections 1.8.1 and 1.8.2 in this chapter). By the elementary renewal theorem (p 432 in [52](2010))*

$$E[d_x] = \frac{1}{r(x)f(x)}. \tag{1.9}$$

The interrenewal times $\{d_{x,n}\}$ *are regenerative cycles of a regenerative process. The regenerative process consists of iid realizations built in the interrenewals. Since there is exactly one level-*x *downcrossing in each regenerative cycle, the renewal reward theorem and formula (1.9) yield*

$$\frac{1}{E[d_x]} = \lim_{t\to\infty} \frac{\mathcal{D}_t(x)}{t},$$

$$r(x)f(x) = \lim_{t\to\infty} \frac{\mathcal{D}_t(x)}{t},$$

gyving formula (1.3) in Theorem above. Formula (1.4) follows by the elementary renewal theorem.

1.4.5 Geometric Sample-Path Level Crossingss

Figures 1.2 and 1.3 show geometric sample-path downcrossings and upcrossings, respectively.

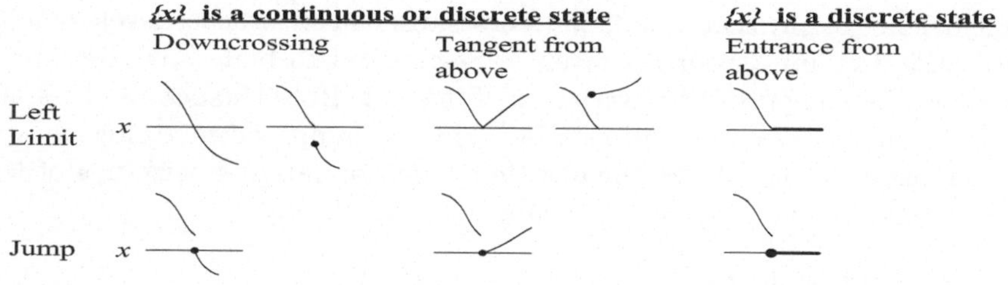

FIGURE 1.2
Sample-path/SP- hits of level x in the state space from above—downcrossings of level x. SP motion has two components—(1) in direction of slope $r(x)$ (2) SP "motion" from left to fight—in direction of time.

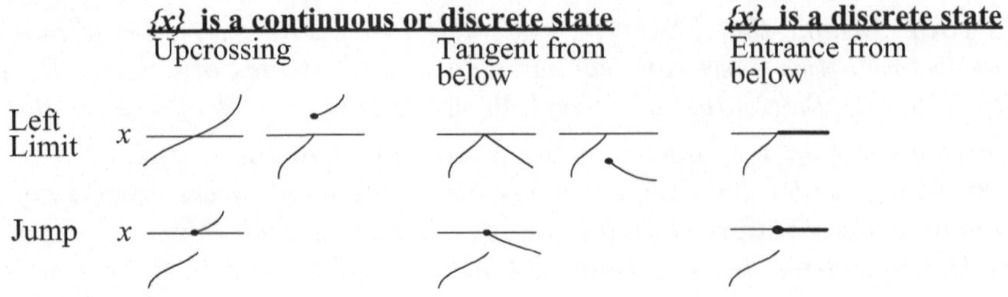

FIGURE 1.3
Sample-path/SP-hits of level x in the state space from below—upcrossings of level $\dot{x}$. SP motion has two components—(1) upward in direction of slope $r(x)$ (2) SP "motion" from left to right—in direction of time.

In addition to the SP hits of level x shown in Figures 1.2 and 1.3, there are other ways that the SP can cross level x, e.g., via SP "vertical jumps".

Generally, an SP "downward vertical jump" downcrosses level x if it begins in state-space set (x, ∞) and ends in state-space set $(-\infty, x)$ An SP upward vertical jump upcrosses level x if it begins in state-space set $(-\infty, x)$ and ends in state-space set (x, ∞).

Vertical jumps which cross level x are important because they combine with the other types of level-x crossings when applying SP "rate-balance" across level x to derive analytic integral equations for the stationary pdfs of the variables of interest. A more complete set of level-crossing types viewed pictorially, is given in Section 2.5, pp. 45–47 in [22](2017).

1.4.6 Rate Balance of a Sample-Path Across a State-space Level and Rate-Balance Equations Using Basic LC Limit Theorems

1.4.6.1 Rate Balance

Rate balance of physical sample-path crossings of a state-space level combines with basic LC limit theorems given in Section 1.4 to build analytic integral equations for the pdfs of the variables of interest. Rate balance can be stated as "the long-run rate of (continuos (left limt) or jump)–downcrosings is equal to the long-run rate of (continuous (left limit) or jump)–upcrosings of level x, for every $x \in S$ (state space). Expressed symbolically:

$$\lim_{t\to\infty} \frac{\mathcal{D}_t^c(x) + \mathcal{D}_t^j(x)}{t} = \lim_{t\to\infty} \frac{\mathcal{U}_t^c(x) + \mathcal{U}_t^j(x)}{t}, x \in \mathbf{S}, \text{with probability 1;} \tag{1.10}$$

$$\lim_{t\to\infty} E\left[\frac{\mathcal{D}_t^c(x) + \mathcal{D}_t^j(x)}{t}\right] = \lim_{t\to\infty} E\left[\frac{\mathcal{U}_t^c(x) + \mathcal{U}_t^j(x)}{t}\right], x \in \mathbf{S}, \tag{1.11}$$

where S is the state space. (For further discussion of rate balance see p. 16 and Sections 2.4.5–2.4.8 on pp. 32–36 in [22] (2017)).

1.4.7 Rate-Balance Equations for Pdfs Using Basic LC Limit Theorems

By applying LC the analyst derives the integral equations to solve for the stationary pdf $\{P_0, f(x)\}_{x\in \boldsymbol{S}}$, by inspection of the sample path.

Example 1.4.1 *Assume all sample-path downcrossings are continuous, and all sample-path upcrossings are vertical jumps. This is the case for the virtual wait in M/M/1 and M/G/1 queues. Formulas (1.1) and (1.10) in Section 1.4 imply*

$$f(x) = \lim_{t\to\infty} \frac{\mathcal{D}_t^c(x)}{t} = \lim_{t\to\infty} \frac{\mathcal{U}_t^{cj}(x)}{t}, x \in \boldsymbol{S}, \tag{1.12}$$

$$f(0^+) = \lim_{t\to\infty} \frac{\mathcal{D}_t^c(0)}{t} = \lim_{t\to\infty} \frac{\mathcal{U}_t^{j}(0)}{t}. \tag{1.13}$$

In formula (1.12) the first equality is due to a basic LC theorem 1.3. The second equality is due to rate balance. Formula (1.12) generates the integral equation for $f(x)$ where P_0 is an unknown constant.

In formula (1.13) the two equalities are similar to those in formula (1.12). Formula (1.13) generates an equation for P_0.

The generated integral equations are in terms of the parameters and pdfs of the input random variables. Building the integral equation for $f(x)$ is demonstrated for the virtual wait in M/M/1 queues in Section 2.1.7 of Chapter 2.

Three main steps in the analysis of a typical stochastic model are shown along the edges of a process triangle in Figure 1.4.

1.5 Exponential Distribution and Poisson Process

The exponential distribution and Poisson process are important in stochastic models. The exponential distribution is continuous and has support $(0, \infty)$. The Poisson process is a "continuous-time discrete-state process" taking integer values in $\{0, 1, 2, ...\}$. A useful connection between the exponential distribution and Poisson process is discussed in Section 1.5.5 in this chapter.

The exponential distribution with rate λ (mean $= 1/\lambda$) is denoted by Exp_λ, where $\lambda > 0$. The pdf $b(x)$, cdf $B(x)$, and complementary cdf (ccdf)

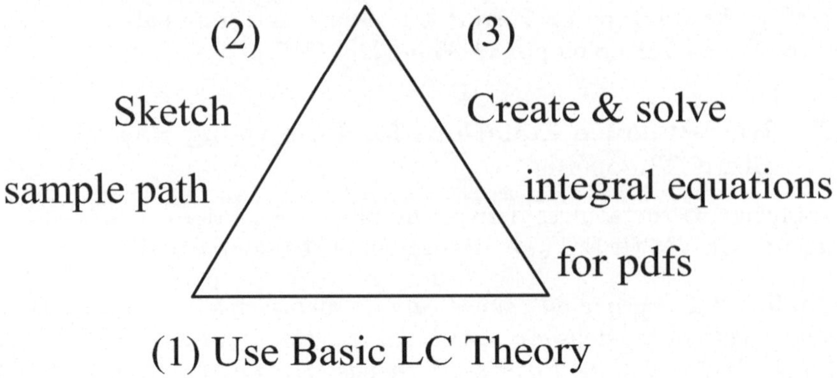

FIGURE 1.4
Three main steps in an LC analysis of a typical stochastic model. Step (3) uses steps (1) and (2) to solve for the pdfs of the random variables of interest.

$\overline{B}(x) := 1 - B(x)$ have formulas

$$\left.\begin{aligned} b(x) &= \lambda e^{-\lambda x}, 0 \le x < \infty; \\ B(x) &= 1 - e^{-\lambda x}, 0 < x < \infty; \\ \overline{B}(x) &= 1 - B(x) = e^{-\lambda x}, 0 \le x < \infty. \end{aligned}\right\} \tag{1.14}$$

The pdf is monotone decreasing from λ towards 0 and is convex for $x \in [0, \infty)$. The cdf $B(x)$ is monotone increasing from 0 towards 1 and is concave, for $x \in (0, \infty)$. The ccdf $\overline{B}(x)$ is monotone decreasing from 1 towards 0 and is convex for $x \in [0, \infty)$. The formulas in (1.14) are plotted in Figure 1.5, where $\lambda = 2$.

Let S_n, n =1, 2, 3..., denote the service times of the customers that arrive to a queueing system, which are iid Exp_μ r.v.s where $\mu > 0$. Let $S_n \stackrel{d}{=} S$ where $S \stackrel{d}{=} \text{Exp}_\mu$. Then,

$$\left.\begin{aligned} P(S \le x) &= 1 - e^{-\mu x}, x > 0, \\ P(S > x) &= e^{-\mu x}, x \ge 0, \\ E[S] &= 1/\mu. \end{aligned}\right\} \tag{1.15}$$

Figure 2.5 in Section 2.1.7 in Chapter 2, is a useful template which gives the crossing rates of the sample path that of level $x > 0$ leading to equation (1.17). That is, if the arrival rate is λ

$$\lim_{t\to\infty} \frac{\mathcal{D}_t^c(x)}{t} = \lambda P_0 \overline{B}(x) + \lambda \int_{y=0}^{x} \overline{B}(x-y) f(y) dy, 0 < x < \infty. \tag{1.16}$$

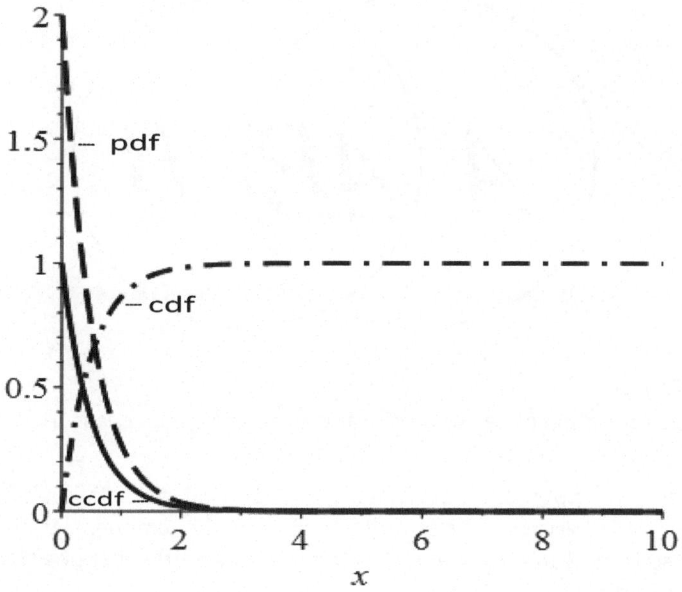

FIGURE 1.5
Exponential distribution (Exp_λ) with $\lambda = 2$: pdf $b(x)$, $0 \le x < \infty$,–dash; cdf $B(x)$, $0 < x < \infty$,–dot-dash; ccdf $\overline{B}(x)$, $0 \le x < \infty$,–solid.

If additionally the service times are $\stackrel{d}{=} \text{Exp}_\mu$, then Equation (1.16) is identical to

$$f(x) = \lambda P_0 e^{-\mu x} + \lambda \int_{y=0}^{x} e^{-\mu(x-y)} f(y)dy, 0 < x < \infty, \qquad (1.17)$$

where $f(x)$ is the continuous part of the virtual-wait pdf. The solution for P_0 is obtained by applying the law of total probability (aka normalizing condition) which states that "the sum of all probabilities is 1 for a single random variable":

$$P_0 + \int_{x=0}^{\infty} f(x)dx = 1. \qquad (1.18)$$

Equation (1.17) is the analytic integral equation for $f(x)$, $x > 0$. Its solution for $f(x)$, $x > 0$ is obtained using standard calculus and algebra, where P_0 is an unknown to be determined. The normalizing condition (1.18) is then used to solve for P_0.

1.5.1 Memoryless Property of the Exponential Distribution

The memoryless property of the exponential r.v. is very useful in stochastic modelling. First we mention conditional probability of one random event given that a second random event occurs, because it leads to the memoryless

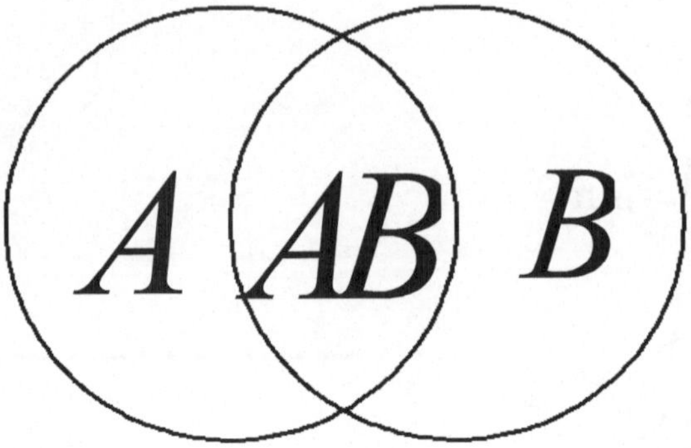

FIGURE 1.6
Venn diagram of: event $\boldsymbol{A}$; conditioning event $\boldsymbol{B}$; joint event $\boldsymbol{AB}$.

property of the exponential r.v. (see pp. 7–16 and p. 294ff in [52](2010) for more theory, properties and examples regarding the exponential distribution.)

Suppose that A and B are different random events with positive probabilities (see Figure 1.6).

The conditional probability that A occurs given that B occurs is given by

$$P(A|B) = \frac{P(A \text{ and } B \text{ both occur})}{P(B \text{ occurs})}$$

or briefly

$$P(A|B) = \frac{P(AB)}{P(B)}. \tag{1.19}$$

If events A and B are statistically independent then

$$P(AB) = P(A) \cdot P(B).$$

Then

$$P(A|B) = \frac{P(AB)}{P(B)} = \frac{P(A)\,P(B)}{P(B)} = P(A).$$

(see p. 217ff in [34](1968) for related theory and examples).

We now apply conditional probability to explain the memoryless property of the exponential distribution. Given that an exponential random variable $X \stackrel{d}{=} \text{Exp}_\lambda$, exceeds an arbitrary given value $t > 0$, the exceedance over t has the same exponential distribution as the original unconditioned exponential random variable X. To see this, let $t > 0$ be the arbitrary value, and let y be a positive number. Then

$$\begin{aligned} P(X > t + y|X > t) &= \tfrac{P(X>t+y \text{ and } X>t)}{P(X>t} \\ &= \tfrac{P(X>t+y)}{P(X>t)} = \tfrac{e^{-\lambda(t+y)}}{e^{-\lambda t}} = e^{-\lambda y} \\ &= P(X > y), y > 0. \end{aligned} \tag{1.20}$$

In explanation of the second equality in formula (1.20), the event $\{X > t + y\}$ is a subset of the event $\{X > t\}$. Therefore $P(X > t+y$ and $X > t) = P(X > t + y)$.

1.5.2 Bayes' Rule

Bayes' rule involving conditional probabilities, is useful in many fields including stochastic models. The simplest form of Bayes' rule is

$$P(A|B) = P(B|A)P(A), \tag{1.21}$$

which is provable from formula (1.19) above (see Figure 1.6).

A more general form of Bayes' rule occurs when the event B must intersect the union of N sets "$\cup_{i=1}^{N} A_i$", where the A_is are "mutually exclusive and exhaustive" (do not overlap with each other, and fill up the entire probability space). Then, event B can occur only if at least one of the events $A_1, ..., A_N$, has a positive probability. Let us assume that $P(A_i) > 0$, $i = 1, ..., N$. Then $B = \cup_{i=1}^{N} (BA_i)$. Applying formula (1.21) gives

$$P(A_i|B) = \frac{P(BA_i)}{P(B)},$$

$$P(A_i|B) = \frac{P(BA_i)}{\sum_{i=1}^{N} P(BA_i)},$$

$$P(A_i|B) = \frac{P(B|A_i)P(A_i)}{\sum_{i=1}^{N} P(B|A_i)P(A_i)}, \quad i = 1, ..., N. \tag{1.22}$$

Formula (1.22) has many real-world applications, e.g., in medical diagnosis. In medicine, we may make the analogy of formula (1.22) as follows: $B :=$ "symptom" of a disease—each A_i is a "potential distinct cause" of the symptom B. Bayes' rule is applied later in this book. e.g., in Section 3.3.4 in Chapter 4. (See p. 114—125 in [34] (1968) for a comprehensive discussion on Bayes' rule.)

1.5.3 Probability Distribution of the Minimum of Two or More Independent Exponential Random Variables

The pdf of the minimum of multiple iid exponential r.v.s is important when analyzing stochastic models where the r.v.s operate simultaneously First consider two iid r.v.s X_1 and X_2 with distributions Exp_{λ_1} and Exp_{λ_2} respectively, and $\lambda_1 \neq \lambda_2$. Let

$$X_{\min} := \min\{X_1, X_2\}.$$

For a fixed value $x > 0$, by independence of X_1 and X_2:

$$\begin{aligned} P\left(X_{\min} > x\right) &= P\left(X_1 > x \text{ and } X_2 > x\right) \\ &= P\left(X_1 > x\right) P(X_2 > x) \\ &= e^{-\lambda_1 x} \cdot e^{-\lambda_2 x} \\ &= e^{-(\lambda_1 + \lambda_2)x} \end{aligned} \tag{1.23}$$

Therefore

$$X_{\min} \overset{d}{=} Exp_{\lambda_1 + \lambda_2}.$$

A similar formula applies to the minimum of any finite number of exponential variates, say $c \geq 2$ Thus if

$$X_{\min} := \min\left\{X_1, X_2, ..., X_c\right\}, c \geq 2.$$

then

$$P\left(X_{\min} > x\right) = e^{-(\lambda_1 + \lambda_2 + \cdots + \lambda_c)x}, c \geq 2 \tag{1.24}$$

If $\lambda_1 = \lambda_2 = \cdots = \lambda_c \equiv \lambda$ then

$$P\left(X_{\min} > x\right) = e^{-c\lambda x}, c \geq 2 \tag{1.25}$$

Formula (1.24) is useful in analyzing multiple-server queues (see Section 4.2 in Chapter 4). It is important in Reliability theory for studying the time until failure of a device operating on multiple components with exponentially distributed lifetimes, e.g., batteries.

1.5.4 Poisson Process

1.5.4.1 Poisson Probability Distribution

A discrete random variable X has a Poisson probebility distribution if $\lambda > 0$ and

$$P(X = n) = \frac{e^{-\lambda}\lambda^n}{n!}, n = 0, 1, 2, ... \tag{1.26}$$

In formula (1.26) "λ" is the mean value of the distribution, i.e., $E\left[X\right] = \lambda$, which is shown using the basic definition (see a book on Calculus):

$$e^{\lambda} \equiv \sum_{n=0}^{\infty} \frac{\lambda^n}{n!}.$$

An important technical point is that λ is a number. Therefore the dimension of λ, denoted by $[\lambda]$, is that $[\lambda] = [1]$ (aka dimensionless)—see Section 1.8.6.

1.5.4.2 Poisson Process Details

Let $N(t)$ denote the number of arrivals during the time period $(0,t)$, $t>0$. The process $\{N(t)\}_{t\geq 0}$ is a Poisson process if $N(t)$ has a Poissin probability distribution with parameter λt, i.e.,

$$P(N(t)=n)=\frac{e^{-\lambda t}\left(\lambda t\right)^{n}}{n!},\ n=0,1,2,... \tag{1.27}$$

The expected number of Pisson events during $(0,t)$ is denoted by $E\left[N(t)\right]$. It is found using the formula for the mean of a discrete r.v. defined on $0,1,...,\infty$, as in (1.26) That is,

$$E\left[N(t)\right]=\sum_{n=0}^{\infty}nP(N(t)=n).$$

From Formula (1.27)

$$E\left[N(t)\right]=\sum_{n=0}^{\infty}n\frac{e^{-\lambda t}\left(\lambda t\right)^{n}}{n!}=\lambda te^{-\lambda t}\sum_{n=1}^{\infty}\frac{\left(\lambda t\right)^{n-1}}{n-1!}=\lambda te^{-\lambda t}e^{\lambda t}=\lambda t. \tag{1.28}$$

where

$$e^{x}=\textstyle\sum_{n=0}^{\infty}\frac{x^{n}}{n!},$$

and "e" is the base of natural logarithms. The rate of the Poisson Process $\{N(t)\}_{t\geq 0}$, e.g., the customer arrival rate to an M/M/1 queue is, by formula (1.27),

$$\lim_{t\to\infty}\frac{N(t)}{t}=\lim_{t\to\infty}\frac{\lambda t}{t}=\lambda=\frac{E\left[N(t)\right]}{t}, \tag{1.29}$$

where the first equality derives from the elementary renewal theorem (see p. 428 in [52] (2010)), and the third equality follows from formula (1.28)—(see Section 2.1 in Chapter 2).

An important technical point is that in the "Poisson process" λ is a "rate". Therefore the dimension of λ, denoted by $[\lambda]$, is $[\lambda]=\left[\frac{1}{Time}\right]$—see Section 1.8.6.

1.5.5 Focus on Connection between Exponential Distribution and Poisson Process

The following fundamental statements hold, connecting the exponential and the Poisson probability distributions: Let $S:=$ time until the first arrival to a queue. For a fixed time $t>0$. Then

$$S\leq t\iff \text{number of arrivals in (0,}t\text{) is }\geq 1.$$

Thus, we obtain the useful equivalence:

$$S\leq t\iff N(t)\geq 1,$$

which translates, in terms of probabilities, to:

$$\left.\begin{aligned} P(S \leq t) = P\left(N(t) \geq 1\right) = 1 - P(N\left(t\right) = 0) \\ P(S \leq t) = 1 - e^{-\lambda t}, t > 0. \end{aligned}\right\} \tag{1.30}$$

when applying (1.27),

For a comprehensive discussion on the exponential distribution and the Poisson process see Chapter 5 in [52] (2010).

1.6 Birth and Death Process

A birth and death process is a special case of continuous-time Markov chain (abbreviated "CTMC"). A comprehensive discussion of CTCMs is given in Chapter 6 in [52](2010), pp. 372–381, 386–389.

For a birth and death process, consider a stochastic model like a queueing system, where items arrive, then get service, and then depart. Assume items arrive at exponential rate $\lambda_n, n = 0, 1, ...,$ when $n :=$ number of items in the system just before an arrival. The number of items in the system goes from n to $n+1$ at rate $\lambda_n, n = 0, 1, ...$ Assume that service times are exponential with rate μ_n, $n = 1, 2, ...,$ when there are n items in the system, then the number of items goes from n to $n - 1$ at exponential ate $\mu_n, n = 1, 2, ...$ At each transition instant (arrival or departure) the number of items in the system changes exactly by "+1 or -1".

Let P_n, $n = 0, 1, 2...,$ denote the stationary probability of n items in the system. A useful result for birth and death processes is

$$P_n = \frac{\lambda_0 \lambda_1 \cdots \lambda_{n-1}}{\mu_1 \mu_{12} \cdots \mu_n} P_0, n = 1, 2, ... \tag{1.31}$$

(see p. 392 in [52](2010)). Formula (1.31) can be used to derive P_0 by applying the normalizing condition

$$\sum_{n=0}^{\infty} P_n = 1. \tag{1.32}$$

In formula (1.32), if the maximum number of units is finite N_{max} then $P_n \equiv 0, n > N_{max}$.

1.7 Stability in Stochastic Models

As an example of a stochastic model consider a queueing system where inter-arrival times and service times are random or uncertain. In this monograph

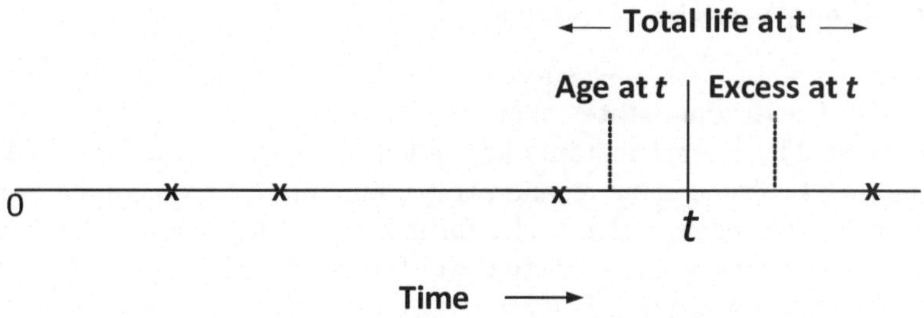

FIGURE 1.7
The "x"s on the horizontal line denote renewal times. For fixed finite time t, figure shows age (backward recurrence time), excess (forward recurrence time), and total life (age + excess). The inter-arrival time between renewals 3 and 4, cover the fixed finite time t.

we will call a queueing system "stable" if the expected customer arrival rate is less than the expected service rate in the system. In a standard $M_\lambda/M_\mu/1$ queue (see Chapter 2) the expected customer arrival rate is λ per unit time and the expected service rate is μ per unit time. Then the queueing system is stable if $\lambda < \mu$.

In the standard M/M/c queue (Section 4.2 in Chapter 4) if the expected customer arrival rate is λ per unit time and the expected maximum service rate is $c\mu$, per unit time, then the queueing system is stable if $\lambda < c\mu$ (when all c servers are busy).

For an interesting presentation of "flow" through a deterministic system and flow through a stochastic system see Section 1.1, pp. 4–7 in [41] (1975).

1.8 Useful Related Concepts when Applying LC

1.8.1 Renewal Process

Consider a sequence of positive iid random variables, $\{X_n\}_{n=1,2,...}$. Let X_n represent the time between events $n-1$ and n in a stochastic model. Let $N(t)$ denote the number of events up to time t. The "counting process" $\{N(t)\}_{t\geq 0}$ is called the "renewal process" generated by the times $\{X_n\}_{n=1,2,...}$ (see pp. 34–35 in [57] (2003) for more details). For example, let $N(t) :=$ number of arrivals to a queue during $(0, t)$. Then $X_n :=$ interarrival time between customers $n-1$ and n. Then, $\{N(t)\}_{t\geq 0}$ is a Poisson process with rate λ iff X_n, $n = 1, 2, ...$, are iid exponentially distributed with mean $1/\lambda$ (see formula (1.30)).

1.8.2 Regenerative Process

A regenerative process has successive random instants when the process "restarts in the original state" that it occupied at time 0, independent of the past history. The restart instants are called "regenerative points." The times between any two successive regenerative points are iid r.v.s, called "regenerative cycles". The regenerative cycles form a "renewal process". An example of a regenerative process is the "virtual wait" in an M/M/1 queue, starting with an arrival when the system is empty. The regenerative points are the arrival instants when arriving customers find the system empty (i.e., when the customer actual wait is zero). For further discussions on regenerative processes see pp. 40–41 in [57] (2003) and pp. 447–457 in [52] (2010).

1.8.3 Elementary Renewal Theorem

Let $M(t)$ denote the expected number of renewals by time t in a renewal process. Let μ denote the average time between renewals. Then

$$\lim_{t\to\infty}\frac{M(t)}{t}=\frac{1}{\mu} \tag{1.33}$$

(See Proposition 7.1 on p. 430 and Theorem 7.1, pp. 432—433 in [52] (2010) for a proof and discussion of the "elementary renewal theorem".)

1.8.4 Renewal Reward Theorem

We state here the "renewal reward theorem" by using the example of the busy cycles in the virtual wait process $\{W(t)\}_{t\geq 0}$ in the standard M/M/1 queue (see Section 2.1 in Chapter 2). Denote the regenerative cycles of $\{W(t)\}_{t\geq 0}$ by $\{\mathcal{C}_n\}_{n=1,2,\ldots}$. Each $C_n := \mathcal{B}_n+\mathcal{I}_n$, where $\mathcal{B}_n :=$ busy period n and $\mathcal{I}_n :=$ idle period n (see Section 1.8.2 on regenerative processes). Let R_t denote the number of times that an "event" $\mathcal{R}$ occurs during a time interval $(0,t)$, where $t>0$. Let $\mathcal{R}_\mathcal{C}$ denote the number of times that the "same event" $\mathcal{R}$ occurs during each regenerative cycle C, having expected value $E\left[\mathcal{R}_\mathcal{C}\right]$. Assume that $0<E\left[\mathcal{C}\right]<\infty$, i.e., $E\left[\mathcal{C}\right]$ is finite (which holds when the queue is stable).

The renewal reward theorem states that

$$\frac{E\left[\mathcal{R}_\mathcal{C})\right]}{E\left[\mathcal{C}\right]}=\lim_{t\to\infty}\frac{\mathcal{R}_t}{t} \text{ with proability 1;} \tag{1.34}$$

$$\frac{E\left[\mathcal{R}_\mathcal{C})\right]}{E\left[\mathcal{C}\right]}=\lim_{t\to\infty}\frac{E\left[\mathcal{R}_t\right]}{t} \tag{1.35}$$

Formula (1.34) implies that the ratio $\frac{E[\mathcal{R}_\mathcal{C})]}{E[\mathcal{C}]}$ is identical to the long-run rate at which event R_t occurs as $t\to\infty$ for every typical sample path of the virtual wait. The probabilistic behavior of the process $\{W(t)\}_{t\geq 0}$ with respect to event

R during each regenerative cycle, for every typical sample path recapitulates the long-run probabilistic behavior of the process with respect to the event R_t as $t \to \infty$. The renewal reward theorem is discussed in further detail in on p. 439ff in [52] (2010), and on p. 41ff in [57] (2003). The renewal reward theorem is considered to be of "utmost importance in [57] (2003)" for stochastic models. It is very useful in many applications of LC.

1.8.5 Lindley Recursions

We use the example of the virtual wait process in a single server queue, to introduce Lindley recursions. Let W_n denote the actual waiting time of arrival customer n before service. Let S_n denote the service time of customer n. Let , $T_{n+1} := \tau_{n+1} - \tau_n$ where τ_n is the arrival instant of customer n, $(n = 0, 1, ...)$. The Lindley recursion for the actual waiting times W_n and W_{n+1} is given by:

$$W_{n+1} = \max\{W_n + S_n - T_{n+1}, 0\}, \; n = 0, 1, 2, ... \tag{1.36}$$

(see [43] (1952)).

Formula (1.36) is important theoretically and computationally, for analyzing r.v.s in queues, inventories, dams, pharmacokinetic models, actuarial risk models, and related stochastic models. The author discovered LC while using Lindley recursions to derive integral equations for the pdfs of r.v.s, during work on his PhD thesis, in 1974 (see [10] (1975)). The benefit of LC is that it bypasses all or most of the required algebra when using Lindley recursions directly to derive the integral equations. LC simplifies the derivation by using a sample-path diagram of the r.v. over time, and a few simple basic level crossing theorems in Section 1.4, which permits writing down the integral equations immediately by inspection.

1.8.6 Dimensional Analysis

Dimensional analysis specifies the dimensions (physical units) of the quantities in a physical measurement, formula, or mathematical equation. Let "L" denote "length", "M" "mass", and "T" "time". Speed and velocity have dimension $\left[\frac{L}{T}\right]$. Conversely a quantity with dimension $\left[\frac{L}{T}\right]$ is the speed or velocity of a point or object. Acceleration is the rate of change (derivative) of speed and has dimension $\left[\frac{L}{T^2}\right]$. Force is equal to Mass $\times$ Acceleration and has dimension $[M]\left[\frac{L}{T^2}\right]=\left[\frac{ML}{T^2}\right]$. Energy is Length $\times$ Force, and has dimension $[L]\,[M]\left[\frac{L}{T^2}\right] = \left[\frac{ML^2}{T^2}\right]$. A number without units is "dimensionless". The probability of an event is dimensionless (a number in $[0, 1]$). Rate has dimension $\left[\frac{1}{T}\right]$; for example the number of entrances into a mall in time period $(0, t)$, since "number of entrances" is number. The dimensions of all terms in an equation must have the same dimension. Knowledge of the dimension can lead to new insights about an equation (e.g., new interpretations).

In LC, dimensional analysis provides a necessary (but not sufficient) condition to test the correctness of equations, especially integral equations (see Exercise 1.9.7). The pdf $f(x)$ of the virtual wait in an M/M/1 queue, has dimension $\left[\frac{1}{T}\right]$ (see Section 2.1.9 in Chapter 2, and Section 1.4, p. 7 in [22](2017)). The importance is that in any equation, all terms must necessarily have the same dimension (i.e., same units), otherwise it has an error. Two useful references for dimensional analysis are: [42](1966), and pp. 213 - 229 in [45](1947).

1.8.7 Review of Statistics

Some basic statistical properties are used at various places in this book (e.g., mean, higher moments, variance, etc.). See pp. 4–37 in Chapter 1 in [53] (2013) for a useful review of Basic Statistics.

1.9 Exercises

Exercise 1.9.1 *In a sample path, the leading point dynamically "SP" can make "vertical" jumps, which take place "instantaneously" (not in Time). Thus the SP can move either "not in Time"; or "in Time" in the state space, depending on the slope at which the SP increases or decreases. Discuss whether this property has relevance only in mathematical models, or also in real-world applications?*

Exercise 1.9.2 *For a birth and death process, use formulas (1.31) and (1.32) to compute P_0. Is there a necessary condition with respect to dimensional analysis such that formula (1.31) must satisfy?*

Exercise 1.9.3 *For a birth and death process, use formula (1.31) and Section 1.8.6 to confirm that the P_ns are "dimensionless" (plain real numbers).*

Exercise 1.9.4 *For an exponential distribution with rate μ, (Exp_{μ},) derive: the meanm and variance. (For a generic r.v. X, the mean (expected value) is denoted by $E[X]$, and the variance by $Var[X] = E\left[(X - E[X])^2\right] =$ "$E\left[X^2\right] - (E[X])^2$". (Note: This requires finding the second moment $E\left[X^2\right]$ $:= \int_{y=0}^{\infty} y^2 f(y)dy = \int_{y=0}^{\infty} y^2 \mu e^{-\mu y} dy = \frac{2}{\mu^2}$, derived using integration by parts. In general, the nth moment of Exp_{μ} is $E[X^n] = \frac{n!}{\mu^n}$.)*

Exercise 1.9.5 *For an exponential distribution with rate μ, (Exp_{μ},) derive the median x_{mdn}, such that $\int_{x_{mdn}}^{\infty} \mu e^{-\mu y} dy = \frac{1}{2}$.*

Exercise 1.9.6 *From formula 1.28 in Section 1.5.4, $E[N(t)] = \lambda t$. Show that the variance of $N(t)$, denoted by $Var(N(t))$, is also given by $Var(N(t)) = \lambda t$. Note that $Var(X) \equiv E[X^2] - (E[X])^2$ for a generic r.v. X. This result is a well-known property of the Poison probability distribution.*

Exercise 1.9.7 *Consider the very celebrated equation "$\boldsymbol{E} = \boldsymbol{M}\boldsymbol{C}^2$" in Physics, where $\boldsymbol{E}$ denotes kinetic energy of a mass; $\boldsymbol{M}$ denotes mass and $\boldsymbol{C}$ denotes the speed of light. Show that the following "necessary condition" holds for the equation "$\boldsymbol{E} = \boldsymbol{M}\boldsymbol{C}^2$": "Both sides must have the same dimension." (Note: This necessary condition "is not" a derivation or proof of the equation. The original derivation using the special theory of relativity is given famously in [31] (1905).). For explanation of the constant speed of light $\boldsymbol{C}$ in the equation, see [28] (2021).*

Exercise 1.9.8 *In Bayes' rule, formula (1.22), consider the connections in the field of Medicine. Let B := "symptom" of an illness", and $\{A_i\}_{i=1}^{N}$:= set of possible diseases or conditions that can cause symptom B. Give an "analogous scenario" in the fields Management Science, Engineering, or Astronomy, etc.*

2

Standard M/M/1 Queue and Variants

This chapter discusses the M/M/1 queue where the first "M" means "Poisson arrivals", and the second "M" means "service times are $\stackrel{d}{=}$ Exp$_\mu$". Section 2.1 treats a standard M/M/1 queue, where there are no state dependencies The other sections in this chapter treat other useful variants, in order to develop techniques for modelling stochastic models using LC: e.g., M/M/1 with service time depending on waiting time; M/M/1 where zero-wait arrivals get exceptional service; M/M/1 with bounded workload.

2.1 Standard M/M/1 Queue

The M$_\lambda$/M$_\mu$/1 queue is a basic stochastic model. Figure 2.1 depicts customers arriving in a Poisson process at rate λ and waiting to be served or being served at rate μ in a single server queue (see Section 1.5.4). Arriving customers are served in order of arrival (FCFS – "first come first served"), and receive iid service times $\stackrel{d}{=} S \stackrel{d}{=}$ Exp$_\mu$. For r.v. S the complementary cdf is $\overline{B}(x) = e^{-\mu x}, x \geq 0$; cdf is $B(x) = 1 - e^{-\mu x}$, $x \geq 0$. We assume is stability, i.e., the "traffic intensity" is $\rho := \lambda \times E[S] = \lambda/\mu < 1$, so that the queue is stable, i.e., in equilibrium (see Sections 1.5 and 1.7). Stability implies that important stationary probability distributions exist as $t \to \infty$, such as the mixed stationary pdf $\{P_0, f(x)\}_{x>0}$ for the virtual wait; and "probability mass function" (pmf) P_n, $n = 0, 1, 2, ..$, where n := number of customers in the system.

Let $N(t)$:= number of customers in the system at time $t > 0$. It is shown using standard methods that $\lim_{t\to\infty} P(N(t) = n)$ is equal to:

$$P_n = \rho^n (1 - \rho), n = 0, 1, .., \tag{2.1}$$

and

$$P_0 = 1 - \rho \tag{2.2}$$

(see pp. 502–504 in [52] (2010) for details on how to derive (2.1) and (2.2).

DOI: 10.1201/9780429297601-2

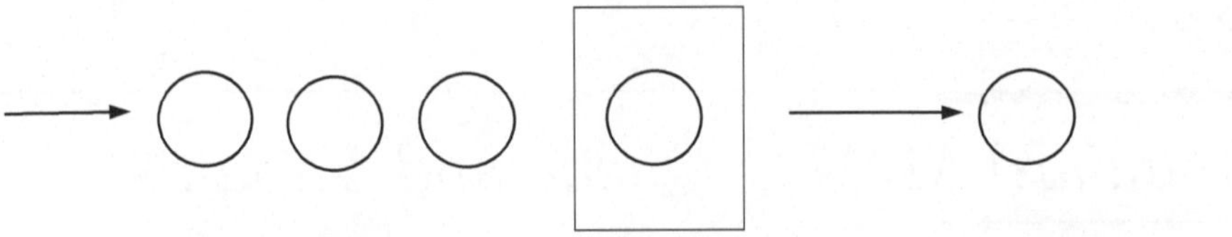

FIGURE 2.1
Sketch of $\text{M}_\lambda/\text{M}_\mu/1$ queue with Poisson arrival rate λ and exponential service rate μ.

2.1.1 Criterion for Existence of Stationary Pdf of Wait

The criterion for existence of $\{P_0, f(x)\}_{x>0}$ is "stability" of the queue. That is, the waiting time returns to 0 with probability 1. Equivalently, $0 < P_0 < 1$. By (2.2), $0 < 1 - \rho < 1 \iff \rho < 1$ as stated in Section 2.1. Sample paths of the virtual wait process return to level 0, with probability 1 ("for all sample paths except for a finite number of possible sample paths).

A useful, intuitive way to state the criterion for stability, which applies generally to M/G/1 queues, where "G" is a general positive r.v is "the expected number of arrivals to the system during a service time is less than 1". This means that "the service rate is faster than the arrival rate".

2.1.2 PASTA Principle–"Poisson Arrivals See Time Averages"

Consider Figure 2.2 in Section 2.1.5. At the arrival instant τ_n of customer n, the corresponding upward jump starts at the level W_n, which is the actual waiting time of arrival n. The sequence of actual waits $\{W_n\}_{n=1,2,...}$ occurs in an arrival stream in the form of a Poison process, and is called "Poisson arrivals". It has been shown by various authors that in queueing systems where there are Poisson arrivals, the limiting probability distribution of $\{W_n\}_{n=1,2,...}$ as $n \to \infty$ is equal to the limiting probability distribution of the virtual wait process $\{W(t)\}_{t\geq 0}$ as $t \to \infty$. This "PASTA principle" is useful since knowledge of either one of the two limiting probability distributions yields immediately the probability distribution of the other (See [36] (2008); [48](2011); [52](2010) for more details on the PASTA,principle and [59](1982) for an article specifically on the topic.)

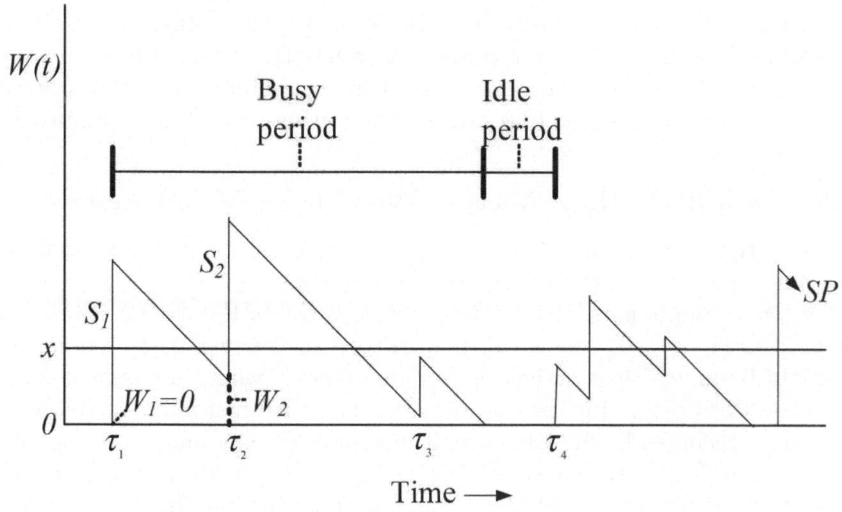

FIGURE 2.2
Sample path of virtual wait in M/M/1 queue. Let S_n := service time,W_n := arrival-point wait at τ_n^-, where τ_n^- := instant just before arrival time of customer $n, n = 1, 2$. The virtual wait at time $\tau_n^+ := W_n + S_n$. Also shows busy and idle periods, and SP := system point.

2.1.3 Rate Balance across State-Space Levels

The LC method focuses on sample-path rate balance across an arbitrary state-space level, say level x, because of the basic level crossing theorems (in Sections 1.4 and 1.4.3) state that sample-path downcrossing rates correspond to very simple mathematical expressions of the continuous part of the pdf $f(x)$ for every $x > 0$ in the state space. The LC method depends on the fact that, the sample-path upcrossing rates correspond to simple mathematical integral transforms of terms within an integral expression for $f(x)$ (see Sections 1.4–1.7 in Chapter 1 of [22](2017)). In the derived integral equations, the probability P_0 appears as a factor of a linear coefficient of $\overline{B}(x)$. We can then use the law of total probability (normalizing condition), e.g., in Equation (1.18) in Section 1.5 to solve for the mixed pdf $\{P_0, f(x)\}_{x>0}$ explicitly.

2.1.4 Virtual Wait in M/M/1 Queue

In a single server queue like $M_\lambda/M_\mu/1$, the virtual wait (virtual waiting time) is a continuos-time, continuous-state stochastic process denoted by $\{W(t)\}_{t\geq\infty}$. The virtual wait measures how long a time an arrival to the system is required to wait until it begins service. In a single server queue the virtual wait is the same as the workload (the sum of all service times of the waiting customers just before the arrival plus the remaining service time of the customer in service). If at least one customer is present in the system the virtual wait and workload are both positive in value; this occurs during busy periods (Figure 2.2). If no customer is present in the system the virtual wait and the workload are both equal to 0; this occurs during idle periods (Figure 2.2).

2.1.5 Sample Path of Virtual Wait in M/M/1 Queue

In this section, we examine properties of sample paths of the virtual wait process and its relation to the actual wait process.

A typical sample path of $\{W(t)\}_{t\geq\infty}$ is depicted in Figure 2.2. The virtual wait process $\{W(t)\}_{t\geq\infty}$ has a slope equal to -1 whenever $W(t) > 0$, because if a would-be arrival had arrived a seconds later, it would necessarily wait a second less until beginning service, provided no other customer arrived within that extra a seconds. Each busy cycle is made up of: (1) a busy period plus (2) an idle period. The state space of $\{W(t)\}_{t\geq\infty}$ is the set $[0, \infty)$ (vertical axis). The stationery mixed pdf of the virtual wait is denoted by $\{P_0, f(x)\}_{x>0}$, where $P_0 = \lim_{t\to\infty} P(W(t) = 0)$ and $f(x)$ is the probability density function on $(0, \infty)$ (we often call P_0 an "atom" and $f(x)$ a "pdf", to simplify context in a discourse). In the standard M/M/1 queue, the limiting pdf is the same as the stationary pdf, denoted by $\{P_0, f(x)\}_{x>0}$.

In the stochastic models treated in this book the stationary pdf and limiting pdf of the key state variable, are the same. In general, if the limiting pdf exists then it is also a stationary pdf. In some stochastic models (not treated in this book) the stationary pdf of a key random variable exists, but its limiting pdf does not exist (see p. 205 in [48] (2011) for a discussion on limiting distributions and stationary distributions).

2.1.6 Focus on Actual Wait in M/M/1 Queue

The actual wait of an arrival is equal to the left limit of the virtual wait process at an arrival instant. Examples of the actual wait are given by $W_1 = 0$ at τ_1^- and W_2 at τ_2^- in Figure 2.2. By the PASTA principle for queues (Section 2.1.2 in this chapter) the stationary pdf of the actual wait is the same as the stationary pdf of the virtual wait.

2.1.6.1 The Slope of the Sample Path of $\{W(t)\}_{t\geq 0}$ Is -1 when $W(t) > 0$

First we quickly explain intuitively why the slope is -1 when $W(t) > 0$. If an arrival who must wait a positive time before service were to arrive instead one second later, then it would have to wait one second less, provided $W(t) > 0$ and no new customer arrives during that one second. causing immediately an upward jump. So during that second, the virtual wait would drop by one second.

Second, we examine further a sample-path diagram in the standard M/M/1 queue. The reason that the slope is equal to -1 when $W(t) > 0$, is as follows.

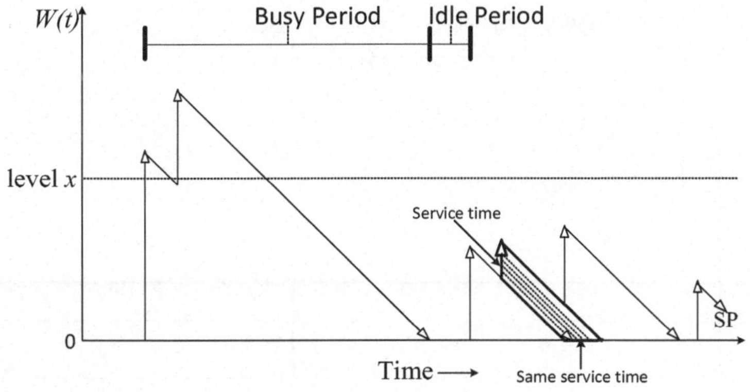

FIGURE 2.3
Features a vertical jump representing an arrival's service time, and the same service time as a horizontal on the time axis, when the arrival is actually in service. This explains why the slope of $W(t)$ equals -1, when $W(t) > 0$. Each arrival that waits a positive time nitiates a similar trapezoidal structure.

The vertical upward jump size at a new customer arrival instant, say t_0, is equal to the arrival's service time, say S. The moment when the arrival begins service is the instant $t_0 + W(t_0)$. The duration of the service when the arrival's service actually takes place is indicated by a horizontal line of equal length S at level 0 (see Figure 2.3). The vertical length S and level-0 horizontal S are opposite sides of a trapezoid, which happens for every arrival to the system who waits a positive time.

2.1.6.2 Customer-initiated Trapezoids and Triangles in Sample Path

Each arrival that waits a positive time initiates such a trapezoid (Figure 2.3). Thus, the slope of the virtual wait diagram is equal to -1 when $W(\cdot) > 0$. The total length of the busy period is equal to the sum of all service times of customers that are served in a busy period. Each arrival that waits zero time for service initiates, similarly, a right-angled isosceles triangle with one vertical side equal to S and a level-0 side equal to S.

2.1.7 Basic LC Limit Theorems and Integral Equation for Stationary Pdf of Wait in M/M/1 Queue

This section applies basic LC theorems stated in Section 1.4 in Chapter 1, to derive an integral equation for the analytic stationary pdf of a random variable, using properties of a physical, typical sample path of the r.v (for further details, see Chapter 1 in [22] (2017)).

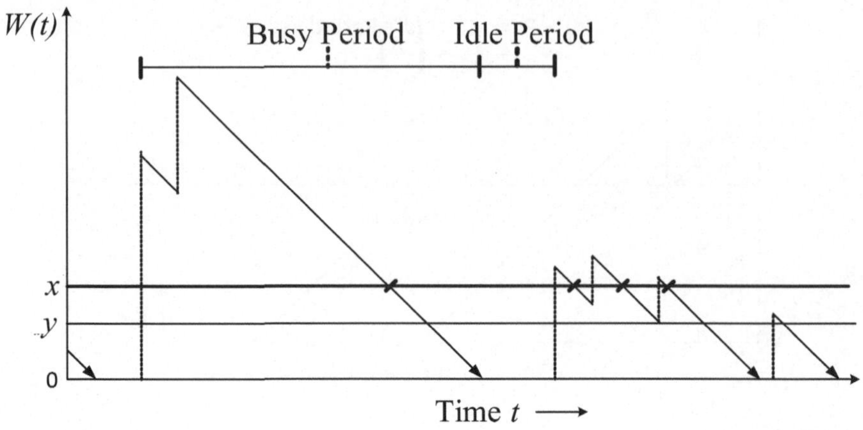

FIGURE 2.4
Sample path of virtual wait $\{W(t)\}_{t\geq 0}$ showing state-space levels (at levels x and y). Also shows SP hits of level 0.

2.1.7.1 Integral Equation for Pdf of the Virtual Wait

By the basic LC theorem stated in formula (1.1) in Section 1.4, the down-crossing rate of level x is

$$f(x) = \lim_{t\to\infty} \frac{\mathcal{D}_t(x)}{t}.$$

We now show intuitively, that the analytic upcrossing rate of level x is given by

$$\lim_{t\to\infty} \frac{\mathcal{U}_t(x)}{t} = \lambda P_0 \overline{B}(x) + \lambda \int_{y=0}^{x} \overline{B}(x-y) f(y)dy, 0 < x < \infty, \tag{2.3}$$

where $\overline{B}(x) = e^{-\mu x}$ is the ccdf of the service time (see formula (1.14) in Section 1.5). By rate balance across level x,

$$\lim_{t\to\infty} \frac{\mathcal{D}_t(x)}{t} = \lim_{t\to\infty} \frac{\mathcal{U}_t(x)}{t}.$$

Therefore, we obtain the analytic basic integral equation for the stationary mixed pdf $\{P_0, f(x)\}_{x>0}$ namely

$$f(x) = \lambda P_0 e^{-\mu x} + \lambda \int_{y=0}^{x} e^{-\mu(x-y)} f(y)dy, 0 < x < \infty, \tag{2.4}$$

where $f(x)$ is the absolutely continuous part of the virtual-wait pdf.

2.1.7.2 Further LC Explanation of Equation (2.4)

In equation (2.4), the left side $f(x)$ is the sample-path downcrossing rate of "continuos" downcrossings of level x with slope -1, by the basic LC theorem given in Section 1.4. On the right side, the term $\lambda P_0 e^{-\mu x}$ is the rate of SP vertical upward jumps starting from level 0, that upcross level x (:= rate of arrivals who wait zero time and have service rates greater than x). The second term $\lambda \int_{y=0}^{x} e^{-\mu(x-y)} f(y)dy$ is the rate of upward jumps staring in a small subinterval $(y, y+dy) \subseteq (0, x)$, that upcross level x, because their service times are greater than $x - y$, summed over all $y \in (0, x)$ as $dy \downarrow 0$. (For definitions of Riemann integral and Riemann-Stieltjes integral see a text book on calculus or on real analysis, e.g., p. 214 in Section 29, in [6])(1976). By rate balance (1.10) across level x, the left and right sides are equal for all $x > 0$ (incidentally, no time-consuming algebra is required).

Figure 2.5 in this Section gives LC explanations of the sample-path crossing rates and the sources of each of the terms of the integral equation using the notation $\overline{B}(x)$ in place of $e^{-\mu x}$ which is used in equation (2.4). This basic integral equation for the mixed pdf of wait holds where the service times are generally distributed.

The solution for P_0 is obtained by applying the law of total probability (aka normalizing condition) which states that the sum of all probabilities for a single random variable is equal to 1. Thus,

$$P_0 + \int_{x=0}^{\infty} f(x)dx = 1, \tag{2.5}$$

or

$$P_0 = 1 - \int_{x=0}^{\infty} f(x)dx. \tag{2.6}$$

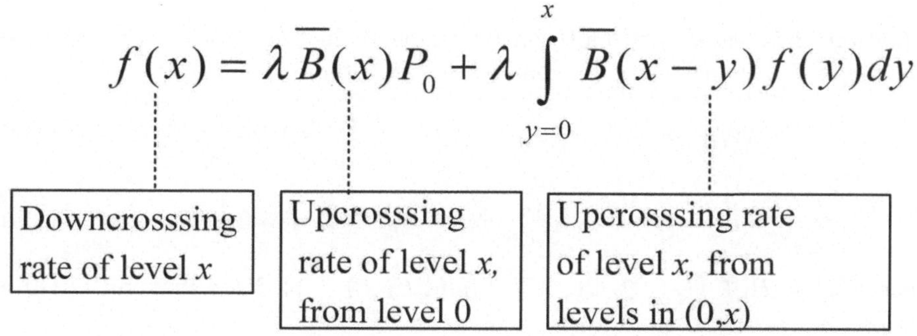

FIGURE 2.5
Integral equation for the stationary pdf of the virtual wait in standard $M_\lambda/M_\mu/1$ queue, where G:= Exp_μ. Shows the sample-path sources of SP downcrossings and SP upcrossings of level x.

This book has many integral equations like that in Figure 2.5. A central theme of LC is to interpret every term in every equation, as an SP (system point) motion relative to state-space levels or boundaries of state-space sets; and vice versa. This connection of SP motions and analytic terms in integral equations, allows an analyst to save time by bypassing many algebraic steps. The thought process also frees the analyst to think of new possibilities about the stochastic model.

2.1.8 Alternative Forms of Integral Equation for PDF of Wait in M/M/1

An example using sample paths and LC concepts to obtain results that otherwise may require considerable algebraic manipulations only, is when developing useful alternative forms of equation (2.4). We derive two different forms of the integral Equation (2.4) for the limiting pdf of wait in the M/M/1 queue (see Figure 2.5 for LC interpretations in the standard form). The alternative forms are useful due to their LC interpretations with respect to the sample-path "dynamic motions" of the SP in the state space. We first rewrite the right-hand side of (2.4) by collecting positive terms and negative, terms separately as follows:

$$f(x) = \lambda(1-(1-e^{-\mu x}))P_0 + \lambda\int_{y=0}^{x}(1-(1-e^{-\mu(x-y)}))f(y)dy$$

$$= \lambda\left(P_0 + \int_{y=0}^{x} f(y)dy\right) - \lambda\left((1-e^{-\mu x})P_0 + \int_{y=0}^{x}(1-e^{-\mu(x-y)})f(y)dy\right). \tag{2.7}$$

Equation (2.7) leads to two alternative forms of the LC equation (2.4), where the integrals are written as Riemann-Stieltjes integrals:

$$f(x) = \lambda F(x) - \lambda\textstyle\int_{y=0}^{x}(1-e^{-\mu(x-y)})dF(y), x \geq 0; \tag{2.8}$$

$$f(x) = \lambda F(x) - \lambda\textstyle\int_{y=0}^{x}F(x-y)dF(x), x \geq 0, \tag{2.9}$$

since $F(x) = P_0 + \int_{y=0}^{x} f(y)dy$, $x \geq 0$ (and $F(\infty) = 1$). Note that in (2.8) and (2.9),

$$dF(y) = f(ydy, y > 0,$$

$$dF(0) = P_0.\cdot$$

Formula (2.9) is known in the queueing literature as the "Takács integral equation" ([56]) (1962). The Takács forms of the equations in formulas (2.8) and (2.9) above, have intuitive LC interpretations, which help to derive them immediately, as follows.

Consider a sample path of the virtual wait (Figure 2.2). Observe a one-to-one correspondence between the set of algebraic terms in equations (2.8) and (2.9), and a set of mutually exclusive and exhaustive sample-path crossings of level x, which are different from those depicted in Figure 2.5.

In equations (2.8) or (2.9) the left side "$f(x)$" is the SP downcrossing rate of level x with slope -1, as usual. On the right side of these equations, the first term "$\lambda F(x)$" is the rate of all vertical SP upward jumps which start below level x, i.e., they start in the state-space interval $[0, x]$. Some of these jumps upcross level x strictly and end above level x; the remaining upward jumps which start in $[0, x]$ end below (do not upcross) level x. The second term on the right side of equations (2.8) and (2.9) subtracts off the rate of upward jumps which end below level x. Therefore the right side is equal precisely to the total rate at which the SP upward jumps strictly upcross level x. Rate balance across level x gives directly equation (2.8). Equation (2.8) yields (2.9) by: (1) applying the transformation $z = x - y$, $dz = -dy$ in the integral term; (2) integrating by parts; replacing variable z by variable y.

Equations (2.8) and (2.9) are useful when analyzing variants of the standard M/D/1,queue where all service times are equal to a common constant $D > 0$, in Section 3.4 in Chapter 3. Equations (2.8) and (2.9) help to derive the stationary cdf $F(x)$ directly. Equations (2.8) and (2.9) also demonstrate the flexibility of identifying the sources of the physical sample-path motions corresponding to the analytic algebraic terms in the integral equations for the pdf of wait. The LC intuitive interpretations of these two equations also suggest how to use LC to develop integral equations for the pdf of r.v.s in other related stochastic models.

2.1.9 Dimensional Analysis for M/M/1

The left side of equation (2.4), i.e., $f(x)$, is the derivative of the cdf $F(x)$, which is a dimensionless quantity because it is a probability (a number in $[0, 1]$) The dimension of the variable x is $[Time]$, since x is a waiting time. The derivative $dF(x)/dx$ is equal to $\lim_{\Delta x \downarrow 0} \frac{F(x+\Delta x)-F(x)}{\Delta x}$, whose numerator is dimensionless (difference of two numbers) and whose denominator has dimension $[Time]$ (Δ is a change in x). Hence the dimension $[f(x)] = [Time]^{-1}$, which is a rate (see Section 1.8.6). It is,a useful observation for LC applications.

The right hand side of (2.4) has the same dimension as the arrival rate of customers to the system λ, whose dimension is $[Time]^{-1}$, because the two coefficients of λ in the equation are readily seen to be dimensionless quantities. Therefore in equation (2.4) both sides have the dimension $[Time]^{-1}$. This matching of dimensions satisfies a necessary condition for equations to be correct, as mentioned in Section 1.8.6 in Chapter 1. Being able to match the dimensions of all terms in an equation is useful, especially when it has many terms. Matching dimensions of terms means only that an equation is not patently false. On the other hand, non-matching dimensions of any terms serves as an alert to the analyst that there is an error in one or more terms of the equation. This situation means that the equation should be checked to find the error(s). An equation satisfying a necessary condition may point to its correctness; but it does not mean that the equation is totally correct. The equation must still be checked for correctness of concepts and algebraic steps, and of the interpretation of sample path motion, as well as for matching dimensions. Any error in the derived model equations, can lead to errors in subsequent inferences about the system.

2.1.10 Solution of Integral Equation for the Stationary PDF and CDF of Wait

The solution of equation (2.4) in Section 2.1.7.1, for $\{P_0, f(x)\}_{x>0}$, uses basic calculus and elementary differential equations. Taking the derivative d/dx on both sides of (2.4) and accounting for differentiation under the integral sign, gives

$$f'(x) = -\mu(\lambda P_0 e^{-\mu x} + \lambda \int_{y=0}^{x} e^{-\mu(x-y)} f(y)dy) + \lambda f(x), \tag{2.10}$$

where $f'(x) := df(x)/dx$. Equation (2.10) is equivalent to the differential equation

$$f'(x) = -\mu f(x) + \lambda f(x). \tag{2.11}$$

(Differentiation under the integral sign, yields the term "$+\lambda f(x)$" on the right-hand side of (2.10).) Differential Equation (2.11) follows from (2.10) by applying the fact in (2.10) that:

$$f(x) = \lambda P_0 e^{-\mu x} + \lambda \int_{y=0}^{x} e^{-\mu(x-y)} f(y)dy.$$

Next we solve the ordinary differential equation in (2.11) now written as

$$f'(x) + (\mu - \lambda)f(x) = 0, x > 0. \tag{2.12}$$

Multiplying by the integrating factor $e^{(\mu-\lambda)x}$ on both sides of (2.12) and then integrating both sides with respect to x yields

$$\frac{d}{dx} f(x)e^{(\mu-\lambda)x} = 0,$$

$$f(x)e^{(\mu-\lambda)x} = A,$$

$$f(x) = Ae^{-(\mu-\lambda)x}, x > 0, \tag{2.13}$$

where A is a constant to be determined. To compute A, first let $x \downarrow 0$, whence $f(0^+) = A$. Second, let $x \downarrow 0$ in Equation (2.13), which gives $f(0) = A.$, and in integral Equation (2.4), which gives $f(0^+) = \lambda P_0$. Thus $A = \lambda P_0$. An LC Interpretation is: "the rate of SP downcrossings into level 0 with slope -1 = SP rate of SP vertical upward jumps out of level 0".

Since $A = \lambda P_0$ we obtain

$$f(x) = \lambda P_0 e^{-(\mu-\lambda)x}, x > 0. \tag{2.14}$$

We then compute P_0 by substituting (2.14) into the normalizing condition (2.5), which yields

$$P_0 + P_0 \int_{x=0}^{\infty} \lambda e^{-(\mu-\lambda)x} dx = 1,$$

$$P_0 + P_0 \frac{\lambda}{\mu - \lambda} = 1,$$

$$P_0 = \frac{\mu - \lambda}{\mu} = 1 - \frac{\lambda}{\mu} = 1 - \rho, \tag{2.15}$$

where $\rho = \lambda/\mu < 1$ is the traffic intensity (defined in line 4 of Section 2.1).

The cdf of wait is

$$F(x) = P_0 + \textstyle\int_{y=0}^{x} \lambda(1-\rho)e^{-(\mu-\lambda)y} dy = 1 - \rho e^{-(\mu-\lambda)x}, \; x \geq 0, \tag{2.16}$$

from equations (2.14) and (2.15).

2.1.11 Expected Waiting Time before Service in M/M/1 Queue

Let W_Q denote the wait before service. From Equation (2.4), we can find $E[W_Q]$ directly, without requiring first to derive the soliton for $\{P_0, f(x)\}_{x>0}$, as follows. Multiply both sides of (2.4) by x, and integrate with respect to $x \in (0, \infty)$, thus obtaining:

$$E[W_Q] = \int_{x=0}^{\infty} x f(x) dx$$

$$= \lambda P_0 \int_{x=0}^{\infty} x e^{-\mu x} dx + \lambda \int_{x=0}^{\infty} x \int_{y=0}^{x} e^{-\mu(x-y)} f(y) dy dx,$$

Integrating the right-most side gives

$$E[W_Q] = \frac{\lambda}{\mu^2}P_0 + \lambda\int_{y=0}^{\infty}\int_{x=y}^{\infty} xe^{-\mu(x-y)}f(y)dxdy$$

$$= \frac{\lambda}{\mu^2}P_0 + \lambda\int_{y=0}^{\infty}\int_{x=y}^{\infty} (x-y+y)\,e^{-\mu(x-y)}f(y)dxdy$$

$$= \frac{\lambda}{\mu^2}P_0 + \lambda\int_{y=0}^{\infty}\left[\int_{x=y}^{\infty}(x-y)\,e^{-\mu(x-y)}dx + y\int_{x=y}^{\infty}e^{-\mu(x-y)}dx\right]f(y)dy$$

$$= \frac{\lambda}{\mu^2}P_0 + \lambda\int_{y=0}^{\infty}\left[\frac{1}{\mu}(1+y)\right]f(y)dy$$

$$= \frac{\lambda}{\mu^2}\left[P_0 + \int_{y=0}^{\infty}(1+y)\,f(y)dy\right]$$

$$= \frac{\lambda}{\mu^2}\left[P_0 + 1 - P_0 + E[W_Q]\right]$$

$$= \frac{\lambda}{\mu^2}\left[1 + E[W_Q]\right]$$

$$E[W_Q] = \frac{\frac{\lambda}{\mu^2}}{\left(1-\frac{\lambda}{\mu}\right)} = \frac{\lambda}{\mu(\mu-\lambda)}. \tag{2.17}$$

Formula (2.17) for $E[W_Q]$ is a special case of the "Pollaczek-Khinchine (P-K) formula" (see formula 3.10 in Section 3.2.3).

2.1.12 Focus on PDF and CDF of System Time

Let σ denote the total time, i.e., actual wait + service, that a customer spends in the system (aka "system time"). Let $f_\sigma(x)$ be its pdf, $F_\sigma(x)$ the cdf, $x > 0$, with ccdf $\overline{F}_\sigma(x)$. Since $\sigma = W_q + S$, (W_q:= actual wait before service), we obtain:

$$\begin{aligned}
&\overline{F}_\sigma(x) = P(\sigma > x) = P(S > x|W_q = 0)\\
&= P_0e^{-\mu x} + \textstyle\int_{y=0}^{x} P(S > x-y|W_q = y)f(y)dy + P(W_q > x)\\
&= P_0e^{-\mu x} + \lambda P_0\textstyle\int_{y=0}^{x} e^{-(\mu-\lambda)y}e^{-\mu(x-y)}dy + \int_{y=x}^{\infty}\lambda P_0e^{-(\mu-\lambda)y}dy\\
&= \tfrac{P_0}{1-\frac{\lambda}{\mu}}e^{-(\mu-\lambda)x} = e^{-(\mu-\lambda)x},\ x > 0.
\end{aligned} \tag{2.18}$$

Thus, the "system time" in $M_\lambda/M_\mu/1$ is $\sigma \stackrel{d}{=} \text{Exp}_{\mu-\lambda}$, i.e.,

$$f_\sigma(x) = (\mu - \lambda)\, e^{-(\mu-\lambda)x}, x > 0,$$

$$F_\sigma(x) = 1 - e^{-(\mu-\lambda)x}, x \geq 0. \tag{2.19}$$

$$\overline{F}_\sigma(x) = e^{-(\mu-\lambda)x}, x \geq 0.$$

(See pp. 72–79 in [22](2017) for further discussions of system time in M/M/1.)

2.1.13 Connecting the Pdf of System Time and Pdf of Waiting Time

We connect the stationary pdf of system time $f_\sigma(x)$ directly with the stationary pdf of waiting time $f(x)$. Consider a sample path of the virtual wait and fix level $x > 0$ (Figure 2.5 in Section 2.1.7.1). We view the SP jumps at arrival instants of customers from the top ends of the jumps, which represent the system times of the corresponding arrivals to the system. Viewing the sample path starting from level "y" where $y > x$, the long-run downcrossing rate of level x is given by

$$\lambda \int_{y=x}^{\infty} e^{-\lambda(y-x)} f_\sigma(y) dy, \tag{2.20}$$

since $e^{-\lambda(y-x)}$ is the probability that the time until the next customer arrival exceeds $y-x$; and $\lambda f_\sigma(y)dy$ is the rate at which jumps due to customer arrivals end within a "dy" neighborhood of level $y > x$. The term $e^{-\lambda(y-x)}$ in (2.20) ensures that no new SP vertical upcrossings of level x occur until at least time interval $y - x$ later. Since the slope of the sample path is -1, the horizontal time interval with length $y - x$ has no new arrivals and associated SP vertical jumps, or SP downcrossings of level x.

By the basic LC theorem (1.1) in Section 1.4, another mathematical expression for the same long-run SP downcrossing rate of x is precisely "$f(x)$"; equation (2.20) then gives the formula connecting $f(x)$ and $f_\sigma(x)$:

$$f(x) = \lambda \int_{y=x}^{\infty} e^{-\lambda(y-x)} f_\sigma(y) dy, \tag{2.21}$$

which can be simplified. Multiplying both sides of Equation (2.21) by $e^{-\lambda x}$, taking d/dx on both sides, and simplifying, gives

$$f_\sigma(x) = f(x) - \frac{f'(x)}{\lambda}, x > 0, \tag{2.22}$$

wherever $f'(x)$ exists.

Equation (2.22) is consistent with formula (2.14) in Section 2.1.10. In $M_\lambda/M_\mu/1$, $f(x) = \lambda P_0 e^{-(\mu-\lambda)x}, x > 0$; substituting into (2.22) gives

$$f_\sigma(x) = (\mu - \lambda)\, e^{-(\mu-\lambda)x}, x > 0,$$
$$F_\sigma(x) = \textstyle\int_{y=0}^{x} f_\sigma(y)dy = 1 - e^{-(\mu-\lambda)x}, x \geq 0, \tag{2.23}$$

which is the same result as in (2.19). Thus, in $M_\chi/M_\mu/1$, the stationary pdf of the system time is $\stackrel{d}{=} \text{Exp}_{\mu-\lambda}$, and

$$E[\sigma] = \frac{1}{\mu - \lambda}.$$

2.1.14 Stationary Probability Distribution of Number in the System

Before deriving the formula for the stationary probability of the number of customers in the system in M/M/1 using LC, we discuss the "Erlang probability distribution", which enters into the computations.

2.1.14.1 Erlang Probability Distribution and Gamma Function

First consider the Erlang(-k, λ) distribution, denoted by $\text{Erl}_{k,\lambda} :=$ probability distribution of the sum of k iid Exp_λ random variables. The pdf of $\text{Erl}_{k,\lambda}$ is given by

$$b(x) = e^{-\lambda x}\frac{(\lambda x)^{k-1}\lambda}{(k-1)!}, x > 0. \tag{2.24}$$

To show that $b(x)$ in (2.24) is a pdf, it must integrate to "1" on $(0, \infty)$:

$$\int_{x=0}^{\infty} b(x)dx = \frac{1}{(k-1)!}\int_{x=0}^{\infty} e^{-\lambda x}(\lambda x)^{k-1}\lambda dx$$

$$= \frac{1}{(k-1)!}\int_{y=0}^{\infty} e^{-y}y^{k-1}dy = \frac{\Gamma(k)}{(k-1)!} = 1, \tag{2.25}$$

where we made the transformation "$y = \lambda x$", and "$\Gamma(k) := (k-1)!$" is the "Gamma function" for integer values of $k \geq 1$, which is given by

$$\Gamma(k) = \int_{y=0}^{\infty} e^{-y}y^{k-1}dy = (k-1)!,\ k = 1, 2, ... \tag{2.26}$$

Formula (2.26) is provable by mathematical induction. For example,

$$\Gamma(1) = \textstyle\int_{y=0}^{\infty} e^{-y}dy = 1 \equiv 0!,$$
$$\Gamma(2) = \textstyle\int_{y=0}^{\infty} e^{-y}ydy = 1 = 1!, \tag{2.27}$$
$$\Gamma(3) = \textstyle\int_{y=0}^{\infty} e^{-y}y^2dy = 2!, \text{ etc.}$$

Mathematical Induction Proof of Formula (2.26)

Assume that formula (2.26) is true for $k =$ an arbitrary $n \geq 1$. Then

$$\Gamma(n) = \int_{y=0}^{\infty} e^{-y} y^{n-1} dy = (n-1)!.$$

The goal of Mathematical Induction is to prove that from this assumption, the formula is true for "all" intgeers ≥ 1. Integratig "by parts" (a tecchnique of integration given in many books on calculus), the ntegral in the last formula, gives

$$\int_{y=0}^{\infty} e^{-y} y^{n-1} dy = \int_{y=0}^{\infty} \frac{e^{-y} y^{n}}{n} dy = (n-1)!$$

$$\int_{y=0}^{\infty} e^{-y} y^{(n+1)-1} dy = n \times (n-1)! = n!,$$

$$\Gamma(n+1) = n!, n \geq 1.$$

Since the induction hypothesis holds for $n = 1$, it holds for all $n = 1, 2, \ldots$. A concise decription of mathematical induction is given on p. 21 of [54] (1967), and comprehensively on pp. 9–20 in [29] (1956).

2.1.15 $\mathrm{M}_\lambda/\mathrm{M}\mu/1$: Probability Distribution of Number in the System in M/M/1

Let N denote the number of units (customers) in the system, i.e., "waiting in the queue before service + in service", in the stable $\mathrm{M}_\lambda/\mathrm{M}_\mu/1$ system (aka, in the "steady state" or "in equilibrium" where $\lambda < \mu$). Denote $P(N = n)$ by P_n, $n = 0, 1, \ldots$. We can derive P_n, $n = 0, 1, \ldots$ by (1) conditioning on the pdf of W_Q (waiting time of customer in the queue before service), or (2) by conditioning on σ (total time in the system), as follows.

(1) Conditioning on W_Q, substitute $b(y)$ given in formula (2.24) in Section 2.1.14.1 for $P(n-1$ arrivals occur during $y|W_Q = y)$. Just after an arbitrary arrival begins service there will be $n-1$ new customers waiting for service,, resulting in n customers in the system with probability

$$P_n = \textstyle\int_{y=0}^{\infty} P(n-1 \text{ arrivals occur during } y)|W_Q = y)\ f(y)dy,$$

giving:

$$P_n = \textstyle\int_{y=0}^{\infty} Erl_{n-1,\lambda} f(y)dy = \int_{y=0}^{\infty} \left(e^{-\lambda y} \frac{(\lambda y)^{n-1}}{(n-1)!} \right) \lambda P_0 e^{-(\mu-\lambda)y} dy$$

$$= P_0 \left(\frac{\lambda}{\mu}\right)^n \textstyle\int_{y=0}^{\infty} e^{-\mu y} \frac{(\mu y)^{n-1}}{(n-1)!} \mu dy = P_0 \rho^n \times 1, n = 0, 1, \ldots, \qquad (2.28)$$

where $\rho = \lambda/\mu < 1$ (noting that the integral in (2.28) is equal to 1 from the result in (2.25) in Section 2.1.14.1 above.

The normalizing condition

$$\sum_{n=0}^{\infty} P_n = 1,$$

holds for $P_n, n = 0, 1, , 2....$ Thus, $P_0 \sum_{n=0}^{\infty} \rho^n = 1$ where $\rho < 1$ (for stability in the M/M/1 queue). So, $P_0 = 1 - \rho$, giving the well-known geometric distribution

$$P_n = P_0\rho^n = (1-\rho)\rho^n, n = 0, 1, \ldots . \tag{2.29}$$

(2) Conditioning on the system time σ: substitute $f_\sigma(y)$ from formula (2.19) into

$$P_n = \int_{y=0}^{\infty} P(n \text{ arrivals occur during } y)|\sigma = y) f_\sigma(y) dy$$

getting

$$P_n = \int_{y=0}^{\infty} \left(e^{-\lambda y}\frac{(\lambda y)^n}{n!}\right)(\mu - \lambda)e^{-(\mu-\lambda)y}dy$$

$$= \left(\frac{\lambda}{\mu}\right)^n (1 - \frac{\lambda}{\mu}) \int_{y=0}^{\infty} e^{-\lambda y}\frac{(\mu y)^n}{n!}\mu dy. \tag{2.30}$$

$$P_n = \left(1 - \frac{\lambda}{\mu}\right)\left(\frac{\lambda}{\mu}\right)^n \cdot 1 = P_0\rho^n, n = 0, 1, \ldots, \tag{2.31}$$

because the integral in (2.30) is equal to 1 $(= \frac{\Gamma(n+1)}{n!})$. From (2.31)

$$P_0 = 1 - \rho = 1 - \frac{\lambda}{\mu}, \tag{2.32}$$

which agrees with formula (2.15) in Section 2.1.10.

2.1.16 Expected Number in the System

The"right-tail probability"of P_n, $n = 0, 1, \ldots$, in formula (2.31) is $P(N \geq n) = \rho^n, n = 0, \ldots$, i.e.,

$$E[N] \equiv \sum_{n=0}^{\infty} nP_n = \sum_{n=1}^{\infty} nP_0\rho^n = \frac{\rho}{1-\rho} = \frac{\lambda}{\mu - \lambda}, \tag{2.33}$$

which agrees with the M/M/1 result $E[N]$ on p. 188 in [41] (1975). The second summation, $\sum_{n=1}^{\infty} nP_0\rho^n$, in formula (2.33) is obtained by changing it to the equivalent:

$$P_0\rho\sum_{n=1}^{\infty} n\rho^{n-1} = P_0\rho\sum_{n=1}^{\infty}\frac{d\rho^n}{d\rho} = P_0\rho\frac{d}{d\rho}\left(\sum_{n=1}^{\infty}\rho^n\right),$$

which then yields the last two terms in formula (2.33).

2.1.16.1 More Classical Way to Derive P_n, $n = 0, 1,...$, in $M_\lambda/M_\mu/1$

We include a classical approach by applying birth and death processes for comparison and completeness (e.g., pp. 49–55 in [36] (2008); Section 6.3, p. 374 and Example 6.14, p.395 in [52](2010); and others).

We can use formula (1.31) in Section 1.6 in Chapter 1 above, repeated here for easy access. The number of customers in the system changes by ± 1 due to Poisson arrivals and exponential service times. Therefore it is a birth and death process. Thus

$$P_n = \frac{\lambda_0\lambda_1\cdots\lambda_{n-1}}{\mu_1\mu_{12}\cdots\mu_n}P_0, n = 1, 2, ..., \tag{2.34}$$

and P_0 is obtained by applying the normalizing condition

$$\sum_{n=0}^{\infty} P_n = 1. \tag{2.35}$$

Using the derived birth-and-death values of P_n, the mixed pdf of the waiting time $\{P_0, f(x)\}_{x>0}$ is then derived by conditioning on N, the number in the system.

In this book, the logical direction is: (1) "waiting time pdf" $\rightarrow$ "the number in the system pmf (probability mass function)". This is the opposite to the birth and death direction, which is (2) "the number in the system pmf" $\longrightarrow$ "waiting time pdf". Both methods are valid. This LC approach in this book is influenced by the original use of "Lindley recursions", when the author was analyzing "Multiple server queues with service time depending on waiting time" (see [43](1952)). (A brief discussion of Lindley recursions is given in Section 1.8.5 in Chapter 1. Also, see pp. 3–4 in [22] (2017).)

As shown in Section 2.1.15, LC derives first the mixed pdf $\{P_0, f(x)\}_{x>0}$ of the waiting time before service W_q, or of the pdf $f_\sigma(x), x > 0$ of the customer total time in the system σ. This is followed by conditioning on the number of arrivals during W_q or during σ, to derive the values of $P_n, n = 0, 1, 2, ...$.

The main advantage of LC is that it avoids a great deal of algebra in deriving integral equations for pdfs of r.v.s of interest, especially in complex models. Another advantage of using LC is that it is more suitable for deriving the pdfs of waiting times, and other random variables (including number in the system) in state-dependent models, which occur, e.g., when service time depends on waiting time; or when service time depends on the number of customers left behind in the waiting line when a service begins, etc.

This book presents LC derivations related to state variables. Classical derivations based only on birth-death processes, generating functions, etc., are given in many classical works, e.g., [52] (2010); [36] (2008), [41] (1975), and others.

Note: LC uses stochastic level-crossing techniques, and jointly includes many companion classical methods as well.

2.1.17 Expected Busy Cycle, Busy Period, and Idle Period in Standard M/M/1 Queue

The consecutive busy cycles in an M/M/1 queue form a renewal process $\{\mathcal{C}_n\}_{n=1,2,..}$, where $C_n \overset{d}{\equiv} C$, and C is a finite r.v.. Starting from time 0, the first renewal is at time C_1; the second renewal is at time $C_1 + C_2$; the nth renewal is at time $\sum_{i=1}^{n} C_i$, $n = 1, 2, ...$ By the elementary renewal theorem (p 432 in [52] (2010))

$$E[\mathcal{C}] = \frac{1}{\text{rate at which renewals occur}},$$

i.e., the rate at which busy cycles start. This occurrence rate is equal to the rate at which a sample path of the virtual wait (workload) makes vertical upward jumps out of level 0, namely λP_0. Thus

$$E[\mathcal{C}] = \frac{1}{\lambda P_0}. \tag{2.36}$$

During the intervals $\{\mathcal{C}_i\}_{i=1,2,...}$, iid random segments of a process are overlaid. The overlaid process is called a regenerative process. For example, consider the virtual wait process in Figure 2.2 in Section 2.1.6, above. Each regenerative cycle C:= "Busy period + Idle period". The regenerative cycles are the same as the underlying inter-renewal times of the renewal process. Expected Busy Period and Expected Idle Period:

In $M_\lambda/M_\mu/1$ the renewal reward theorem given by formula (1.34) in Section 1.8.4 in Chapter 1 is useful for obtaining $E[\mathcal{B}]$ and $E[\mathcal{I}]$. Thus

$$\frac{E[\mathcal{B}]}{E[\mathcal{C}]} = \lim_{t\to\infty} \frac{\text{time server is busy in } (0,t)}{t} = 1 - P_0,$$

which is the long-run proportion of time that the queue is busy with at least one customer in the system. Applying formula (2.36) gives

$$E[\mathcal{B}] = \frac{1 - P_0}{\lambda P_0} = \frac{\frac{\lambda}{\mu}}{\lambda P_0} = \frac{1}{\mu(1 - \frac{\lambda}{\mu})} = \frac{1}{\mu - \lambda}. \tag{2.37}$$

Similarly for the idle period

$$\frac{E[\mathcal{I}]}{E[\mathcal{C}]} = P_0,$$

and

$$E[\mathcal{I}] = \frac{P_0}{\lambda P_0} = \frac{1}{\lambda}. \tag{2.38}$$

Formula (2.38) can be inferred also by by applying the memoryless property of Exp_λ, which is the distribution of the time interval between arrivals to the system. This is because the idle period begins when the virtual wait (workload) hits level 0 from above, say at time t. Hence, by the memoryless property the length of the idle period after time t is $\stackrel{d}{=} \text{Exp}_\lambda$, which has mean $= 1/\lambda$.

2.1.18 Expected Number Served in a Busy Period

We can also use the renewal reward theorem to derive the expected number of customers served in a busy period. Let $N_\mathcal{B}$ denote the number served during $\mathcal{B}$. Then

$$\frac{E[N_\mathcal{B}]}{E[\mathcal{C}]} = \lim_{t\to\infty} \frac{\text{number of arrivals to the system in time } t}{t} = \frac{\lambda t}{t} = \lambda,$$

because all arrivals to the standard M/M/1 system get served (e.g., there are no blocked customers). Thus

$$E[N_\mathcal{B}] = \frac{\lambda}{\lambda P_0} = \frac{1}{P_0}, \tag{2.39}$$

where $E[\mathcal{C}] = 1/(\lambda P_0)$.

Formula (2.39) illustrates two important concepts: (1) the usefulness of the renewal reward theorem and (2) the importance of the probability P_0.

2.1.19 Expected Sojourn above and Expected Sojourn below a Level in the Virtual Wait in the $\text{M}_\lambda/\text{M}_\mu/1$ Queue

in various stochastic models, it is of theoretical interest and useful in practice, to determine the expected duration of a sojourn of the sample path of a key r.v., above or below a state-space level $x > 0$. Denote a sojourn "above level x" by "a_x", and the sojourn below level x by "b_x". Here, we derive the expected values $E[a_x]$ and $E[b_x]$, $x > 0$ for the "virtual wait" in the $\text{M}_\lambda/\text{M}_\mu/1$ queue. See Figure 2.2 above.

2.1.19.1 Expected Sojourn above Level x

Since the service times are $\stackrel{d}{=} Exp_\mu$, all sample-path vertical jump sizes are $\stackrel{d}{=} Exp_\mu$. Hence every sample-path upcrossing of a fixed level $x \geq 0$ has an "excess" above x which is also $\stackrel{d}{=} Exp_{\mu}$, by the memoryless property of the exponential distribution (see Section 1.5.1 in Chapter 1). Thus,

$$a_x \stackrel{d}{=} \mathcal{B},$$
$$E\left[a_x\right] \stackrel{d}{=} E\left[\mathcal{B}\right] = \tfrac{1}{\mu-\lambda}, x \geq 0, \tag{2.40}$$

where $E\left[\mathcal{B}\right] = \frac{1}{\mu-\lambda}$ is given in formula (2.37). Note that formula (2.40) implies that $E\left[a_x\right]$ is a constant, independent of level $x \geq 0$. In Section 2.1.19.2, we shall derive the same result for $E\left[a_x\right]$ using a method that shows the mechanism which may be appropriate for deriving $E\left[a_x\right]$, as well as $E\left[b_x\right]$, in more complex stochastic models.

2.1.19.2 Time between Two Successive Downcrossings of Level x

Let "d_x" denote the time between two successive SP "downcrossings of level x", $x \geq 0$. The SP downcrossing rate of level $x \geq 0$, is $\lim_{t\to\infty} \frac{\mathcal{D}_t(x)}{t}$. By the elementary renewal theorem (see Section 1.8.3 in Chapter 1)

$$E\left[d_x\right] = \frac{1}{f(x)} = \frac{1}{\lambda P_0 e^{-(\mu-\lambda)x}}, \tag{2.41}$$

where $P_0 = 1 = \frac{\lambda}{\mu}$ (see formula (2.14) in Section 2.1.10). Let $\{d_{x,n}\}_{n=1,2,\cdots}$ denote the sequence of inter-downcrossing times at level x. Then $\{d_{x,n}\}_{n=1,2,}$ is a regenerative process where each cycle is $\stackrel{d}{=} d_x$ (see Section 1.8.2 in Chapter 1)

By the renewal reward process (see Section 1.8.4 in Chapter 1),

$$\frac{E\left[a_x\right]}{E\left[d_x\right]} = 1 - F(x),$$

where $1 - F(x) := P(W(t) > x)$. Moreover,

$$F(x) = P_0 + \textstyle\int_{y=0}^{x} \lambda(1-\rho)e^{-(\mu-\lambda)y}dy = 1 - \rho e^{-(\mu-\lambda)x},\ x \geq 0.$$

Therefore, since $\rho = \frac{\lambda}{\mu}$,

$$\begin{aligned} E\left[a_x\right] &= \tfrac{1-F(x)}{\lambda P_0 e^{-(\mu-\lambda)x}} \\ &= \tfrac{\frac{\lambda}{\mu} e^{-(\mu-\lambda)x}}{\lambda P_0 e^{-(\mu-\lambda)x}} \\ &= \tfrac{\frac{\lambda}{\mu}}{\lambda P_0} \\ &= \tfrac{1}{\mu\left(1-\frac{\lambda}{\mu}\right)} \\ &= \tfrac{1}{\mu-\lambda}, \end{aligned} \tag{2.42}$$

which agrees with formula (2.40) above.

2.1.19.3 Expected Sojourn below Level x

The SP inter-downcrosing time d_x of level x, is:

$$d_x = a_x + b_x$$

Thus, from formulas (2.41) and (2.40),

$$E[d_x] = E[a_x] + E[b_x],$$

$$E[b_x] = E[d_x] - E[a_x]$$

$$= \frac{1}{\lambda P_0 e^{-(\mu-\lambda)x}} - \frac{1}{\mu-\lambda} \tag{2.43}$$

$$= \frac{1}{\lambda(1-\frac{\lambda}{\mu})e^{-(\mu-\lambda)x}} - \frac{1}{\mu-\lambda}$$

$$E[b_x] = \frac{1}{\mu-\lambda}\left(\frac{\mu}{\lambda}e^{+(\mu-\lambda)x} - 1\right).$$

From formula (2.43),

$$E[b_x] > E[a_x], x > 0.$$

When $x = 0$, $E[b_0] = \frac{1}{\lambda}$ ($= E$[idle period] $= E[\mathcal{I}]$–see formula (2.38 in Section 2.1.17)). Also $E[a_0] = E[\mathcal{B}] = \frac{1}{\mu-\lambda}$. Then, since $0 < \lambda < \mu$ for stability in the $M_\lambda/M_\mu/1$ queue,

$$E[a_0] < E[b_0] \text{ if } 0 < \lambda < \tfrac{\mu}{2},$$
$$E[a_0] > E[b_0] \text{ if } \tfrac{\mu}{2} < \lambda < \mu. \tag{2.44}$$

2.2 Exercises

Exercise 2.2.1 *In formula (2.22) for the stationary pdf of the system time, check whether the necessary condition holds with respect to dimensional analysis.*

Exercise 2.2.2 *Use formula (2.34) to compute* P_0.

Exercise 2.2.3 *Explain why the sample path of the virtual wait process* $\{W(t)\}_{t\geq 0}$ *in the standard M/M/1 queue has a slope equal to* -1 *when* $W(t) > 0$.

Exercise 2.2.4 *Each arrival to an M/M/1 queue who waits a positive time before beginning service, initiates a trapezoid within the sample path. What is the analogous geometric stricture initiated within the sample path by arrivals who wait zero time before beginning service.*

Exercise 2.2.5 *In equation (2.4) in Section 2.1.7 explain why the terms* $P_0 e^{-\mu x}$ *and* $\int_{y=0}^{x} e^{-\mu(x-y)} f(y)dy$ *are dimensionless.*

Exercise 2.2.6 *In formula (2.16) in Section 2.1.10 show that cdf* $F(0) = 0$ *and* $F(x)$ *is a continuous increasing function with limit* 1 *as* $x \to \infty$*. Also show that* $F(x)$ *is a concave function on* $(0, \infty)$ *(i.e.,second derivative is neagative).*

Exercise 2.2.7 *In equation (2.16) in Section 2.1.10 prove the validity of the third term.*

Exercise 2.2.8 *In equation (2.33) in Section 2.1.16 show the algebraic steps to prove the equality:* $P_0 \rho \sum_{n=1}^{\infty} \rho^{n-1} = \frac{\rho}{1-\rho}$.

Exercise 2.2.9 *In equation (2.18) in Section 2.1.12 provide details of the integration and algebraic steps.*

Exercise 2.2.10 *Provide the details which prove that Equation (2.22) follows from Equation (2.21) in Section 2.1.13.*

Exercise 2.2.11 *Show that equation (2.9) follows by integration of equation (2.8) by parts.*

Exercise 2.2.12 *Verify the inequalities in foormula (2.44) regarding* $E[a_0]$ *and* $E[b_0]$.

2.3 M/M/1 with Service Time Depending on Waiting Time

Consider a variant of the standard M/M/1 queue where the Poisson arrival rate is λ and the service time for each customer is $S \stackrel{d}{=} \text{Exp}_\mu$ and the service rate μ is a function of the arrival's waiting time before service begins, which is denoted by W_Q . That is, the conditional probability $P(S \leq y | W_Q = y) = 1 - e^{-\mu(y)x}$, $x > 0$, where S is the arrival's service time. The function $\mu(y)$ is the service-time parameter which depends on the arriving customer's waiting time y before starting service. Denote the mixed stationary pdf of W_Q, by $\{P_0, f(x)\}, x > 0$ where $P_0 =$ stationary $P(W_Q = 0)$ and $f(x)$ is absolutely continuous on $(0, \infty)$. From the basic LC theorem stated in in Section 1.4 in Chapter 1, the sample-path downcrossing rate of level $x > 0$ is equal to $f(x)$, with probability 1 (see Figure 2.7 in Section 2.3.1). The SP exit rate out of level 0 is λP_0 (which possibly upcrosses level x).

The sample-path upcrossing rate of level x is due only to vertical jumps representing service times of arriving customers, which start at a level $W_Q = y, 0 \leq y < x$. The conditional probability that such a jump upcrosses level x is

$$P(S > x - y | W_Q = y) = e^{-\mu(y)(x-y)}, 0 \leq y < x,$$

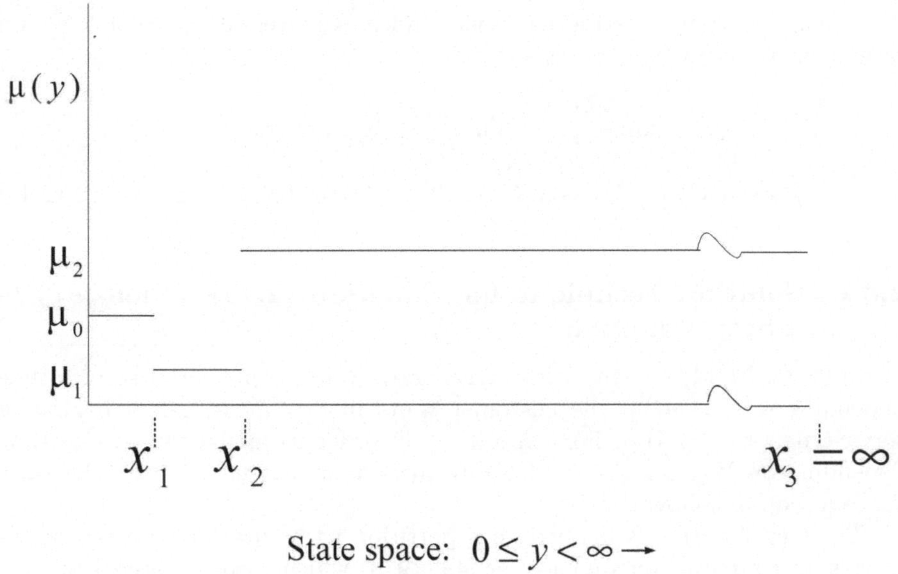

FIGURE 2.6
Step function $\mu(y)$,$y \geq 0$, in formula (2.46) showing the exponential service rates μ_0, μ_1, μ_2 over the state-space intervals $\lfloor 0, x_1)$, (x_1, x_2), (x_2, x_3), respectively, where $x_3 = \infty$.

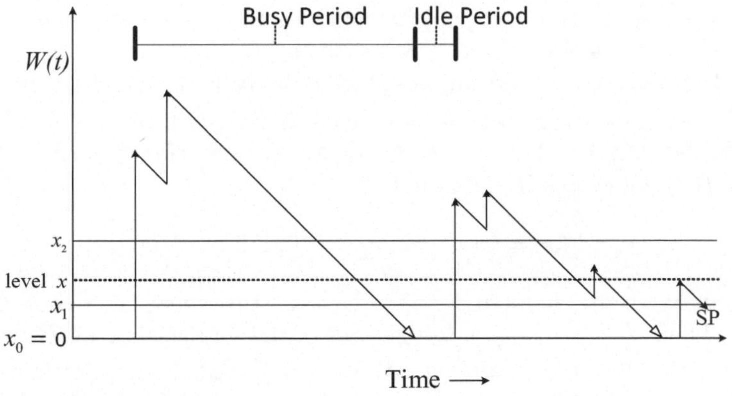

FIGURE 2.7
Sample path of virtual wait when service time S depends on actual wait W_Q. In this example, the state-space partition creates three intervals: $[x_0, x_1)$, (x_1, x_2), (x_2, ∞).

due to exponentially distributed service times. An integral equation based on rate balance across level x, is:

$$\lim \frac{\mathcal{D}_t(x)}{t} = \lim \frac{\mathcal{U}_t(x)}{t}, 0 \le x < \infty,$$

$$f(x) = \lambda P_0 e^{-\mu(0)x} + \lambda \int_{y=0}^{x} e^{-\mu(y)(x-y)} dy, 0 \le x < \infty. \tag{2.45}$$

2.3.1 Solution Technique for Equation (2.45) When $\mu(y)$ Is a Step Function

Consider an M/M/1 queue with arrival rate λ and exponential service time depending on how long the customer waits in the queue before beginning service, denoted by W_Q. For exposition, in order to model the service time depending on W_Q, let the service rate $\mu(y)$ be a step function on the state space $[0, \infty]$, as follows.

The step function is defined on a partition with three state-space subintervals. (for a more general case see [49](1973) which used a classical Lindley recursion–not LC; and [22](2017) using LC). Thus,

$$\mu(y) = \begin{cases} \mu_0, \ x_0 \le y < x_1; \\ \mu_1, \ x_1 < y < x_2, \\ \mu_2, \ x_2 < y < x_3 = \infty, \end{cases} \tag{2.46}$$

where $x_0 = 0$. Assume that the μ_js are all different rates.

The step function in (2.46) corresponds to partitioning the state space by $0 = x_0 < x_1 < x_2 < x_3 = \infty$, which separates the state space into three contiguous subintervals: $I_0 = [0, x_1)$, $I_1 = (x_1, x_2), I_2 = (x_2, \infty)$ (Figure 2.6 and Figure 2.7). Formula (2.46) implies that an arrival whose actual waiting time W_Q is in the interval I_j receives a service time $S_j \stackrel{d}{=} \text{Exp}_{\mu_j}$, $j = 0, 1, 2$.

The mixed pdf of $\{W(t)\}_{t \ge 0}$ is $\{P_0, f(x)\}_{x \ge 0}$. For notational clarity, define that part of $f(x)$ when x is in interval I_j by

$$f_j(x) = f(x), x \in I_j, j = 0, 1, 2.$$

Let the actual wait, denoted by W_q, have a stationary cdf denoted by $F(x), x \ge 0$. Then $\frac{d}{dx}F(x) = f(x)$. Let the cdf of the service time be $B_j(x) := P(S_j \le x) = 1 - e^{-\mu_j x}, x \ge 0$, having pdf $b_j(x) = \frac{\partial}{\partial x}B_y(x) = \mu_j e^{-\mu_j x}, x > 0$, whenever an arrival "sees" $W_Q \in I_j$. A sample path of $\{W(t)\}_{t \ge 0}$ resembles that for the standard M/M/1 queue, except that the sample-path jump size (= arrival-point service time) depends on what the sample-path level is at the start of the upward jump (the actual wait). (See Figure 2.7.)

Consider a fixed state-space level $x \geq 0$ in a sample path of $\{W(t)\}_{t\geq 0}$. The downcrossing rate of x is $f(x)$, by the basic LC Theorem in Section 1.4. The balance equations for $f_0(x)$, $f_1(x)$, $f_2(x)$, where the right-hand sides are the corresponding upcrossing rates of level x, are given by:

$$f_0(x) = \lambda P_0 e^{-\mu_0 x} + \lambda \int_{y=0}^{x} e^{-\mu_0(x-y)} f_0(y)dy,\ 0 < x < x_1,$$

$$f_1(x) = \lambda P_0 e^{-\mu_0 x} + \lambda \int_{y=0}^{x_1} e^{-\mu_0(x-y)} f_0(y)dy + \lambda \int_{y=x_1}^{x} e^{-\mu_1(x-y)} f_1(y)dy,$$
$$x_1 < x < x_2,$$

$$f_2(x) = \lambda P_0 e^{-\mu_0 x} + \lambda \int_{y=0}^{x_1} e^{-\mu_0(x-y)} f_0(y)dy + \lambda \int_{y=x_1}^{x_2} e^{-\mu_1(x-y)} f_1(y)dy$$
$$+\lambda \int_{y=x_2}^{x} e^{-\mu_2(x-y)} f_2(y)dyx, x_2 < x < x_3 = \infty. \tag{2.47}$$

Assume there are exactly two subintervals in the state-space partition. In formula (2.47) P_0 := stationary P(virtual wait $= 0$). To derive the value of P_0, apply the normalizing condition (aka law of total probability)

$$P_0 + \int_{x=0}^{x_1} f_0(x)dx + \int_{x=x_1}^{\infty} f_1(x)dx = 1. \tag{2.48}$$

In (2.47) the term $\lambda P_0 e^{-\mu_0 x}$ is the upcrossing rate of x caused by SP jumps at arrival instants when the system is empty. The terms $\lambda \int_{y=x_j}^{x_{n+1}} e^{-\mu_n(x-y)} f_j(y)dy$ are the upcrossing rates of level x by SP jumps at arrival instants when the virtual wait $\{W(t)\}_{t\geq 0}$ is at state-space levels $y \in (xj, x_{n+1}), n = 0, 1$. The terms $\lambda \int_{y=x_n}^{x} e^{-\mu_2(x-y)} f_n(y)dyx$ are the upcrossing rates of level x by SP jumps at arrival instants when $\{W(t)\}_{t\geq 0}$ is at state-space levels $y \in (x_n, x), n = 2$. Rate balance across level x yields the integral equation for $f_n(x), n = 0, 1, 2$.

To solve for $f_n(x), n = 0, 1, 2$, use a "stair-step" procedure: first solve for $f_0(x)$, $0 < x < x_1$; then solve for $f_1(x), x_1 < x < x_2$ using the continuity of the total pdf of wait $f(x)$ at x_1 to utilize the result obtained for $f_0(x)$; then solve for $f_2(x)$, $x_2 < x < x_3 = \infty$ using the continuity of the total pdf of wait $f(x)$ at x_2, to utilize the result obtained for $f_1(x)$. Finally solve for P_0 using the normalizing condition. Continuity of $f(x)$ is guaranteed because there are no atoms (points with positive probability) in $(0, \infty)$. The only atom is at level 0, having probability $P_0 > 0$.

2.3.2 Solution of Equation (2.45) When $\mu(y)$ Is Defined on Two Subintervals $I_0 = [0, x_1)$, $I_1 = (x_1, \infty)$

This example illustrates a derivation of the total mixed pdf $\{P_0, f(x)\}_{x\geq 0}$. For exposition let the state-space partition have only two points, 0 and $x_1 > 0$, with state-space intervals: $I_0 = [0, x_1)$, $I_1 = (x_1, \infty)$. A potential real-world application may be to patients arriving in a Poisson process at a walk-in clinic for service. If arriving patients wait in the queue before service less than time x_1 they receive an exponential service time with rate μ_0; if they wait in the queue longer than time x_1 they receive an exponential service rate μ_1 $(\neq \mu_0)$. This model requires only two main state-space intervals: I_0 and I_1 with corresponding partial pdfs $f_0(x)$, $0 < x < x_1$, and $f_1(x)$, $x_1 < x < \infty$.

2.3.2.1 Solution for $f_0(x)$

The pdf $f_0(x)$ satisfies the first equation in formula (2.47). By using the same steps as in the solution of equation (2.4) in Section 2.1.7 in Chapter 2, we obtain

$$f_0(x) = \lambda P_0 e^{-(\mu_0 - \lambda)x}, 0 < x < x_1. \tag{2.49}$$

2.3.2.2 Solution for $f_1(x)$

An integral equation for $f_1(x), x_1 < x < \infty$, leads to a more involved formula than (2.49) because $f_1(x)$ requires knowledge of equation (2.49), the limiting value $f_0(x_1^-)$ from below, and the limiting value $f_1(x_1^+)$ from above.

Consider the second integral equation in formula (2.47) for $f_1(x), x_1 < x < x_2 = \infty$, repeated here for ease of reference. For $x_1 < x < \infty$:

$$F_1(x) = \lambda P_0 e^{-\mu_0 x} + \lambda \int_{y=0}^{x_1} e^{-\mu_0(x-y)} f_0(y)dy + \lambda \int_{y=x_1}^{x} e^{-\mu_1(x-y)} f_1(y)dy,$$

Substituting the formula for $f_0(y)$ given in (2.49) yields

$$f_1(x) = \lambda P_0 e^{-\mu_0 x} + e^{-\mu_0 x}\lambda^2 P_0 \int_{y=0}^{x_1} e^{\lambda y} dy + \lambda \int_{y=x_1}^{x} e^{-\mu_1(x-y)} f_1(y)dy,$$

$$f_1(x) = \lambda P_0 e^{-\mu_0 x} + \lambda e^{-\mu_0 x} P_0 \lambda A_0 + \lambda \int_{y=x_1}^{x} e^{-\mu_1(x-y)} f_1(y)dy,$$

$$f_1(x) = \lambda P_0 e^{-\mu_0 x}(1 + \lambda A_0) + \lambda \int_{y=x_1}^{x} e^{-\mu_1(x-y)} f_1(y)dy, \tag{2.50}$$

where the constant A_0 is given by

$$A_0 = \int_{y=0}^{x_1} e^{\lambda y} dy = \frac{e^{\lambda x_1} - 1}{\lambda}. \tag{2.51}$$

A_0 has dimension $[Time]$ and λA_0 is dimensionless.

Taking $\frac{d}{dx}$ in formula (2.50) (accounting for differentiation under the integral), gives

$$\begin{aligned} f_1'(x) &= -\mu_0 \lambda P_0 e^{-\mu_0 x}(1 + \lambda A_0) - \mu_1 \lambda \\ &\quad \times \int_{y=x_1}^{x} e^{-\mu_1(x-y)} f_1(y)dy + \lambda f_1(x), 0 < x < \infty. \end{aligned} \tag{2.52}$$

Substituting from the integral on the right side of (2.50) into the immediately previous equation, results in

$$f_1'(x) = -\mu_0 \lambda P_0 e^{-\mu_0 x} (1 + \lambda A_0) - \mu_1 \left(f_1'(x) - \lambda P_0 e^{-\mu_0 x} (1 + \lambda A_0) \right) + \lambda f_1(x). \tag{2.53}$$

or

$$f_1'(x) + (\mu_1 - \lambda) f_1(x) = (-\mu_0 \lambda P_0 (1 + \lambda A_0) + \mu_1 \lambda P_0 (1 + \lambda A_0)) e^{-\mu_0 x},$$

where A_0 is is given in (2.51). This formula reduces to the first order differential equation

$$f_1'(x) + (\mu_1 - \lambda) f_1(x) = (\mu_1 - \mu_0) \lambda (1 + \lambda A_0) P_0 e^{-\mu_0 x}, \tag{2.54}$$

where it is assumed that $\lambda < \mu_1$. Multiplying throughout by the integration factor $e^{(\mu_1 - \lambda)x}$ and expressing the left side as a derivative, gives

$$\frac{d}{dx}(f_1(x) e^{(\mu_1 - \lambda)x}) = (\mu_1 - \mu_0) \lambda (1 + \lambda A_0) P_0 e^{(-\mu_0 + \mu_1 - \lambda)x}.$$

Integrating both sides with respect to x gives

$$f_1'(x) = \frac{(\mu_1 - \mu_0) \lambda (1 + \lambda A_0)}{(-\mu_0 + \mu_1 - \lambda)} P_0 e^{-\mu_0 x} + C_1 e^{-(\mu_1 - \lambda)x}, x_1 < x < \infty,$$

where C_1 is a constant. To determine C_1 we apply the continuity of the total pdf of the virtual waiting time $f(x)$, $x > 0$, including in particular when $x = x_1$. Continuity at state-space level x_1 gives

$$\lim_{x \uparrow x_1} f_0(x_1) = \lim_{x \downarrow x_1} f_1(x_1^+)$$

$$f_0(x_1^-) = f_1(x_1^+), \tag{2.55}$$

or

$$\lambda P_0 e^{-(\mu_0 - \lambda)x_1} = \frac{(\mu_1 - \mu_0) \lambda (1 + \lambda A_0)}{(-\mu_0 + \mu_1 - \lambda)} P_0 e^{-\mu_0 x_1} + C_1 e^{-(\mu_1 - \lambda)x_1},$$

which gives

$$C_1 = \frac{\lambda P_0 e^{-(\mu_0 - \lambda)x_1} - \frac{(\mu_1 - \mu_0)\lambda(1 + \lambda A_0)}{(-\mu_0 + \mu_1 - \lambda)} P_0 e^{-\mu_0 x_1}}{e^{-(\mu_1 - \lambda)x_1}}$$

$$= \lambda P_0 e^{+(\mu_1 - \mu_0)x_1} - \frac{(\mu_1 - \mu_0) \lambda (1 + \lambda A_0)}{(-\mu_0 + \mu_1 - \lambda)} P_0 e^{(-\mu_0 + \mu_1 - \lambda)x_1}$$

or

$$C_1 = \lambda P_0 e^{+(\mu_1 - \mu_0)x_1} \left(1 - \frac{(\mu_1 - \mu_0)(1 + \lambda A_0)}{(-\mu_0 + \mu_1 - \lambda)} e^{-\lambda x_1} \right) = C_2 P_0, \tag{2.56}$$

where

$$C_2 = \lambda e^{(\mu_1-\mu_0)x_1}\left(1 - \frac{(\mu_1-\mu_0)(1+\lambda A_0)}{(-\mu_0+\mu_1-\lambda)}e^{-\lambda x_1}\right); \tag{2.57}$$

C_2 has dimension $[Time]^{-1}$, and A_0 is given in formula (2.51).

2.3.2.3 Solution for P_0

Applying (2.48) gives

$$P_0\left[\begin{array}{c} 1+\int_{x=0}^{x_1}\lambda e^{-(\mu_0-\lambda)x}dx \\ +\int_{x=x_1}^{\infty}\left(\frac{(\mu_1-\mu_0)\lambda(1+\lambda A_0)}{(-\mu_0+\mu_1-\lambda)}e^{-\mu_0 x}+C_2e^{-(\mu_1-\lambda)x}\right)dx\end{array}\right]=1\,,$$

$$P_0 = \frac{1}{1+\int_{x=0}^{x_1}\lambda e^{-(\mu_0-\lambda)x}dx+\int_{x=x_1}^{\infty}\left(\frac{(\mu_1-\mu_0)\lambda(1+\lambda A_0)}{(-\mu_0+\mu_1-\lambda)}e^{-\mu_0 x}+C_2e^{-(\mu_1-\lambda)x}\right)dx},$$

$$P_0 = \frac{1}{1+\lambda\left(\frac{1-e^{-(\mu_0-\lambda)x_1}}{\mu_0-\lambda}\right)+\frac{(\mu_1-\mu_0)\lambda(1+\lambda A_0)e^{-\mu_0 x_1}}{(-\mu_0+\mu_1-\lambda)\mu_0}+\frac{C_2e^{-(\mu_1-\lambda)x_1}}{\mu_1-\lambda}}. \tag{2.58}$$

For P_0 to be positive, the inequality $\lambda < \mu_1$ must hold (see right-most term in denominator of (2.58).

A mild check, which shows a necessary condition on (2.58) is obtained by letting $x_1 \to \infty$, and assuming that $\lambda < \mu_0$. Then (2.58) becomes

$$P_0 = \frac{\mu_0-\lambda}{\mu_0} = 1-\frac{\lambda}{\mu_0},$$

which is equal, as it should be, to the probability of a zero wait in a standard $M_\lambda/M_{\mu_0}/1$ queue (see formula 2.15 in Section 2.1.10).

2.4 Exercises

Exercise 2.4.1 *Explain why the dimension of the constant A_0 in formula (2.51) is [Time].*

Exercise 2.4.2 *Explain why differentiation under the integral sign in Equation (2.52) results in the term "$+\lambda f_1(x)$".*

Exercise 2.4.3 *Derive equation (2.53) from Equation (2.52).*

Exercise 2.4.4 *Explain why the pdf $f(x), 0 < x < \infty$, in formula (2.55), is continuous on $(0,\infty)$. [Hint: Are there any points in the state space, other than $\{0\}$, which have positive probabilities? See Section 3.4.1 below with regard to level D in sample paths of the virtual wait in the M/D/1 queue.]*

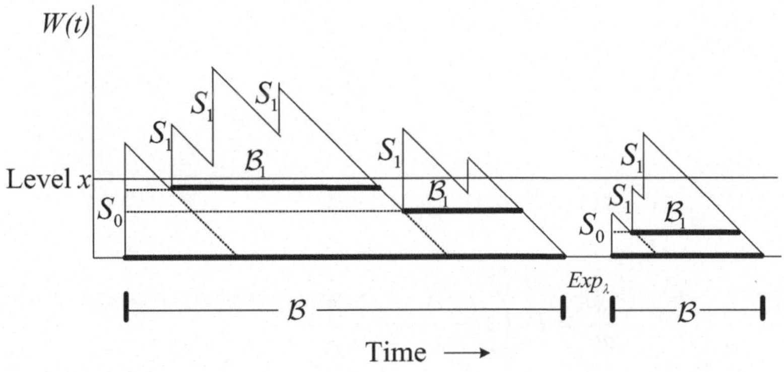

FIGURE 2.8
The $\mathcal{B}$s are busy periods initiated by S_0s in $M_\lambda/M/1$, where zero-waits receive service times $\stackrel{d}{=} S_0$. In the figure, the first S_0 has two "first-descendant" arrivals and corresponding $\mathcal{B}_1$s. The second S_0 has one first- descendant arrival, and one corresponding $\mathcal{B}_1$.

2.5 M/M/1 Where Zero-wait Arrivals Receive Special Service

A particular case of M/M/1 with service time depending on wait, which has many useful applications, is a model where the initial customer of each busy period receives special service (also called exceptional service). It is a special case of the model in Section 2.3.2. In that example, we now set $0 = x_0 < x_1 = \infty$ as the state-space partition. Let zer0-wait customers have service times $S_0 = \text{Exp}_{\mu_0}$ and customers who must wait have service times $S_1 = \text{Exp}_{\mu_1}$, where $\mu_1 \neq \mu_0$ Let the corresponding pdf of wait be $\{P_0, f(x)\}_{0<x<\infty}$ (see Figure 2.8)

The integral equations in formula (2.47) reduce to

$$f(x) = \lambda P_0 e^{-\mu_0 x} + \lambda \int_{y=0}^{x} e^{-\mu_1 (x-y)} f(y)dy,\ 0 < x < \infty, \tag{2.59}$$

The solution of Equation (2.59) for $\{P_0\ , f(x)\}_{x>0}$, is given below in Section 2.5.3. Before solving for$\{P_0\ , f(x)\}_{x>0}$, in (2.59) we can get P_0 by integrating both sides with respect to x over $(0, \infty)$; using the "law of total probability" (aka "normalizing condition")

$$P_0 + \int_{x=0}^{\infty} f(x)dx \ = 1,$$

which gives

$$1 - P_0 = \frac{\lambda}{\mu_0} P_0 + \lambda \int_{y=0}^{\infty} f(y) \int_{x=y}^{\infty} e^{-\mu_1 (x-y)} dx dy$$

$$1 - P_0 = \frac{\lambda}{\mu_0} P_0 + \frac{\lambda}{\mu_1} (1 - P_0)$$

$$1 - \frac{\lambda}{\mu_1} = P_0 \left(1 - \frac{\lambda}{\mu_1} + \frac{\lambda}{\mu_0}\right)$$

$$P_0 = \frac{1 - \frac{\lambda}{\mu_1}}{1 - \frac{\lambda}{\mu_1} + \frac{\lambda}{\mu_0}} = \frac{1 - \rho_1}{1 - \rho_1 + \rho_0}, \tag{2.60}$$

where the traffic intensities are: $\rho_n = \frac{\lambda}{\mu_n} > 0, n = 0, 1$. Formula (2.60) is very important in queueing theory.

2.5.1 Condition for Stability in M/M/1 Where Zero-Wait Arrivals Receive Special Service

A necessary condition for stability is "$\rho_1 < 1$" (guarantees $P_0 > 0$ and $\{0\}$ is a positive recurrent state, i.e., sample paths return to level 0 in a finite time). If $\rho_1 = 1$ then $P_0 = 0$, implying that the system is not in equilibrium. Thus, $\rho_1 \neq 1$ and with probability 1 (a.s.):

$$P_0 = \frac{1}{1 + \frac{\rho_0}{1-\rho_1}} = \frac{1 - \rho_1}{1 - \rho_1 + \rho_0}. \tag{2.61}$$

In (2.61), P_0 is a probability, in $(0, 1)$, only if $\frac{\rho_0}{1-\rho_1} > 0 \iff 0 < \rho_1 < 1$. Formula (2.61) is classical and well tested (For further discussion of this model where zero-wait arrivals receive special service, see [22] (2017).)

2.5.2 Expected Busy Period in M/M/1 Where Zero-Wait Arrivals Receive Special Service

Customers that wait zero time receive service time $S_0 \stackrel{d}{=} \text{Exp}_{\mu_0}$. Customers that must wait a positive time receive service time $S_1 \stackrel{d}{=} \text{Exp}_{\mu_1}$ (see Figure 2.8). Here, we determine $E[\mathcal{B}]$ where $\mathcal{B} :=$ busy period when initial service time $\stackrel{d}{=} S_0$, and all other service times of arrivals within a busy period are $\stackrel{d}{=} S_1$. The idle periods are $\stackrel{d}{=} \text{Exp}_\lambda$ by the memoryless property of the customer interarrival times to the system (see Section 1.5.1).

In Figure 2.8, the "$\mathcal{B}$"s are busy periods in $\text{M}_\lambda/\text{M}_{\mu_0}/1$ where zero-waits receive service time $\stackrel{d}{=} S_0$ ($\neq S_1$). The "$\mathcal{B}_1$"s are busy periods in $\text{M}_\lambda/\text{M}_{\mu_1}/1$

with all service times $\stackrel{d}{=} S_1$, initiated by what we call here "first descendant arrivals of an initial S_0". First descendant arrivals may occur at any time during or after the actual S_0, within the $\mathcal{B}$ which it initiates. First descendant arrivals of an S_0 initiate busy periods of type $\mathcal{B}_1$ whose left-projected horizontal base level-0 lines intersect the vertical state-space interval $(0, S_0)$. We give two methods to evaluate $E[\mathcal{B}]$, in Sections 2.5.2.1 and 2.5.2.2, respectively.

2.5.2.1 Method 1 for Expected Busy Period in M/M/1 Where Zero-wait Arrivals Receive Special Service

The busy period B is given by the initial service time S_0 plus the sum of the busy periods $B_{1,i}$, $n = 1, ..., A_{S_0}$

$$\mathcal{B} = S_0 + \sum_{i=1}^{A_{S_0}} \mathcal{B}_{1,i}, \tag{2.62}$$

where A_{S_0} = number of arrivals, including the "first descendant arrivals", during S_0 (see Figure 2.8). The sub-busy periods $\{\mathcal{B}_{1,i}\}_{n=1,2,...}$ are iid r.v.s which are $\stackrel{d}{=} \mathcal{B}_1$ in a standard $\mathrm{M}_\lambda/\mathrm{M}_{\mu_1}/1$ queue with service time S_1 (see Figure 2.8). Because all $\mathcal{B}_{1,i}$s are $\stackrel{d}{=} \mathcal{B}_1$, regardless of their number, they are "independent of A_{S_0}". More formally,

$$P(\mathcal{B}_{1,i} = x | A_{S_0} = n) = P(\mathcal{B}_{1,i} = x) \text{ for all } x > 0, \text{ and all } n = 1, 2, ...$$

Thus, the expected value of the "sum of the r.v.s" on the right side of formula (2.62), is given by:

$$E[\mathcal{B}] = E[S_0] + \lambda E[S_0] E[\mathcal{B}_1])$$

$$E[\mathcal{B}] = E[S_0] + \lambda E[S_0] \tfrac{E[S_1]}{1-\lambda E[S_1]} \tag{2.63}$$

$$E[\mathcal{B}] = \tfrac{E[S_0]}{1-\rho_1} = \tfrac{E[S_0]}{P_{0,1}},$$

where $P_{0,1} := P(\text{wait} = 0)$ in the standard "$\mathrm{M}_\lambda/\mathrm{M}_{\mu_1}/1$ queue with common service time. S_1", where we applied "Wald's equation".

To understand "Wald's equation", (see p. 47ff in [51] (2004); or Problem 13 on p. 486 in Chapter 7 in [52] (2010)).

2.5.2.2 Method 2 for Expected Busy Period in M/M/1 Where Zero-wait Arrivals Receive Special Service

We can apply an LC-based result for the expected "busy period" denoted by $E[\mathcal{B}]$, by using the "renewal reward theorem" (see Section 1.8.4) in Chapter 1 for the expected "busy cycle," denoted by $E[\mathcal{C}]$), in M/G/1, where more genrally "$\mathrm{G} \stackrel{d}{\neq} \mathrm{Exp}_\lambda$".

The derivation of $E[\mathcal{C}]$ is similar to that in the M/M/1 queue given in (2.69). Using P_0 in (2.37) gives formula (2.63) in Chapter 3, as follows:

$$E[\mathcal{B}] = \frac{1-P_0}{\lambda P_0} = \frac{1 - \frac{1-\lambda E[S_1]}{1-\lambda E[S_1]+\lambda E[S_0]}}{\lambda \frac{1-\lambda E[S_1]}{1-\lambda E[S_1]+\lambda E[S_0]}} = \frac{E[S_0]}{1-\lambda E[S_1]} = \frac{E[S_0]}{1-\rho_1} = \frac{E[S_0]}{P_{0,1}}.$$

Once we have $E[\mathcal{B}]$ using either Method 1 or Method 2, we can derive the expression for "P_0" directly using it. Thus

$$E[\mathcal{C}] = E[\mathcal{B}] + E[\mathcal{I}],$$

where $\mathcal{I}$;= "Idle period"$= \frac{1}{\lambda}$. Hence,

$$P_0 = \frac{E[\mathcal{I}]}{E[\mathcal{C}]} = \frac{\frac{1}{\lambda}}{\frac{1}{\lambda} + E[\mathcal{B}]} = \frac{\frac{1}{\lambda}}{\frac{1}{\lambda} + \frac{E[S_0]}{1-\rho_1}} = \frac{1-\rho_1}{1-\rho_1+\rho_0}.$$

2.5.3 Solution for Pdf of Wait in the M/M/1 Queue where Zero-Waits Receive Special Service

We now derive the stationary mixed pdf $\{P_0, f(x)\}_{\dot{x}>0}$ when $S_0 \stackrel{d}{=} \text{Exp}_{\mu_0}$, and $S_1 \stackrel{d}{=} \text{Exp}_{\mu_1}$. The corresponding service-time cdfs are $B_0(x) = 1 - e^{-\mu_0 x}$, and $B_1(x) = 1 - e^{-\mu_1 x}$, as in Equation (2.59). Applying the differential operator "$\langle D+\mu_0\rangle \langle D+\mu_1\rangle$" to both sides of Equation (2.59), which is equivalent to differentiating with respect to x and adding $\mu_n f(x)$, for $n = 0, 1$, in succession, yields a "second order differential equation":

$$\langle D+\mu_1-\lambda\rangle \langle D+\mu_0\rangle f(x) = 0, \tag{2.64}$$

with solution

$$f(x) = ae^{-(\mu_1-\lambda)x} + be^{-\mu_0 x}, \; x > 0, \tag{2.65}$$

provided $\mu_0 \neq \mu_1 - \lambda$ (Note: if the special relation "$\mu_0 = \mu_1 - \lambda$" were to hold, then the differential equation (2.64) would have a different solution for $f(x)$; (see a text on differential equations, e.g., pp. 106–113 in [9] (1969)). Quantities a, and b in (2.65) are constants obtained by using (2.65) and the two independent initial conditions:

$$f(0) = \lambda P_0, \text{ and } f'(0) = -\mu_0 \lambda P_0 + \lambda f(0). \tag{2.66}$$

The two constants, and P_0 are given by:

$$a = \frac{-\lambda^2 P_0}{\mu_1-\mu_0-\lambda}, \quad b = \frac{\lambda(\mu_1-\mu_0)P_0}{\mu_1-\mu_0-\lambda}, \quad P_0 = \frac{1-\rho_1}{1-\rho_1+\rho_2}, \tag{2.67}$$

where $\rho_i = \lambda/\mu_i$, $i = 1, 2$. (Note: The book [22] (2017) gives an alternative solution technique to derive $f(x)$, $x > 0$.)

2.5.4 Expected Busy Period When Service Times Are Exponential

We can check some of the properties obtained above in M/M/1 where zero-waits receive special service, using the solution for $\{P_0, f(x)\}_{x>0}$. From equation (2.63),

$$E[\mathcal{B}] = \frac{\frac{1}{\mu_0}}{1 - \frac{\lambda}{\mu_1}} = \frac{\mu_1}{\mu_0(\mu_1 - \lambda)}. \tag{2.68}$$

A check on $E[\mathcal{B}]$ in (2.68), is that the model of M/M/1 where zero-waits receive special service, reduces to the standard M/M/1 queue when $\mu_0 = \mu_1 = \mu < \lambda$. Then, $E[\mathcal{B}] = 1/(\mu - \lambda)$, which agrees with $E[\mathcal{B}]$ in the standard $M_\lambda/M_\mu/1$ queue, given in formula (2.37) in Section 2.1.17.

2.5.5 Sojourn above Level x When Service Times Are Exponential

Analysts of a system may be interested in how long a time that the virtual waiting time spends above a certain level, each time the waiting-time process upcrosses that level. Suppose that the level of interest is $x > 0$. Let γ_x denote the "excess of the vertical jump" over level x, given that $\{W(t)\}_{t\geq 0}$ upcrosses level x. Due to the "memoryless property" of the exponential distribution, the excess $\gamma_x \stackrel{d}{=} S_0 \stackrel{d}{=} \text{Exp}_{\mu_0}$ whenever the vertical jump upcrpssing is the "initial jump" of a busy period, i.e., when $S_0 \stackrel{d}{=} \text{Exp}_{\mu_0}$ (which we call here a "type-0 jump"); or the excess $\gamma_x \stackrel{d}{=} S_1 \stackrel{d}{=} \text{Exp}_{\mu_1}$ whenever the jump upcrossing occurs from "within" a busy period (which we call here a "type-1 jump"). Thus a_X will be equal to a busy period, whose initial jump size is either $\gamma_x \stackrel{d}{=} \text{Exp}_{\mu_0}$; or, the excess will be $\gamma_x \stackrel{d}{=} \text{Exp}_{\mu_1}$. In either case, all other jump sizes that begin during a_x are $\stackrel{d}{=} \text{Exp}_{\mu_1}$ (see Figure 2.8). Thus,

$$E[a_x] = p_0(x)\frac{1/\mu_0}{1 - \frac{\lambda}{\mu_1}} + p_1(x)\frac{1/\mu_1}{1 - \frac{\lambda}{\mu_1}}, \tag{2.69}$$

where $p_i(x) := P$(a vertical jump upcrossing is due to the service-time jump of type-i), $i = 0, 1$. Moreover

$$p_0(x) + p_1(x) = 1.$$

Thus, if $\frac{1}{\mu_0} < \frac{1}{\mu 1}$ then $E[a_x] < E[\mathcal{B}_1]$, where $\mathcal{B}_1$ is the busy period in a standard $M_\lambda/M_{\mu_1}/1$ queue. If $\frac{1}{\mu_0} > \frac{1}{\mu 1}$ then $E[a_x] > E[\mathcal{B}_1]$. It is shown in [22] (2017), that the probabilities $p_0(x)$ and $p_1(x)$ are given by

$$p_0(x) = \frac{P_0(x)e^{-\mu_0 x}}{f(x)/\lambda}, \quad p_1(x) = \frac{\int_{y=0}^{x} e^{-\mu_1(x-y)} f(y)dy}{f(x)/\lambda},$$

where the atationary pdf $\{P_0, f(x)\}_{\dot{x}>0}$ is given for P_0 in (2.60), and for $f(x)$ in (2.65).

2.6 Exercises

Exercise 2.6.1 *List at least three real-world potential applications of a queue where zero-wait customers get special service.*

Exercise 2.6.2 *In formula (2.60), give an analytic or practical condition such that* $\rho_0 > 1$*, and the queue is still in equilibrium.*

Exercise 2.6.3 *Give two real-world examples where it may be important to use the expected sojourn time above a level* x*,* $E[a_x]$*, given in formula (2.69).*

Exercise 2.6.4 *Check that the solution of the differential Equation (2.64) is that given in formula (2.67), by consulting, if necessary, [9] (1969) or other text on differential equations.*

Exercise 2.6.5 *In formula (2.62),* $\mathcal{B} = S_0 + \sum_{i=1}^{A_{S_0}} \mathcal{B}_{1,i}$*, explain why it is necessary that the r.v.* A_{S_0} *is independent of the r.v.s* $\mathcal{B}_{1,i}$*,* $j = 1, .., A_{S_0}$*, in order to derive formula (2.63).*

Exercise 2.6.6 *In Figure 2.8, clarify the meaning of "first-descendant arrival".*

Exercise 2.6.7 *Check the algebraic steps giving the resuts for* a*,* b*, and* P_0*, in formula (2.67).*

Exercise 2.6.8 *Give the dimensions of all terms in formulas (2.66) and (2.67), and their reasons for being so.*

Exercise 2.6.9 *Check the dimension of* $E\left[a_x\right]$ *in formula (2.69) in Section 2.5.5.*

2.7 M/M/1 with Bounded Workload

Briefly, queues with bounded workload, and related models with barriers on their key random variable, have many real-world applications (e.g., [20] (2019) and references therein).

2.7.1 Variant 1: M/M/1 Model

All customers join the system, and all actual waiting times (up to start of service) are in state-space interval $[0, K)$. Each arrival gets either full service S, or a truncated service if S would cause $\{W_K(t)\}_{t \geq 0}$ to exceed K (as deemed

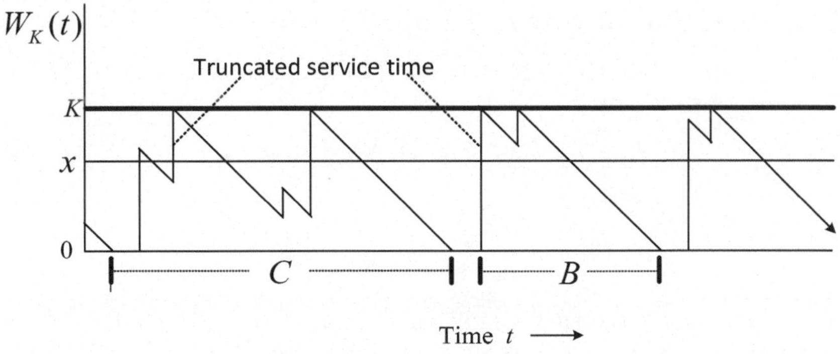

FIGURE 2.9
Sample path of $\{W_K(t)\}_{t\geq 0}$ in M/M/1 with bounded workload. $\mathcal{C}_K :=$ busy cycle, $\mathcal{B}_K :=$ busy period.

by a system manager, or other means). We define the service time S_K (due to the level-K barrier) in terms of the "prior" brought service S, as follows.

If a customer has an actual wait $y \in [0, K)$ then

$$S_K = \min(S, K - y). \tag{2.70}$$

Let the brought (aka "prior") service time $S \stackrel{d}{=} \text{Exp}_\mu$. Then the arrival's brought ccdf of service is $\overline{B}(x) = e^{-\mu x}, x > 0$.

Consider a sample path of $\{W_K(t)\}_{t\geq 0}$ (Figure 2.9). Let the stationary mixed pdf of wait be $\{P_{K,0}, f_K(x)\}_{x>0}$. Applying sample-path rate balance across level x gives immediately Equation (2.71) for $f_K(x)$ below, where the left and right sides are the sample-path downcrossung and upcrossing rates, respectively. Equation (2.72) is the normalizing condition, to evaluate $P_{0.}$.

$$f_K(x) = \lambda P_{K,0}\overline{B}(x) + \lambda \int_{y=0}^{x} \overline{B}(x-y) f_K(y)dy, 0 < x < K, \tag{2.71}$$

$$P_{K,0} + \int_{y=0}^{K} f_{K,0}(x)dx = 1. \tag{2.72}$$

If the barrier $K \in (0, \infty)$ then the mixed pdf $\{P_{K,0}, f_K(x)\}_{x>0}$ exists for all values of $\lambda > 0$ (see Section 2.1 in [24]). Also, [24] shows that $E[S_K] = (1 - P_{K,0})/\lambda$, equivalently $P_{K,0} = 1 - \lambda E[S_K]$. (Interestingly, this is similar to $P_0 = 1 - \lambda E[S]$ in the standard no-barrier M/M/1 queue in steady state.)

2.7.2 Solution of Integral Equation in Variant 1

In the $M_\lambda/M_\mu/1$ model $\overline{B}(x) = e^{-\mu x}$. The solution of equations (2.71) and (2.72) are, respectively:

$$f_K(x) = \lambda P_{K,0} e^{-(\mu-\lambda)x}, 0 < x < K, \tag{2.73}$$

$$P_{K,0} = \frac{\mu - \lambda}{\mu + e^{-(\mu-\lambda)K}}. \tag{2.74}$$

For a mild check (necessary condition) on formulas (2.73) and (2.74), let us assume that $\lambda < \mu$; under that condition the no-barrier M/M/1 queue is stable (reaches equilibrium). Let $K \to \infty$. Then $P_{K,0} \to 1-\lambda/\mu = P_0$, $f_K(x) \to \lambda P_0 e^{-(\mu-\lambda)x,}$, and the state-space interval $(0, K) \to (0, \infty)$. This result is anticipated, since the mixed pdf

$$\left\{1 - \lambda/\mu, \lambda P_0 e^{-(\mu-\lambda)x,}\right\}_{x>0}$$

is equal to the stationary mixed pdf of wait in the standard no-barrier $M_\lambda/M/_\mu/1$ queue, given in formulas (2.14) and (2.15) in Section 2.1.10.

2.7.3 Variant 2: M/M/1 Model

Upon arrival, customers are cleared immediately from the system if their total time in the system would exceed K, as deemed by a system manager. We assume that the workload $W_K(t^-)$ and the service time S of a would-be time-t arrival are known to the "system manager" by some means. A time-t arrival joins the system only if $W_K(t^-) + S < K$. We define the service time S_K due to the level-K barrier in terms of S as follows. If a customer must wait $y \geq 0$ then

$$S_K = 0 \quad \text{if } y + S > K.$$

$$S_K = S \quad \text{if } y + S \leq K, \tag{2.75}$$

Customers that are allowed to join receive full service S, and depart upon completing service. In this system,some arrivals get no service.

Consider a sample path of $\{W_K(t)\}_{t\geq 0}$ (Figure 2.10). We obtain via LC the integral equation for $f_K(x)$, $0 < x < K$,

$$f_K(x) = \lambda P_{K,0} \left(\overline{B}(x) - \overline{B}(K)\right) + \lambda \int_{y=0}^{x} \left(\overline{B}(x-y) - \overline{B}(K-y)\right) f_K(y)dy, \tag{2.76}$$

with normalizing condition

$$P_{K,0} + \int_{y=0}^{K} f_K(x)dx = 1. \tag{2.77}$$

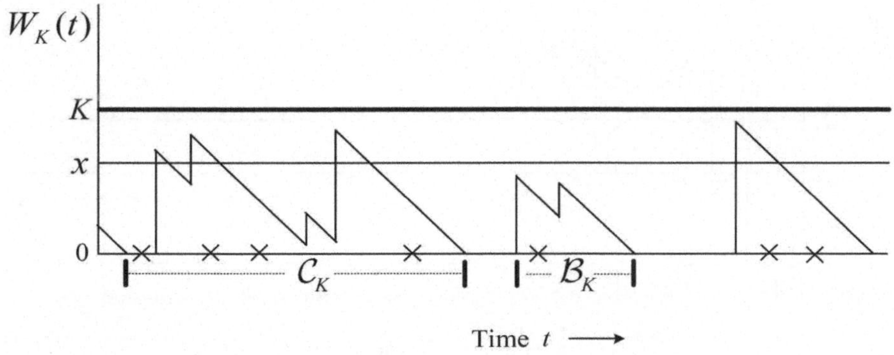

FIGURE 2.10
Variant 2. Sample path of $\{W_K(t)\}_{t\geq 0}$ in M/M/1 with bounded workload. '×' indicates arrivals who balk because their *wait*+ $S > K$. $\mathcal{C}_K$:= busy cycle, $\mathcal{B}_K$:= busy period.

In (2.76), the term $\overline{B}(x) - \overline{B}(K) := P(x < S < K)$, and the term $\overline{B}(x - y) - \overline{B}(K - y) := P(x - y < S < K - y)$. Equation (2.77) is the normalizing condition to compute P_0.

In the $\mathrm{M}_\lambda/\mathrm{M}_\mu/1$ queue with $\overline{B}(x) = e^{-\mu x}$, we obtain the solution of (2.76) for$\{P_{K,0}, f_K(x)\}_{x\in(0,K)}$ from previous work using LC, as a special case of the M/M/c queue with bounded system time. Thus, in Example 1, p. 44 in [25](1981), we set number of servers = 1, which results in

$$f_K(x) = \lambda e^{\rho\beta} P_{K,0} e^{\mu(\rho-1)x}(1 - \beta e^{\mu x})e^{-\mu\beta e^{\mu x}}, 0 < x < K, \tag{2.78}$$

$$P_{K,0} = \frac{1}{1 + \lambda e^{\rho\beta}\int_{x=0}^{K} e^{\mu(\rho-1)x}(1 - \beta e^{\mu x})e^{-\mu\beta e^{\mu x}}dx}, \tag{2.79}$$

where $\rho = \lambda/\mu$, $\beta = e^{-\mu K}$. The solution in Variant 2 is much more complex than the solution for the pdf and P_0 in Variant 1, given above in formulas (2.73) and (2.74).

Formulas (2.78) and (2.77) check with the single-server Markovian result obtained in [35] using a different method of analysis.

If $K \to \infty$ then $\beta \downarrow 0$. Additionally, if $\lambda < \mu$ then (2.78) becomes

$$f(x) = \lambda P_0 e^{-(\mu-\lambda)x}, x > 0, \qquad P_0 = 1 - \frac{\lambda}{\mu},$$

as in the standard no-barrier M/M/1 queue. This shows secondarily, that two quite different solutions can converge to the solution for the same M/M/1 queue as their model parameters change appropriately (in this case the workload barrier tends to ∞.

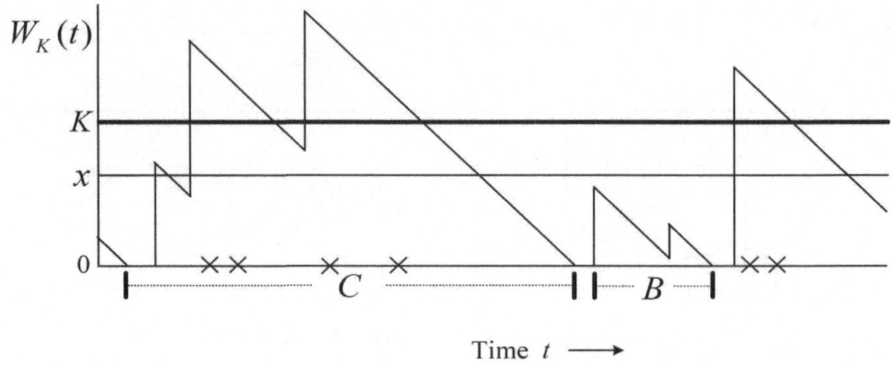

FIGURE 2.11
Variant 3. Sample path of $\{W_K(t)\}_{t\geq 0}$ in M/M/1 with threshold at level K. Symbol "×" indicates arrivals that balk because $W_K(\cdot) > K$ at their arrival instants. $\mathcal{C}_K$:= busy cycle, $\mathcal{B}_K$:= busy period.

2.7.4 Variant 3: M/M/1 Model

All arrivals that "see" a wait $< K$ join the system and receive full service S. Some of these service tines will cause jumps that upcross level K (Figure 2.11). (In variant 3 we call level K a "threshold" rather than a "barrier", because sample-path conditions of the workload "switch at level K". The effective state space is $[0, \infty) = [0, K) \cup [K, \infty)$.) Arrivals that "see" a waiting time $> K$, are blocked from joining the system, and are cleared immediately, thus causing no change of motion on the sample path of the virtual wait (workload) (see Figure 2.11).

We define the service time S_K due to the level-K threshold (threshold because the state space, in terms of S as follows. If a customer must wait y then

$$S_K = S \quad \text{if } y \in [0, K),$$

$$S_K = 0 \quad \text{if } y \in [K, \infty),$$

which may be written as $S_K = S \cdot I_{[0,k)}\ (W_\iota) + 0.I_{[k,\infty)}\ (W_\iota)$, where $I_{\boldsymbol{A}}(\cdot)$ is the characteristic function of set A.

The absolutely continuous mixed pdf of wait consists of two separate parts, defined on state-space interval $(0, K)$ and $[K, \infty)$, respectively. We denote the mixed pdf of wait as $\{P_{K,0}, f_{K,i}(x)\}_{i=0,1}$ where $P_{K,0} := P(\text{workload}= 0)$, the domain of $f_{K,0}(x)$ is $(0, K)$ and the domain of $f_{K,1}(x)$ is $[K, \infty)$. Using LC we can write integral equations for $f_{K,i}(x)$, $i = 0, 1$, by inspection of Figure

2.11, as follows:

$$f_{K,0}(x) = \lambda P_{K,0}\overline{B}(x) + \lambda \int_{y=0}^{x} \overline{B}(x-y) f_{K,0}(y)dy, x \in (0,K),$$

$$f_{K,1}(x) = \lambda P_{K,0}\overline{B}(x) + \lambda \int_{y=0}^{K} \overline{B}(x-y) f_{K,0}(y)dy, x \in [K,\infty),$$

$$P_{K,0} + \int_{y=0}^{K} f_{K,0}(x)dx + \int_{x=K}^{\infty} f_{K,1}(x)dx = 1. \quad (2.80)$$

We infer from Figure 2.11 and the fact that level K is not a tangent of the sample path, that $f(x)$ is continuous when $x = K$. That is, the continuity condition at K is:

$$f_{K,1}(K^+) = f_{K,0}(K^-). \quad (2.81)$$

This is equivalent to "$\lim_{t\to\infty} D_t(x)/t = \lim_{t\to\infty} D_t(x^-)/t$, and there are no SP tangents at level K". (Contrast this continuity property with the sample-path condition at level D in the M/D/1 queue where there is a discontinuity. See Section 3.4.1, regarding the discontinuity of the pdf of the virtual wait at level D in the M/D/1 queue. Contrast the discontinuity at level D, with the continuity at level K in M/M/1 with a workload barrier, Variant 3, due to different sample-path structure of the two models.

2.7.4.1 Expected Sojourn above Level K

Let $\gamma_K :=$ excess of a jump over level K, $a_K :=$ sojourn above level K. Then $a_K = \gamma_K$, and $E[a_K] = E[\gamma_K]$. Let $F_{\gamma_K}(z) := P(\gamma_K \le z)$, $z > 0$. Two different expressions for $\lim_{t\to\infty} U_t(K+z)/t$ are

$$(1 - F_{\gamma_K}(z))\, f_{K,0}(K^-)$$

$$\text{and} \quad \lambda P_{K,0}\overline{B}(K+z) + \lambda \int_{y=0}^{K} \overline{B}(K+z-y) f_{K,0}(y)dy.$$

In the first expression $f_{K,0}(K^-)$ is the upcrossing rate (also the downcrossing rate) of level K, and $1 - F_{\gamma_K}(z)$ is the upcrossing rate of level $K+z$ "given" the SP upcrosses level K. The second term is the upcrossing rate of level $K+z$ due to upward jumps that start in $[0,K)$. Thus

$$1 - F_{\gamma_K}(z) = \frac{\lambda P_{K,0}\overline{B}(K+z) + \lambda \int_{y=0}^{K} \overline{B}(K+z-y) f_{K,0}(y)dy}{f_K, 0(K^-)}$$

$$= \frac{\lambda P_{K,0}\overline{B}(K+z) + \lambda \int_{y=0}^{K} \overline{B}(K+z-y) f_{K,0}(y)dy}{\lambda P_{K,0}\overline{B}(K) + \lambda \int_{y=0}^{K} \overline{B}(K-y) f_{K,0}(y)dy}$$

and

$$E\left[a_K\right] = E]\gamma_K] = \int_{z=0}^{\infty}\left(1 - F_{\gamma_K}(z)\right)dz$$
$$= \int_{z=0}^{\infty}\left[\frac{\lambda P_{K,0}\overline{B}(K+z)+\lambda\int_{y=0}^{K}\overline{B}(K+z-y)f_{K,0}(y)dy}{\lambda P_{K,0}\overline{B}(K)+\lambda\int_{y=0}^{K}\overline{B}(K-y)f_{K,0}(y)dy}\right]dz. \tag{2.82}$$

2.7.5 Variant 3: Solution for Pdf of Wait

Setting $\overline{B}(x) = e^{-\mu x}$ in (2.80), and solving by converting to differential equations, gives

$$\left.\begin{aligned} f_{K,0}(x) &= \lambda P_{k,0}e^{-(\mu-\lambda)x}, x \in (0,K), \\ f_{K,1}(x) &= \lambda P_{k,0}e^{-(\mu x-\lambda K)}, x \in [K,\infty), \\ P_{K,0} &= \frac{1}{1+\frac{\lambda}{\mu-\lambda}(1-e^{-(\mu-\lambda)K})+\frac{\lambda}{\mu}e^{-(\mu-\lambda)K}}. \end{aligned}\right\} \tag{2.83}$$

In (2.83) if $x > (\lambda K)/\mu$ then $\mu x - \lambda K > 0$ and $\int_{x=K}^{\infty} f_{K,1}(x)dx$ is finite. If additionally $\lambda < \mu$ then as $K \to \infty$ the denominator of $P_{K,0} \to \frac{1}{1+\lambda/(\mu-\lambda)}$ $= 1 - \lambda/\mu$, which is P_0 in the no-threshold M/M/1 queue. Also $f_{K,0}(x) \to \lambda P_0 e^{-(\mu-\lambda)x}$, $x \in (0,\infty)$ which is $f(x)$, $x > 0$ in the no-threshold M/M/1 queue.

From (2.82), $E[a_K] = \int_{z=0}^{\infty} e^{-\mu z}dz = 1/\mu$. is M/G/1 queue which deals with multiple Poisson inputs, where each Poisson stream has its own exponential service. It turns out that this G is distributed as a linear combination of exponential distributions (which is not a plain exponential random variable).

2.8 Exercises

Exercise 2.8.1 *Give at least two potential real-world applications of tVariant 1, 2, and 3 in the workload-barrier M/M/1 queue , in Section 2.7.*

Exercise 2.8.2 *For Variant 1 of the M/M/1 queue, show that in formula (2.73) $f(x)$ is constant on the state-space interval $(0,K)$, and thus has a uniform distribution on $(0,K)$, denoted by "$U_{(0,K)}$". State why this uniform distribution on $(0,K)$ is consistent with $P_{K,0} = 0$ in formula (2.74).*

Exercise 2.8.3 *Explain formula (2.70)*

Exercise 2.8.4 *Give a potential real-world application of Variant 1 of M.N/1 in Section 2.7.2.*

Exercise 2.8.5 *Give a potential real-world application of Variant 2 of M.N/1 in Section 2.7.3.*

Exercise 2.8.6 *Give a potential real-world application of Variant 3 of M.N/1 in Section 2.7.4.*

3

M/G/1 Queues

In "M/G/1" queues, the letter "G" may represent different service mechanisms. "G" may represent the type of random variable for the service-time distribution of the arriving customers. For example service time may depend on waiting time in the queue. Or, servivice time may depend on the number of customers left waiting when the arrival starts service, etc. In general, G is distributed differently from a pure Exp_{μ} random variable. Also, "G" may be a discrete random variable such as a "fixed amount of time" $D > 0$ for each arrival. Or, "G" may be a continuous random variable, like a uniform r.v. on an interval $(0, a)$, $a > 0$, denoted by "$\mathrm{U}_{(0,a)}$", $a > 0$. "G" may be a linear combination of several Exp_{μ}s, e.g., a "generalized hyper-exponential" distribution (see [8]; [60] (2012)). (Note: If G := Exp_{μ}, the notation used is $\mathrm{M}_{\lambda}/\mathrm{M}_{\mu}/1$, as in Chapter 2.

Note that "M" stands for "Markovian"–see pp. 237, 238, 239, 456, in [52] (2010) for the "Markovian Property." Briefly, the Markov proerty means that the future probabilistic behaviour given knowledge of the exact present state of a peocess, is idependent of the past states prior to the present sate–see p. 29 in [40] 1975 for further explanation.

In Sections 3.2–3.3.2, the first M/G/1 model we consider in this Chapter, "G" represents a mixture of several independent Poisson arrival strems with distinct "general" service times, which are distributed differently from pure exponential probability distributions. Section 3.3.3, then treats the case where "G" represents a mixture of several independent Poisson arrival streams with distinct pure exponential probability distributions.

3.1 Integral Equation for Pdf of Virtual Wait in General M/G/1 Queue

Let the stationary pdf of the virtual wait be $\{P_{0,}, f(x)\}_{x>0.}$. Let the cdf of Service Time S, be $B(x), x > 0$, and the ccdf be $\overline{B}(x), x \dot{\geq} 0$. The integral equation for $f(x), x > 0$, is

$$f(x) = \lambda P_0 \overline{B}(x) + \lambda \int_{y=0}^{x} \overline{B}(x-y) f(y) dy, \tag{3.1}$$

DOI: 10.1201/9780429297601-3

$$P_0 + \int_0^\infty f(x)dx = 1. \tag{3.2}$$

We will use formulas (3.1) and (3.2) in this chapter, e.g., in Section 3.3.2.

3.2 M/G/1 with Independent Multiple Poisson Inputs

Customers arrive at a "single-server" system in N independent Poisson streams at rates λ_i, $i = 1, ..., N$, where $\lambda = \sum_{i=1}^{N} \lambda_i$. A physical picture of the system is similar to Figure 2.1 in Section 2.1.1, except that here, the large circles in the figure represent randomly distributed so-called "type-i" customers arriving at rate λ_i, and having prior service service with general probability distribution "G". That is, the "G"s may be Exp_{μ_i}s, or may be another type of r.v. such as $U_{(0,a)}$s (uniform r.v. on $(0, a)$).

Denote the service times of customers that arrive at rate λ_i by S_i with cdf $B_i(x)$, complementary cdf (aka "ccdf") $\overline{B}_i(x)$, $x \geq 0$, and pdf $b_i(x)$, $x > 0$, $i = 1, ..., N$. The service discipline is FCFS (service in order of arrival). Denote the stationary mixed pdf of the wait before service. W_q by $\{P_0, f(x)\}_{x>0}$ and its cdf of wait by $F(x)$, $x \geq 0$, i.e., $F(x) = P_0 + \int_0^\infty f(x)dx$.

The service time of an arbitrary arrival is S_i with probability λ_i/λ. An argument for this fact follows. Let $A_{\mathcal{C},i}$ denote the number of arrivals of the Poisson stream i, during a busy cycle C of the virtual wait process. By the renewal reward theorem

$$\frac{E[A_{\mathcal{C},i}]}{E[\mathcal{C}]} = \lim_{t\to\infty} \frac{\text{number of type-}i\text{ arrivals in }(0,t)}{t} = \frac{\lambda_i t}{t} = \lambda_i,$$

$$E[A_{\mathcal{C},i}] = E[\mathcal{C}]\,\lambda_i = \frac{\lambda_i}{\lambda P_0}$$

where the term "λ_i" is due to the arriving Poisson streams being mutually independent (see Chapter 5 in [52] (2010), and $E[\mathcal{C}] = \frac{1}{\lambda P_0}$). This,

$$E[A_{\mathcal{C},i}] = \frac{\lambda_i}{\lambda P_0} = \frac{\lambda_i}{\lambda} \times \frac{1}{P_0},\ i = 1, ..., N,$$

where $P_0 = P(\text{wait} = 0)$. Moreover, the expected total number of "arrivals in a busy cycle" C is equal to "$\frac{1}{P_0}$" (same as expected "number served in a busy period"). Assuming the arrival times are mixed according to their arrival rates during a busy cycle, the probability that an arrival is type i, is equal to $\frac{\lambda_i}{\lambda}$, $i = 1, ..., N$ (see Section 5.3.4 on p. 319ff in [52] (2010)).

By the foregoing argument, the system is. in effect, an M/G/1 queue with a total arrival rate "$\lambda = \sum_{i=1}^{N} \lambda_i$". The total prior service time S has a "mixture" probability distribution: given by

$$S = \begin{cases} S_1 \text{ with probability } \frac{\lambda_1}{\lambda}, \\ \quad \cdots \\ S_N \text{ with probability } \frac{\lambda_N}{\lambda}. \end{cases} \tag{3.3}$$

The random variable S has mixture components S_i and mixture "weights" $\lambda_i/\lambda > 0$, $i = 1, ..., N$, where

$$\sum_{i=1}^{N} \lambda_i/\lambda = 1.$$

The expected value of the service time S is:

$$E[S] = \sum_{i=1}^{N} \frac{\lambda_i}{\lambda} E[S_i] = \frac{1}{\lambda} \sum_{i=1}^{N} \rho_i, \tag{3.4}$$

using the notation "$\rho_i = \lambda_i E[S_i]$", $i = 1, ..., N$.

For a standard M/G/1 queue, a straightforward way to show that $P_0 = 1 - \lambda E[S]$, is to integrate both sides of integral equation (3.1) on $(0, \infty)$ above, and apply formula (3.2). This procedure utilizes:the facts that $\int_{x=0}^{\infty} f(x)dx = 1 - P_0$, and $E[S] \equiv \int_{x=0}^{\infty} \overline{B}(x)dx$. Simplifying the double integral on the right side; followed by some algebra, gives P_0. as:

$$P_0 = 1 - \lambda E[S] = 1 - \sum_{i=1}^{N} \rho_i. \tag{3.5}$$

Explicitly, the foregoing algebraic steps are:

$$\begin{aligned}
f(x) &= \lambda P_0 \overline{B}(x) + \lambda \int_{y=0}^{x} \overline{B}(x-y) f(y)dy, \\
1 - P_0 &= \int_{x=0}^{\infty} f(x)dx = \lambda P_0 \int_{x=0}^{\infty} \overline{B}(x) + \lambda \int_{x=0}^{\infty} \int_{y=0}^{x} \overline{B}(x-y) f(y)dy, \\
1 - P_0 &= \lambda P_0 E[S] + \lambda \int_{y=0}^{\infty} \int_{x=y}^{x} \overline{B}(x-y) f(y)dxdy, \\
1 - P_0 &= \lambda P_0 E[S] + \lambda E[S] \int_{y=0}^{\infty} f(y)dxdy, \\
1 - P_0 &= \lambda P_0 E[S] + \lambda E[S](1 - P_0) = \lambda E[S], \\
P_0 &= 1 - \lambda E[S].
\end{aligned} \tag{3.6}$$

3.2.1 Stability in M/G/1 Queue with Multiple Poisson Inputs

The M/G/1 queue is stable iff every "typical" sample path of the virtual wait process $\{W(t)\}_{t\geq 0}$ returns to state $\{0\}$. This occurs iff $P_0 > 0$. From formula (3.5) stability holds iff

$$\sum_{i=1}^{N} \rho_i < 1. \tag{3.7}$$

Inequality 3.7 is equivalent to:

$$E[\text{number of arrivals in a service time}] < 1. \tag{3.8}$$

That is, the overall arrival rate of customers to the system is less than the overall service rate.

The validity of (3.7) follows from:

$$E\left[\text{number of arrivals during a service time}|S = S_i\right] = \lambda E\left[S_i\right],$$

the proportion of arrivals having $S = S_i$ is $\frac{\lambda_i}{\lambda}$; the N Poisson processes are independent (see Proposition 5.2 on p. 320ff in [52] (2010))

Therefore

$$E\left[\text{number of arrivals during a service time}\right] = \sum_{i=1}^{N} \lambda E\left[S_i\right] \times P(S = S_i)$$

$$= \sum_{i=1}^{N} \lambda E\left[S_i\right] \times \frac{\lambda_i}{\lambda} = \sum_{i=1}^{N\lambda_i} \lambda_i E\left[S_i\right] = \sum_{i=1}^{N} \rho_i,$$

and

$$E\left[\text{number of arrivals during a service time}\right] < 1 \iff \sum_{i=1}^{N} \rho_i < 1.$$

3.2.2 Example of Integral Equation for Pdf of Wait When $N = 2$

For clarity of exposition, assume that $N = 2$. Sample paths of the virtual wait $\{W(t)\}_{t\geq 0}$ resemble those of the standard M/G/1 queue, except that the jump size due to an arrival is $\stackrel{d}{=} S_i$ with probability λ_i/λ, $i = 1, 2$, and the jump sizes have ccdf $\overline{B}_i(x)$, $x > 0$, $i = 1, 2$. Jumps $\stackrel{d}{=} S_i$ occur at Poisson rate λ_i, $i = 1, 2$. By the basic LC Theorem in Section 1.4 in Chapter 1, for a fixed state-space level $x > 0$, the SP downcrossing rate is the $f(x)$ component of the mixed stationary pdf $\{P_0, f(x)\}_{x>0}$. The SP upcrossing rate of level x for an individual stream of type i arrivals is

$$\lambda_i P_0 \overline{B}_i(x) + \lambda_i \int_{y=0}^{x} \overline{B}_i(x-y) f(y) dy, i = 1, 2,$$

"Balancing" the total SP downcrossing and upcrossing rates of level x for both customer types, yields the integral equation for $f(x)$,

$$f(x) = \sum_{i=1}^{2} \lambda_i \left[P_0 \overline{B}_i(x) + \int_{y=0}^{x} \overline{B}_i(x-y) f(y) \right] dy;$$

or, factoring out the term "λP_0":

$$f(x) = \lambda P_0 \left(\sum_{i=1}^{2} \frac{\lambda_i}{\lambda} \overline{B}_i(x) \right) + \lambda \int_{y=0}^{x} \left(\sum_{i=1}^{2} \frac{\lambda_i}{\lambda} \overline{B}_i(x-y) \right) f(y) dy. \quad (3.9)$$

Integral equation (3.9) is in the form of the analogous integral equation depicted in Figure 2.5 in Section 2.1.7.1 for the pdf of wait in a "standard" M/G/1 queue with "$\lambda \equiv \sum_{i=1}^{N} \lambda_i$", and "$\overline{B}(x) \equiv \sum_{i=1}^{N} (\lambda_i/\lambda) \overline{B}_i(x)$, $\dot{x} > 0$". See also integral equation (3.1) and formula (3.2).

3.2.3 Expected Wait before Service in M/G/1 Queue with Multiple Poisson Inputs: The Pollaczek-Khinchine (P-K) Formula

Consider the second moment of the prior service time S, denoted by "$E\left[S^2\right] \equiv \sum_{i=1}^{N} \frac{\lambda_i}{\lambda} E\left[S_i^2\right]$". This is the Pollaczek-Khinchine (P-K) formula (see formula (3.63) on p. 75 in [22] (2017). Also, pp. 220–225 in [36] (2008)) gives the expected wait before service as:

$$E\left[W_q\right] = \frac{\lambda E\left[S^2\right]}{2(1 - \lambda E\left[S\right])} = \frac{\lambda \sum_{i=1}^{N} \left(\frac{\lambda_i}{\lambda}\right) E\left[S_i^2\right]}{2(1 - \sum_{i=1}^{N} \rho_i)} = \frac{\sum_{i=1}^{N} \lambda_i E\left[S_i^2\right]}{2P_0}. \quad (3.10)$$

where P_0 is given in formula (3.5).

Alternatively, we can obtain $E\left[W_q\right]$ in (3.10) directly from integral equation (3.9) on multiplying both sides by x, then integrating both sides over $x \in (0, \infty)$, changing the order of integration in the double integral, and doing some algebra. For an example of this calculation, a similar sequence of algebraic steps for $E\left[W_q\right]$ in the M/M/1 queue is carried out above in Section 2.1.11 of Chapter 2.

3.3 Exercises

Exercise 3.3.1 *Carry out the algebraic steps for deriving $E[W_q]$ in (3.10) directly from integral equation (3.9) in M/G/1 Queue with Multiple Poisson Inputs.*

3.3.1 Little's Law and Expected Number in M/G/1 Queue with Multiple Poisson Inputs

3.3.1.1 Little's Law

Let $\boldsymbol{W}$:= "average wait in a queue before service", in equilibrium. Let $\boldsymbol{L}$:= "average number in the queue before service, in equilibrium. Let λ := Poisson arrival rate to the queue. Little's law states:

$$\boldsymbol{L} = \boldsymbol{\lambda W} \tag{3.11}$$

(see [44] (1981)).

Let L_q:= expected number of units in a queue before service, in equilibrium. Using formula (3.10), and applying Little's law in formula (3.11) gives

$$L_q = \lambda E[W_q] = \frac{\lambda \sum_{i=1}^{N} \lambda_i E[S_i^2]}{2P_0}. \tag{3.12}$$

Denote the expected number of type i units in the queue by $L_{q,i}$. Let the wait of an arbitrary type i customer be $W_{q,i}$, and let the wait of an arbitrary customer be W_q. Then $W_{q,i} \stackrel{d}{=} W_q$, because the waiting time of any arrival depends only on the current workload at the arrival instant. Thus $E[W_{q,i}] = E[W_q]$, $i = 1, ..., N$. By formula (3.11) ,

$$L_{q,i} = \lambda_i E[W_{q,i}] = \lambda_i E[W_q] = \frac{\lambda_i \sum_{i=1}^{N} \lambda_i E[S_i^2]}{2P_0}, \; i = 1, ..., N. \tag{3.13}$$

As a mild (necessary) check, the terms "$L_{q,i}$" and of "$\frac{\lambda_i \sum_{i=1}^{N} \lambda_i E[S_i^2]}{2P_0}$" in formula (3.21) all terms must be dimensionless numbers (see Section 1.8.6 in Chapter 1).

3.3.2 Expected Busy Period in M/G/1 Queue with Multiple Poisson Arrivals

Applying similar derivations as the derivations of formulas (2.37) in Section 2.1.17 and (3.5) in Section 3.2, the expected busy period is given by:

$$E[\mathcal{B}] = \frac{1 - P_0}{f(0)} = \frac{1 - P_0}{\lambda P_0} = \frac{\sum_{i=1}^{N} \rho_i}{\lambda \left(1 - \sum_{i=1}^{N} \rho_i\right)}. \tag{3.14}$$

Example 3.3.1 *Suppose that each of the N arrival streams has the same probability of next staring service, i.e.,*

$$S = S_i \text{ with probability } \frac{1}{N}, i = 1, ..., N,$$

$$E[S] = \sum_{i=1}^{N} \frac{1}{N} E[S_i],$$

where it is assumed that $E[S]$ *is a convex combination of* $E[S_i]$, $i = 1, ...,$ N *(coefficients are positive and sum to 1). In formula (3.14),* $\lambda_i/\lambda \equiv 1/N$, $i = 1, ..., N$. *Thus,*

$$\lambda_i = \frac{\lambda}{N}, i = 1, ..., N,$$

$$\rho_i \equiv \lambda_i E[S_i] = \frac{\lambda}{N} E[S_i], \ i = 1, ..., N,$$

Hence,

$$\sum_{i=1}^{N} \frac{\lambda_i}{\lambda} = \sum_{i=1}^{N} \frac{1}{N} = 1,$$

$$\sum_{i=1}^{N} \rho_i = \sum_{i=1}^{N} \lambda_i E[S_i] = \sum_{i=1}^{N} \frac{\lambda}{N} E[S_i].$$

$$= \lambda \sum_{i=1}^{N} \frac{1}{N} E[S_i] = \lambda E[S].$$

In this case, the multiple Poisson input model reduces to a standard M/G/1 queue with arrival rate λ *and* $E[S] = \frac{1}{N}\sum_{i=1}^{N} E[S_i]$. *From formula (3.14),*

$$E[\mathcal{B}] = \frac{\frac{\lambda}{N}\sum_{i=1}^{N} E[S_i]}{\lambda\left(1 - \sum_{i=1}^{N} \rho_i\right)} = \frac{\sum_{i=1}^{N} \frac{1}{N} E[S_i]}{1 - \sum_{i=1}^{N} \rho_i} = \frac{E[S]}{P_0},$$

which is the formula for $E[\mathcal{B}]$ *in the "standard M/G/1 queue".*

3.3.3 M/M/1 Queue with Multiple Poisson Inputs with Exponentially Distruted Service Times

The previous Section 3.2 gives results for M/G/1 queues with multiple Poisson arrivals, where "G" represents the prior probability distribution of an arrival's service time. In this section, the arrival stream has multiple Poisson arrivals, and specialize the prior service times "G" exponential distributions. Let $\{P_0, f(x)\}_{x>0}$ denote the mixed pdf of the virtual wait. To outline a solution technique for integral Equation (3.9), we assume here that the service times for arrival stream i are $\stackrel{d}{=} \text{Exp}_{\mu_i}$, $i = 1, ..., N$. In the terminology of Section 3.2, $\overline{B}_i(x) = e^{-\mu_i x}$, $i = 1, 2, .., N$. Then (3.9) becomes

$$f(x) = \sum_{i=1}^{N} \lambda_i \left[P_0 e^{-\mu_i x} + \int_{y=0}^{x} e^{-\mu_i(x-y)} f(y) dy \right], \ x > 0. \tag{3.15}$$

We can apply the differential operator $\langle D + \mu_1 \rangle ... \langle D + \mu_N \rangle$ to Equation (3.15), to derive and solve analytically an Nth order differential equation for $f(x)$ with constant coefficients, using "initial conditions" to obtain the constants of integration.

The differential operator $\langle D + \mu_i \rangle$ is "commutative" with respect to exponential functions of the form $e^{\alpha x+\beta}$, where α and β are constants, i.e., for any permutation $(i_1, i_2, ..., i_N)$ of the numbers $(1, 2, ..., N)$. Hence,

$$\langle D + \mu_1 \rangle \cdots \langle D + \mu_N \rangle e^{\alpha \alpha x+\beta} = \langle D + \mu_{i_1} \rangle \cdots \langle D + \mu_{i_N} \rangle e^{\alpha x+\beta}. \tag{3.16}$$

For example, if $N = 2$, then

$$\begin{aligned}
&\langle D + \mu_1 \rangle \langle D + \mu_2 \rangle e^{\alpha x+\beta} \\
&= \langle D + \mu_1 \rangle (\alpha e^{\alpha x+\beta} + \mu_2 e^{\alpha x+\beta}) \\
&= \langle D + \mu_1 \rangle (\alpha + \mu_2) e^{\alpha x+\beta} \\
&= (\alpha + \mu_2)\alpha e^{\alpha x+\beta} + \mu_1(\alpha + \mu_2) e^{\alpha x+\beta} \\
&= (\alpha + \mu_2)(\alpha + \mu_1) e^{\alpha x+\beta} \\
&= (\alpha + \mu_2) \langle D + \mu_1 \rangle e^{\alpha x+\beta} \\
&= \langle D + \mu_2 \rangle \langle D + \mu_1 \rangle e^{\alpha x+\beta}.
\end{aligned} \tag{3.17}$$

The calculation in (3.17) shows that commutativity of the differential operators on an exponential function, is valid when $N = 2$. Using mathematical induction, shows commutativity is valid for all $N \geq 2$. The commutativity property simplifies the conversion of an integral equation into a differential equation, when the active kernel of any term in the integral has an exponential form like "$e^{-\mu_i(x-y)}$" in integral Equation (3.15).

3.3.3.1 *E*[Number in Queue], *E*[Wait], and *E*[Number in Queue]

If $S_i \stackrel{d}{=} \text{Exp}_{\mu_i}$ then the second moment is

$$E\left[S_i^2\right] = \int_{x=0}^{\infty} x^2 \mu_i e^{-\mu_i x} dx = \frac{2}{\mu_i^2}. \tag{3.18}$$

When formula (3.18) substituted into (3.10) and (3.12) respectively, then

$$E[W_q] = \frac{\sum_{i=1}^{N} \frac{\lambda_i}{\mu_i^2}}{1 - \sum_{i=1}^{N} \frac{\lambda_i}{\mu_i}}, \tag{3.19}$$

$$L_q = \frac{\lambda \sum_{i=1}^{N} \frac{\lambda_i}{\mu_i^2}}{1 - \sum_{i=1}^{N} \frac{\lambda_i}{\mu_i}}. \tag{3.20}$$

where $L_q := E$[number of customers waiting for service]. Similarly, substituting into (3.13), gives $L_{q,i}$,$= E$[number of "type-i customers" waiting for service] as:

$$L_{q,i} = \frac{\lambda_i \sum_{i=1}^{N} \frac{\lambda_i}{\mu_i^2}}{1 - \sum_{i=1}^{N} \frac{\lambda i}{\mu_i}}, i = 1, ..., N. \tag{3.21}$$

3.3.3.2 M/M/1 Queue with Two Customer Types

In order to illustrate the solution of integral Equation (3.15), we consider two distinct customer types, and derive the stationary pdf $\{P_0, f(x)\}$. Asume $N = 2$ in (3.15), and apply to both sides, the differential operator "$\langle D+\mu_1\rangle\langle D+\mu_2\rangle$" (see the sample calculation in formula (3.17)). This application results in a second order differential equation for $f(x)$:

$$\left\langle D^2 + (\mu_1 + \mu_2 - \lambda)D + (\mu_1\mu_2 - \mu_1\lambda_2 - \mu_2\lambda_1)\right\rangle f(x) = 0, \tag{3.22}$$

with solution

$$f(x) = ae^{R_1 x} + be^{R_2 x}, \tag{3.23}$$

where R_i, $i = 1, 2$ are the roots for z in the "characteristic equation" (see definition of characteristic equation on p. 107 in [9](1969))

$$z^2 + (\mu_1 + \mu_2 - \lambda)z + \mu_1\mu_2 - \mu_1\lambda_2 - \mu_2\lambda_1 = 0. \tag{3.24}$$

Both roots of equation (3.24) are negative since the product $R_1 \times R_2 = \mu_1\mu_2 - \mu_1\lambda_2 - \mu_2\lambda_1 > 0$ (absolute term, equivalent to $1 - \rho_1 - \rho_2 > 0$, which is the stability condition for this queue), and the sum (coefficient of z) $R_1 + R_2 = -(\mu_1 + \mu_2 - \lambda) < 0$. The constants a and b are determined by applying two independent initial conditions: $f(0) = a + b$ and $f'(0) = R_1 a + R_2 b$, obtained from (3.23) and also from (3.15), resulting in two equations for a and b, namely:

$$a + b = \lambda P_0,$$

$$R_1 a + R_2 b = -(\mu_1\lambda_1 + \mu_2\lambda_2)P_0 + \lambda f(0) = -(\mu_1\lambda_1 + \mu_2\lambda_2 - \lambda^2)P_0. \tag{3.25}$$

Thus $f(x)$ is given by formula (3.23), where a and b are the solutions of the two equations in (3.25):

$$\begin{aligned} a &= -P_0(R_2\lambda - \lambda^2 + \lambda_1\mu_1 + \lambda_2\mu_2)/(R_1 - R_2), \\ b &= P_0(R_1\lambda - \lambda^2 + \lambda_1\mu_1 + \lambda_2\mu_2)/(R1 - R2), \end{aligned} \tag{3.26}$$

and

$$\left.\begin{aligned} &P_0 = 1 - \tfrac{\lambda_1}{\mu_1} - \tfrac{\lambda_2}{\mu_2}, \\ &R_1 = \tfrac{-\boldsymbol{B}}{2} + \tfrac{\sqrt{\boldsymbol{B}^2 - 4\boldsymbol{AC}}}{2}, \quad R_2 = \tfrac{-B}{2} - \tfrac{\sqrt{\boldsymbol{B}^2 - 4\boldsymbol{AC}}}{2}, \\ &\boldsymbol{A} = 1,\ \boldsymbol{B} = \mu_1 + \mu_2 - \lambda,\ \boldsymbol{C} = \mu_1\mu_2 - \mu_1\lambda_2 - \mu_2\lambda_1. \end{aligned}\right\} \tag{3.27}$$

Example 3.3.2 *Consider a numerical example with* $N = 2$, $\lambda_1 = 1.0$, $\lambda_2 = .0.5$, $\mu_1 = 3$, $\mu_2 = 2$. *Then* $P_0 = 0.4167$, $R_1 = -1.0$, $R_2 = -2.5$, $a = 0.555555$, $b = 0.069444$, *and*

$$f(x) = 0.555555\ e^{-1.0x} + 0.069444\ e^{-2.5x}, x > 0.$$

A computational check shows that $F(\infty) = 1$, *and* $f(0) = \lambda P_0$, *i.e.,*

$$\begin{aligned} F(\infty) &= P_0 + \int_{x=0}^{\infty} f(x)dx \\ &= 0.4167 + \int_{x=0}^{\infty} (0.555555\ e^{-1.0x} + 0.069444\ e^{-2.5x})dx = 1, \end{aligned}$$

$$f(0) = a + b = \lambda P_0 = 0.625.$$

3.3.4 Expected Sojourn above Level x in M/M/1 Queue with Multiple Poisson Arrival Types

Assume that N independent Poisson straems of customers arrive at different rates λ_i , $i = 1, ...N$. Denote the sojourn time above a fixed level x by a_x with expected value $E[a_x]$. Let $P_i^{\mathcal{U}}(x) :=$ $P(\text{arrival type } i | \{W(t)\}_{t\geq 0} \text{ upcrosses } x)$, $i = 1, ...N$. By Bayes' rule (seee equation (1.21) in Section 1.5.2 in Chapter 1)

$$P_i^{\mathcal{U}}(x) = \frac{P(\{W(t)\}_{t\geq 0} \text{ upcrosses } x | \text{arrival type } i) \times P(\text{arrival type } i)}{P(\{W(t)\}_{t\geq 0} \text{ upcrosses } x)}.$$

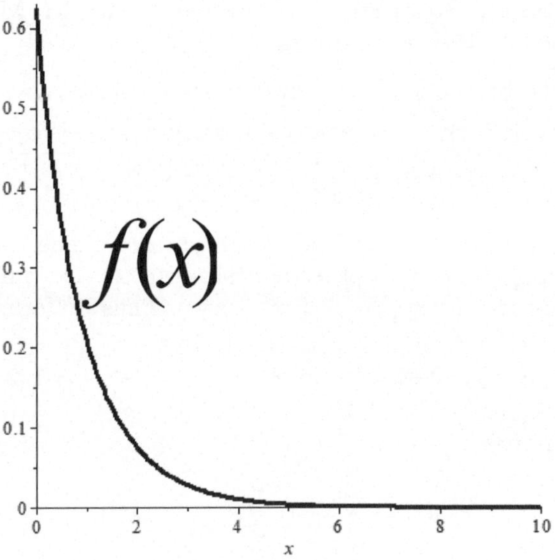

FIGURE 3.1
Pdf of $f(x)$ in formula (3.23) in Example 3.3.2.

In integral equation (3.15) in Section 3.3.3, apply formula (1.22) to obtain:

$$P_i^{\mathcal{U}}(x) = \frac{\left(P_0 e^{-\mu_i x} + \int_{y=0}^{x} e^{-\mu_i(x-y)} f(y)dy\right) \times (\lambda_i/\lambda)}{\sum_{i=1}^{N}\left[P_0 (\lambda_i/\lambda) e^{-\mu_i x} + \int_{y=0}^{x} (\lambda_i/\lambda) e^{-\mu_i(x-y)} f(y)dy\right]}$$

$$= \frac{P_0 (\lambda_i/\lambda) e^{-\mu_i x} + \int_{y=0}^{x} (\lambda_i/\lambda) e^{-\mu_i(x-y)} f(y)dy}{f(x)/\lambda}. \tag{3.28}$$

Let $\gamma_{S_i}(x)$:= excess over level x due to a type-i upcrossing of x. The sojourn $a_x \overset{d}{=} \mathcal{B}_{(i)}$ with probability $p_i^{\mathcal{U}}(x)$, where $\mathcal{B}_{(i)}$ is a busy period where "zero-waits get service $S_0 \overset{d}{=} \gamma_{S_i}(x)$". In $\mathcal{B}_{(i)}$, positive-waits get service time S_i (see formula (3.3) in Section 3.2). Applying (2.63) in Section 2.5.2.1 in Section 2.5.2.1 in Chapter 2, gives

$$E[a_x] = \frac{E[\gamma_S(x)]}{1 - \sum_{i=1}^{N} \rho_i}. \tag{3.29}$$

3.3.4.1 Expected Sojourn above Level x in M/M/1 with Two Types of Poisson Inputs

Consider an M_λ/M/1 queue with two types of arrivals with service times $S_i \overset{d}{=} \mathrm{Exp}_{\mu_i}$, $i = 1,2$. By the memoryless property, the excess over level x on upcrossing x by a type-i arrival, is $\gamma_{S_i}(x) \overset{d}{=} \mathrm{Exp}_{\mu_i}$, $i = 1,2$, for all $x \geq 0$. If the upcrossing of level x which initiates the sojourn a_x is due to a type-i arrival then $a_x \overset{d}{=} \mathcal{B}_i$, a "busy period" in an M/M/1 queue with initial service time "$S_0 = \mathrm{Exp}_{\mu_i}$", and with other nonzero-wait service times in a_x being a randomized mixture of Exp_{μ_1} and Exp_{μ_2} r.v.s. Thus by formula (3.29),

$$E\left[a_x|\text{excess} = \gamma_i(x)\right] = \frac{\frac{1}{\mu_i}}{1 - \sum_{i=1}^{2} \rho_i}, \; i = 1,2,$$

Unconditioning gives

$$E\left[a_x\right] = \sum_{i=1}^{2} \frac{\frac{1}{\mu_i}}{1 - \sum_{i=1}^{2} \rho_i} \times P_i^{\mathcal{U}}(x). \tag{3.30}$$

From (3.28) the probability of a type-i upcrossing of level x is

$$P_i^{\mathcal{U}}(x) = \frac{P_0\left(\frac{\lambda_i}{\lambda}\right) e^{-\mu_i x} + \int_{y=0}^{x} \left(\frac{\lambda_i}{\lambda}\right) e^{-\mu_i(x-y)} f(y)dy}{\frac{f(x)}{\lambda}}, \; i = 1,2.$$

3.3.4.2 Numerical Example: M/M/1 with Two Customer Types

Using the input values and pdf $f(x)$, $x > 0$ in Example 3.3.2 in Section 3.3.3.2, computation gives the values of $P_1^{\mathcal{U}}$ and $P_2^{\mathcal{U}}$ as:

$$P_1^{\mathcal{U}}(x) = \frac{0.27777\, e^{-3.x}(0.5\, e^{0.5x} + e^{2.x})}{0.55555\, e^{-x} + 0.06944\, e^{-2.5x}}, \tag{3.31}$$

$$P_2^{\mathcal{U}}(x) = 1 - p_1^{\mathcal{U}}(x), \; x \geq 0. \tag{3.32}$$

Applying formula (3.30) in Section 3.3.4.1, gives

$$E\left[a_x\right] = P_1^{\mathcal{U}}(x)\frac{1}{1.25} + P_2^{\mathcal{U}}(x)\frac{1}{0.83333},$$

where $P_2^{\mathcal{U}}(x)$ and $P_1^{\mathcal{U}}(x)$ are given in formulas (3.31) and (3.32) respectively. From (3.14), $E[\mathcal{B}] = 0.9333$ for the M/G/1 busy period with multiple Poisson inputs A plot, shows clearly that $E[a_x] > E[\mathcal{B}]$ for all $x > 0$, and $E[a_0] = E[\mathcal{B}]$;also $\lim_{x\to\infty} E[a_x] = 1$ (Figure 3.3). Moreover, $\lim_{x\to\infty} P_i^{\mathcal{U}}(x) = 0.5, i = 1,2$ (Figure 3.2). These observations are readily verified analytically. The increasing concave growth of $E[a_x]$ as x increases is due to the fact that $E[\gamma_x] = 1/\mu_i$ independent of x, implying that the evolution of $E[a_x]$ on $(0,\infty)$ is determined by the $P_i^{\mathcal{U}}(x)$s, which do change as x increases. (In examples where $E[\gamma_x]$ depends on x, e.g., non-exponential prior service times, the properties of $E[a_x]$ as x increases would be different.)

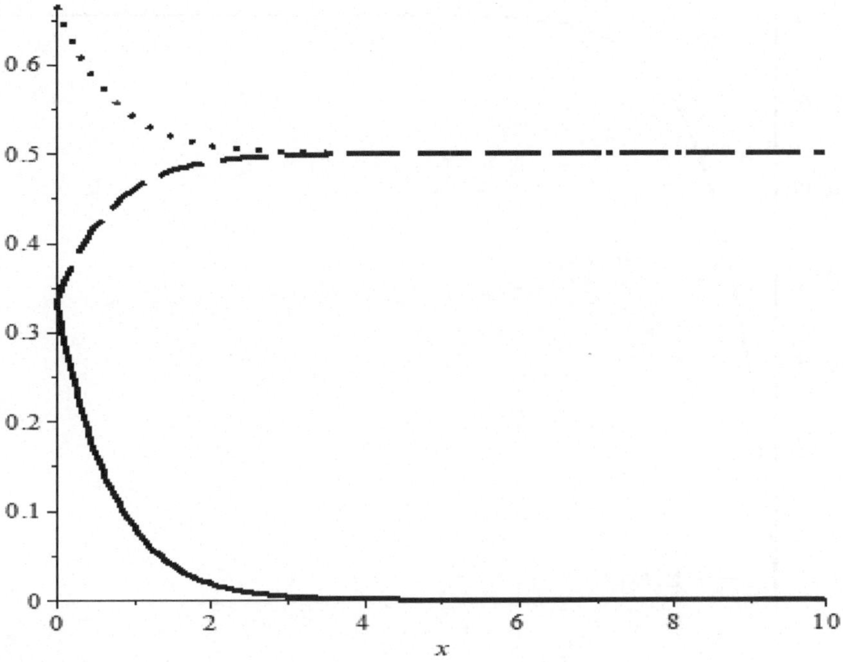

FIGURE 3.2
Changes in probabilities $P_i^{\mathcal{U}}(x)$, $i = 1, 2$ and in $P_i^{\mathcal{U}}(x) - P_2^{\mathcal{U}}(x)$, as x increase on $[0, 10)$ in Example 3.3.4.2.
In figure, $P_i^{\mathcal{U}}(x)$–**dots**; $P_2^{\mathcal{U}}(x)$–**dashes;** $P_i^{\mathcal{U}}(x)(x) - P_2^{\mathcal{U}}(x)(x)$–**solid**.

3.4 M/D/1 Queue

The standard M/D/1 queue is a classical model in queueing theory, first analyzed by A.K. Erlang in 1909 [33](1909). Here, we use LC to derive: the stationary cdf "$F(x)$, $x \geq 0$" of the customer wait before service, and the stationary mixed pdf "$\{P_0, f(x)\}_{x>0}$". Note that $f(x) = dF(x)/dx$, $x > 0$, wherever the derivative exists.

Definition 3.1 *The "pmf" is the "probability mass function" which gives the exact probabilities on the discrete states in the state space of a stochastic model.*

We also obtain the stationary pmf (see Definition 3.1) of the "number of customers" in the system P_n, $n = 0, 1, 2, ...$, and related quantities. A typical sample path of the virtual wait is shown in Figure 3.4.

In the standard M/D/1 queue, the customer arrival stream is Poisson at rate λ. The service time for each customer is deterministic, i.e., $S = D > 0$.

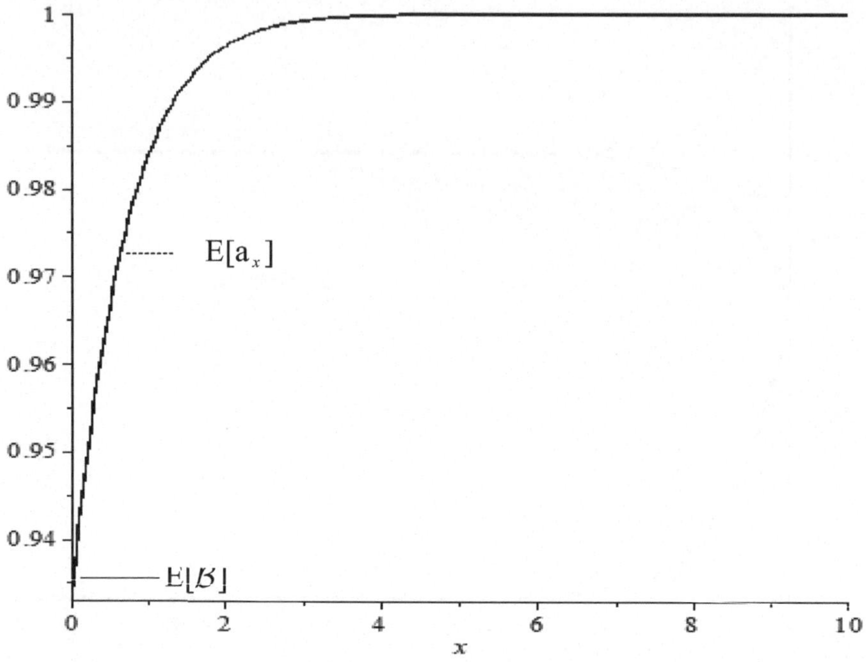

FIGURE 3.3
$E[a_x]$ in numerical Example in M/M/1 with two types of arrivals.

The traffic intensity is $\rho = \lambda E[S] = \lambda D < 1$, implying stability. The "virtual" wait is denoted by $\{W(t)\}_{t\geq 0}$ (Figure 3.4), and the "actual" waiting times are denoted by $\{W_n\}_{n=1,2,...}$. Denote the cdf of actual wait $P(W_n \leq x)$, by $F_n(x)$, $x \geq 0$ and $\lim_{n\to\infty} F_n(x) = F(x)$, $x \geq 0$. Due to the PASTA principle for Poisson arrivals (e.g. Section 2.1.2 in Chapter 2; also [56])

$$F(x) \equiv \lim_{t\to\infty} P(W(t) \leq x) = \lim_{n\to\infty} P(W_n \leq x), x \geq 0.$$

The connection between $\{W(t)\}_{t\geq 0}$ and $\{W_n\}_{n=0,1,...}$ ensures that a study of the virtual wait yields considerable information about both the virtual-wait process and actual-wait process. (See Figure 2.2 in Section 2.1.6 in Chapter 2.)

Define the pdf of the virtual wait $f(x)$, $x > 0$, to be right continuous, and for notational convenience, define $f(0) = f(0^+)$ (which adds zero probability to $F(0)$). The stationary probability of a zero wait is

$$P_0 = F(0) = 1 - \rho = 1 - \lambda D.$$

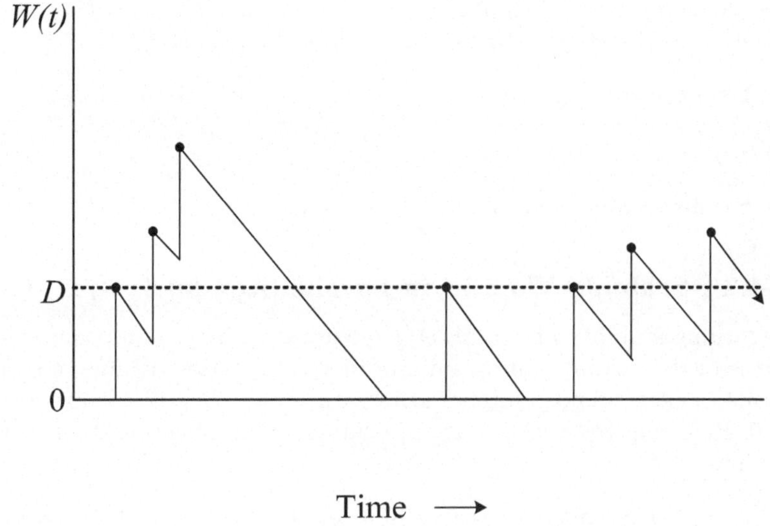

FIGURE 3.4
Sample path of the virtual wait $\{W(t)\}_{t\geq 0}$ in M/D/1 queue. Small black circles at peaks indicate right continuity. The horizontal dotted line at level D is tangent to the sample path just after each arrival instant when the system is empty. The tangent at D causes a discontinuity at level D in the pdf; $f(D^-) > (D^+)$ since $\lim_{t\to\infty} \frac{\mathcal{D}_t(D^-)}{t} > \lim_{t\to\infty} \frac{\mathcal{D}_t(D^+)}{t}$.

The limiting mixed pdf $\{P_0, f(x)\}_{x>0}$ is related,to the cdf $F(x)$, by

$$F(x) = P_0 + \int_{y=0}^{x} f(y)dy, x \geq 0, \tag{3.33}$$

$$F(\infty) = P_0 + \int_{y=0}^{\infty} f(x)dx = 1. \tag{3.34}$$

3.4.1 Properties of Cdf and Pdf of Wait in the Standard M/D/1 Queue

Proposition 3.2, states two properties of the cdf of wait "$F(x)$, $x \geq 0$", and three properties of the mixed pdf "$\{P_0, f(x)\}, x > 0$". A proof of Proposition 3.2 using LC, is given in Proposition 4.9, p. 133ff, in Chapter 3 of [22] (2017).

Proposition 3.2 *For the M/D/1 queue: (1-cdf) the cdf $F(x), x \geq 0$ has exactly one atom, which is at level $x = 0$; (2-cdf) The cdf $F(x), x \geq 0$ is continuous for every $x > 0$; (1-pdf) the mixed pdf of wait $\{P_0, f(x)\}_{x>0}$ has one atom, which is at level 0; (2-pdf) the pdf of wait $f(x)$ has a single downward-jump* *discontinuity of size* λP_0 *at* $x = D$*; (3-pdf) the pdf of wait* $f(x)$ *is continuous for all* $x > 0$, $x \neq D$ *(i.e., for all* $x > 0$, *except at level* $x = D$*).*

Proof *The foregoing stated properties can be seen intuitively by examining Figure 3.4. For more details, see Proposition 4.9, p. 133ff, in Chapter 3 of [22] (2017).*

See also Proposition 3.3 in Section 3.4.2.

3.4.2 Stability Condition for the Standard M/D/1 Queue

Regarding stability of the M/D/1 queue, state $\{0\}$ is an atom, since sample paths of the virtual wait spend a positive proportion of time in state $\{0\}$ *a.s.* –(*a.s.* means "almost surely" aka "with probability 1"). Thus, $P_0 = 1 - \lambda D > 0$. Referring to Section 2.1.1 in Chapter 2, and Section 3.2.1 in Chapter 3, we have

$$E[\text{number of arrivals in a service time}] = \lambda E[S] = \lambda D$$

Hence, for a stationary pdf to exist, a necessary and sufficient condition is $\lambda D < 1$. The state space $S = [0, \infty)$ has no other atoms, since the proportion of time the SP spends in each state $x > 0$, is 0.

Proposition 3.3 *(1)* $F(x), x \geq 0$, *has a jump discontinuity at* $x = 0$ *of size* P_0*; (2)* $F(x)$ *is continuous for all* $x > 0$.

Proof *(1)* $F(x)$ *has a discontinuity at* $x = 0$, *since* 0 *is an atom having probability* $F(0) = P_0$. *(2) Fix* $x > 0$ *in the state space. Then* x *is not an atom (Proposition 3.2 part 1); therefore level* x *is not a point of increase in probability. Thus,* x *is a point of continuity of* $F(\cdot), x > 0$.

3.4.3 Integral Equation for Pdf of Wait in Standard M/D/1 Queue

Applying the alternative form of the basic LC integral Equation (2.9) in Chapter 2 with $B(x-y) = 0$ if $x - y < D$ and $B(x-y) = 1$ if $x - y \geq D$, write immediately an equation for $f(x)$ in terms of $F(x)$ and $F(x-D)$, which is in fact, a differential equation for the cdf $F(x)$ since $f(x) = F'(x)$. Thus,

$$f(x) = \lambda F(x) - \lambda F(x-D), x > 0, \tag{3.35}$$

where $F(x) = P_0 + \int_{y=0}^{x} f(y)dy$, and $F(x-D) = P_0 + \int_{y=0}^{x-D} f(y)dy$.

3.4.3.1 LC Interpretation of Equation (3.35)

To explain (3.35) via LC, consider a sample path of $\{W(t)\}_{t \geq 0}$ (Figure 3.4). In formula (3.35) the left side $f(x)$ is the SP downcrossing rate of level x. SP vertical upward jumps occur at rate λ, which are all of size D.

On the right side of (3.35), the first term "$\lambda F(x)$" (having dimension $[Time]^{-1}$) is the rate of SP jumps that begin in state-space set $[0, x]$. The second term, "$-\lambda F(x-D)$" subtracts off the rate of vertical jumps that begin in $[0, x-D]$ and must end below level x, because jumps starting below $x - D$ cannot upcross x due to their fixed size D. Thus the right side is the upcrossing rate of x. Rate balance across level x then yields analytic equation (3.35).

Remark 3.4 *Another way to look at equation (3.35), is as follows. On the right side,*

$$\lambda F(x) - \lambda F(x-D) = \lambda\left(F(x) - F(x-D)\right) = \lambda\left(P(x-D < W_Q \leq x)\right).$$

Thus, $\lambda\left(F(x) - \lambda F(x-D)\right)$ = rate at which jumps, all of size D, that begin in the state space interval $(x-D, x]$, and must upcross level x. (Note that continuity of $F(x), x > 0$, necessitates that $F(x^-) = F(x)$. Thus, we can use the half-open interval $(x-D, x]$ or the open interval $(x-D, x)$ in equation (3.35) without affecting the count of upcrossings of level x.)

Remark 3.5 *The properties in Proposition 3.2, and equation (3.35) are readily inferred intuitively upon considering a sample path (Figure 3.4), and applying LC interpretations of sample path motion with respect to state-space levels. Such intuitive insights often lead immediately to formal analytic proofs such as in Proposition 3.2.*

3.4.4 Analytic Solution for Cdf and Pdf of Wait in Standard M/D/1 Queue

3.4.4.1 Cdf of Wait in Standard M/D/1 Queue

We give the solution of (3.35), for completeness. For $x \in [0, D)$, $F(x-D) \equiv 0$; thus $f(x) = \lambda F(x)$, or the differential equation

$$F^{'}(x) - \lambda F(x) = 0,$$

having solution

$$F(x) = A_0 e^{\lambda x},\ x \in [0, D),$$

where A_0 is a constant. Letting $x \downarrow 0$, gives the constant

$$A_0 = F(0) = P_0 = 1 - \rho.$$

Thus

$$F(x) = P_0 e^{\lambda x}, x \in [0, D).$$

For $x \in [D, 2D)$, formula (3.35) is equivalent to the differential equation

$$F^{'}(x) - \lambda F(x) = -\lambda P_0 e^{\lambda(x-D)}, x \in [D, 2D).$$

Multiplying both sides by the integrating factor $e^{-\lambda(x-D)}$ and then integrating both sides over $[D, x)$ yields the solution up to a constant

$$F(x) = -P_0\lambda(x-D)e^{\lambda(x-D)} + A_1 e^{\lambda(x-D)}, x \in [D, 2D).$$

The constant A_1 is determined from the continuity of $F(x), x > 0$ (Proposition 3.3). Thus $F(D^-) = F(D)$, or $A_1 = P_0 e^{\lambda D}$ resulting in the solution

$$F(x) = P_0\left(-\lambda(x-D)e^{\lambda(x-D)} + e^{\lambda x}\right), x \in [D, 2D).$$

Mathematical induction on formula (3.35) (induction details are not presented here) yields the classical formula for the cdf of wait, originally derived in [33].

$$F(x) = P_0 \sum_{i=0}^{m} (-\lambda)^i \frac{(x-iD)^i}{i!} e^{\lambda(x-iD)}, \ x \in [m, (m+1)D), \ m = 0, 1, 2, ... \tag{3.36}$$

A more compact form of (3.36) is (e.g., p. 385 in [36]) (2008),

$$F(x) = P_0 \sum_{i=0}^{\lfloor x/D \rfloor} (-\lambda)^i \frac{(x-iD)^i}{i!} e^{\lambda(x-iD)}, x \geq 0, \tag{3.37}$$

where $\lfloor \alpha \rfloor :=$ greatest integer $\leq \alpha$.

3.4.4.2 Pdf of Wait in Standard M/D/1 Queue

The pdf $f(x)$ can be obtained by differentiating $F(x)$ with respect to x. More simply, we obtain $f(x)$ by substituting (3.36) into (3.35) giving

$$f(x) = \lambda P_0 e^{\lambda x}, 0 < x < D,$$

and for $x \in [mD, (m+1)D)$, $m = 0, 1, 2, ...$, (where $\sum_{i=0}^{-1}(\cdots) = 0$),

$$f(x) = \lambda P_0 \left[\sum_{i=0}^{m} (-\lambda)^i \frac{(x-iD)^i}{i!} e^{\lambda(x-iD)} - \sum_{i=0}^{m-1} (-\lambda)^i \frac{(x-(i+1)D)^i}{i!} e^{\lambda(x-(i+1)D)} \right] \tag{3.38}$$

$$\begin{aligned} f(x) &= \lambda P_0(-\lambda)^m \frac{(x-mD)^m}{m!} e^{\lambda(x-mD)} \\ &+ \lambda P_0 \sum_{i=0}^{m-1} \frac{(-\lambda)^i}{i!} \left[(x-iD)^i e^{\lambda(x-iD)} - (x-(i+1)D)^i e^{\lambda(x-(i+1)D)} \right]. \end{aligned} \tag{3.39}$$

The pdf $f(x)$ in (3.38) has a discontinuity at $x = D$ (Proposition, more subtlety, 3.2). That is

$$f(D^-) = \lambda P_0 e^{\lambda D},$$

$$f(D^-) - f(D) = \lambda P_0,$$

illustrating that $f(x)$ has a prominent vertical downward jump of size λP_0 at level $x = D$. Moreover $f(x)$ is continuous for all other $x > 0$ (see Figure 3.5). In Figure 3.5 it is interesting that there is a "concave wave" in $f(x)$ for the interval $x \in [D, 2D)$; the waviness dampens on the interval $x \in [2D, 3D]$. The waviness gradually decrease, but is still discernible in the next three waves. The cdf $F(x)$, for the same example, is given in formula (3.36) and plotted in Figure 3.6, where the continuity of $F(x)$, $x > 0$ is evident, and more subtly than for $f(x)$, $x > 0$, in Figure the discontinuity of $\frac{d}{dx}F(x)$ at $x = D$, is evident.

Remark 3.6 *An LC examination of a typical sample path of* $\{W(t)\}_{t\geq 0}$ *(Figure 3.4) suggests that the sample-path properties at level D are are reflected in the analytical properties of* $f(x)$ *and* $F(x)$ *shown in the Figure 3.5 and Figure 3.6.*

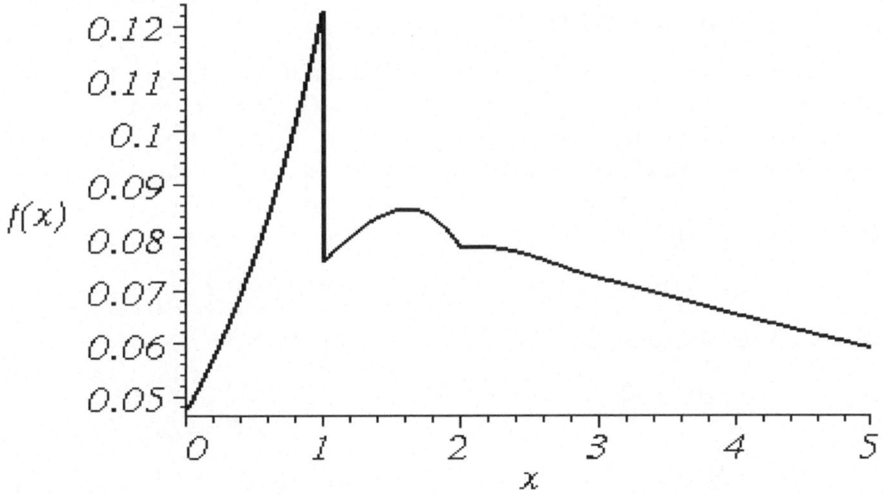

FIGURE 3.5
PDF $f(x)$ of wait in M/D/1: $\lambda = 0.95$, $D = 1$, $\rho = 0.95$ (high traffic). Shows discontinuity and downward jump of size λP_0 at $x = D$; and extreme waviness in right neighborhood $[D, 2D)$.

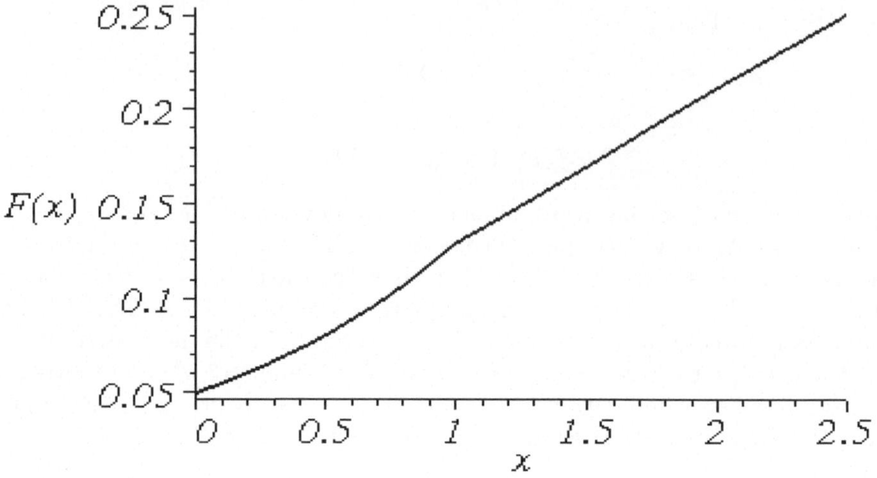

FIGURE 3.6
CDF $F(x)$ of wait in M/D/1: $\lambda = 0.95$, $D = 1$. Shows continuity of $F(x), x > 0$; and decrease in slope of $F(x)$ at $x = D$.

3.4.5 Probability Distribution (Pmf) of Number in System in Standard M/D/1 Queue

Let N be the number of customers in the system at an arbitrary time point and let W_Q (≥ 0) be the wait before service. Let $P_n := P(N = n)$. Let a_n, := $P(n$ units in system "just before an arrival"). Let $d_n := P(n$ units in system "just after a departure"). Poisson arrivals implies: $a_n = P_n = d_n$, $n = 0, 1, 2, ...$

Arrivals "see" n customers in the system iff $W_Q \geq 0$ and $W_Q \in ((n-1)D, nD]$, $n = 0, 1, 2, ...$). Thus

$$a_n = F(nD) - F((n-1)D) = P_n = d_n, n = 0, 1, 2, ...$$

Because all service times are of size D, from formula (3.36),

$$P_0 = F(0) - F(-D) = F(0)$$

$$P_1 = F(D) - F(0) = P_0 e^{\lambda D} - P_0 = P_0(e^{\lambda D} - 1)$$

$$P_2 = F(2D) - F(D) = P_0 e^{\lambda D}(-\lambda D + e^{\lambda D} - 1)$$

$$\cdots$$

$$P_n = F(nD) - F((n-1)\,D),\ n = 0, 1, 2, ... \tag{3.40}$$

(See p. 501, Proposition 8.1 in [52], which proves that $a_n = d_n$.)

Summing the values $P_0 + \cdots + P_n$ in (3.40) above, and cancelling $\pm$ terms, gives the cdf of the number in the system N, namely:

$$P(N \leq n) = \sum_{i=0}^{n} P_i = F(nD), \ n = 0, 1, 2, ..., \tag{3.41}$$

where $F(nD)$ is the cdf of the virtual wait at level nD.

3.4.6 Expected Busy Cycle and Expected Busy Period in Standard M/D/1 Queue

Let $\mathcal{C}$ denote the "busy cycle", and let $\mathcal{B}$ denote the "busy period". Then:

$$E[\mathcal{C}] = \frac{1}{\lambda P_0} = \frac{1}{\lambda(1 - \lambda D)}$$

$$E[\mathcal{B}] = \frac{1 - P_0}{\lambda P_0} = \frac{\lambda D}{\lambda(1 - \lambda D)} = \frac{D}{1 - \lambda D}. \tag{3.42}$$

As a check on formula (3.42), let $\mathcal{I} :=$ idle period, which is $\stackrel{d}{=} \text{Exp}_\lambda$. Then $\mathcal{C} = \mathcal{B} + \mathcal{I}$, aher $\mathcal{I} \stackrel{d}{=} Exp_\lambda$. Thus,

$$E[\mathcal{C}] = E[\mathcal{B}] + E[\mathcal{I}] = E[\mathcal{B}] + \frac{1}{\lambda},$$

$$\frac{1}{\lambda(1 - \lambda D)} = E[\mathcal{B}] + \frac{1}{\lambda},$$

$$E[\mathcal{B}] = \frac{1}{\lambda(1 - \lambda D)} - \frac{1}{\lambda},$$

$$= \frac{1}{\lambda}\left(\frac{1}{1 - \lambda D} - 1\right) = \frac{1}{\lambda}\left(\frac{1 - 1 + \lambda D}{1 - \lambda D}\right),$$

$$E[\mathcal{B}] = \frac{D}{1 - \lambda D}. \tag{3.43}$$

Formulas (3.45) and (3.42) are identical.

3.4.6.1 M/D/1 where Zero-Wait Customers Get Exceptional Service

Queues where zero-wait customers get "special" (aka "exceptional") service are important in queueing theory. Examples are where a copy machine, or printer, loses time in order to be turned on, which adds service time to the first arrival to use the device. Assume that "zero-wait" customers receive a fixed service time D_0, and that all "positive-wait" customers receive a constant service time D_1, where $D_1 \neq D_0$. Let $\mathcal{B}_0 :=$ busy period of the queue. Using similar reasoning as in the M/M/1 queue where zero-wait customers get exceptional service, i.e., Figure 2.8 and formula (2.63) in Section 2.5.2 in Chapter 2, we obtain

$$E[\mathcal{B}_0] = D_0 + \lambda D_0 E[\mathcal{B}_1], \tag{3.44}$$

where $\mathcal{B}_1 :=$ busy period in a standard M/D/1 queue where all of the service times are equal to a constant D_1. Substituting $E[\mathcal{B}]$ from formula (3.42), into formula (3.44) leads to

$$\begin{aligned} E[\mathcal{B}_0] &= D_0 + \lambda D_0 \frac{D_1}{1-\lambda D_1} \\ &= D_0 \left(1 + \frac{\lambda D_1}{1-\lambda D_1}\right) \\ &= \frac{D_0}{1-\lambda D_1}, \end{aligned} \tag{3.45}$$

which is analogous to formula (2.63).

The expected busy cycle is given by:

$$\begin{aligned} E[\mathcal{C}_0] &= E[\mathcal{B}_0] + E[\text{idle period}] \\ &= \frac{D_0}{1-\lambda D_1} + \frac{1}{\lambda} \\ &= \frac{\lambda(D_0 - D_1) + 1}{\lambda(1-\lambda D_1)}. \end{aligned}$$

Let P_{00} denote the probability of a zero wait in the M/D/1 queue where zero-waits receive special service By the elementary renewal theorem, that is,

$$E[\mathcal{C}_0] = \frac{1}{\text{rate at which busy cycles begin}},$$

$$E[\mathcal{C}_0] = \frac{1}{\lambda P_{00}}.$$

Hence

$$P_{00} = \frac{1}{\lambda E[\mathcal{C}_0]} = \frac{1}{\frac{\lambda(D_0 - D_1)+1}{1-\lambda D_1}} = \frac{1-\lambda D_1}{1-\lambda D_1 + \lambda D_0}, \tag{3.46}$$

which is analogous to formula (2.60) in Section 2.5 in Chapter 2. Formula (3.46) indicates that the queue is stable iff $\lambda D_1 < 1$; i.e., iff the traffic intensity for the regular-served customers is less than 1.

3.4.7 Workload-Barrier M/D/1 Queue

In this M/D/1 queue, every arriving customer brings the same "prior service time" = "$D > 0$". Let arrivals be accepted into the system only if their prior service D does not cause the arrival-time workload to exceed a boundary,at level $K > D$. Otherwise, the arrivals are "blocked and cleared" immediately from the system. This M/D/1 model is analogous to Variant 1 in the workload-barrier M/M/1 queue given in Section 2.7.1 in Chapter 2. On arrival (at time t, say), the common "prior" service time D is transformed into a "posterior service time" which is:

$$S_K = \begin{cases} D \text{ with probability } P\left(0 \le W(t) \le K - D\right), \\ 0 \text{ with probability } P\left(K - D < W(t) \le K\right). \end{cases} \tag{3.47}$$

3.4.8 Expected Service Time in a Workload-Barrier M/D/1 Queue

From formula (3.47)

$$E\left[S_K\right] = P\left(0 \le W(t) < K - D\right) \times D + P\left(K - D < W(t) \le K\right) \times 0,$$

$$E\left[S_K\right] = P\left(0 \le W(t) < K - D\right) \times D. \tag{3.48}$$

From equation (3.48), we need to determine $P\left(0 \le W(t) < K - D\right)$ in order to determine $E\left[S_K\right]$. We will now derive the expected service time $E\left[S_K\right]$, when the workload barrier is $K \in (D, 2D)$, since we assume that $K > D$. The purpose of this derivation is to fix ideas and provide clarity to the calculations. (Note that if $K \in (nD, (n+1)D)$ where $n \ge 2$, the analysis would be similar.) The quantities of interest are: the expected "posterior service time" $E[S_K]$; $P_{K,0} := P(\text{wait before service} = 0)$; and the cdf or pdf of wait on three contiguous disjoint intervals inside the state-space interval $(0, K)$, namely: (1) $(0, K - D)$, (2) $(K - D, D)$, and (3) (D, K). All arrivals to the system when the workload,$W(t)$ is within the state-space set $(K - D, D) \cup (D, K)$ $(= (K - D, K))$ are blocked from entering the system because their prior service times (D) would cause the workload to overshoot the barrier at level K. Such arrivals would contribute zero amount to the arrival-time workload. All arrivals when the workload is in $[0, K - D)$ would be accepted into the system for service, and would add an amount D to the arrival-time workload.

Denote the stationary mixed pdf of workload $\{W(t)\}_{t\ge 0}$ by $\{P_{K,0}, h_{K}(x)\}_{0<x<K}$, where $P_{K,0} := \lim_{t\to\infty} P(W(t) = 0)$. We derive the continuous parts of the mixed pdf, "$h_{K}(x)$", on the three exclusive (non-overlapping) intervals mentioned above (see formulas (3.49), (3.50), and

(3.51)). Their derivations are detailed after formula (3.56).

$$h_K(x) = \lambda P_{K,0} e^{\lambda x}, 0 < x < K - D, \tag{3.49}$$

$$h_K(x) = \lambda P_{K,0} e^{\lambda(K-D)}, K - D < x < D, \tag{3.50}$$

$$h_K(x) = \lambda P_{K,0} \left(e^{\lambda(K-D)} - e^{\lambda(x-D)}\right), D < x < K. \tag{3.51}$$

Note that the middle partial pdf $h_K(x)$, $K - D < x < D$, is constant on the state-space subinterval $(K - D, D)$; it is continuous at level $K - D$ since $h_K((K - D)^-) = h_K((K - D)^+)$; and it has a "discontinuity at level D" of size "$\lambda P_{K,0}$", since $h_K(D^-) - h_K(D^+) = \lambda P_{K,0}$. This discontinuity causes a downward jump in the pdf at $x = D$). Substituting from formulas (3.49)–(3.51) into the normalizing condition

$$P_{K,0} + \int_0^K h_K(x)dx = 1,$$

and solving for $P_{K,0}$ yields

$$P_{K,0} = \left[\lambda D e^{\lambda(K-D)} + 1\right]^{-1} \tag{3.52}$$

The stationary continuous part, $h(x)$, $0 < x < K$, of the mixed pdf of $\{W(t)\}_{,t\geq 0}$ is plotted in Figure 3.7, showing the partial pdfs $h_i(x), i = 1, 2, 3$.

The system is assumed to be stable (in equilibrium). We now apply Little's law to the server, thought of as a subsystem of the queueing system (see Section 3.3.1 in Chapter 3). Little's Law states that: the expected number in the server is $\boldsymbol{L} = \boldsymbol{\lambda W}$, where "$\boldsymbol{W}$" := expected service time $E[S_K]$ ([44]) (1961). Thus,

$$\boldsymbol{L} = 0 \cdot P_{K,0} + 1 \cdot (1 - P_{K,0}) = \lambda\, \boldsymbol{W} = \lambda E[S_K],$$

$$E[S_K] = \frac{1 - P_{K,0}}{\lambda}. \tag{3.53}$$

Substituting (3.50) above, into formula (3.53), gives

$$E\left[S_K\right] = \frac{D e^{\lambda(K-D)}}{\lambda D e^{\lambda(K-D)} + 1}. \tag{3.54}$$

It is readily shown from equation (3.49) that

$$E\left[S_K\right] = D\left(P_{K,0} + \int_{x=0}^{K-D} h_K(x)dx\right) \tag{3.55}$$

$$= D \times P(\textit{an arrival is accepted}) < D, \tag{3.56}$$

where the last inequality follows since $0 < P(\textit{an arrival is accepted}) < 1$.

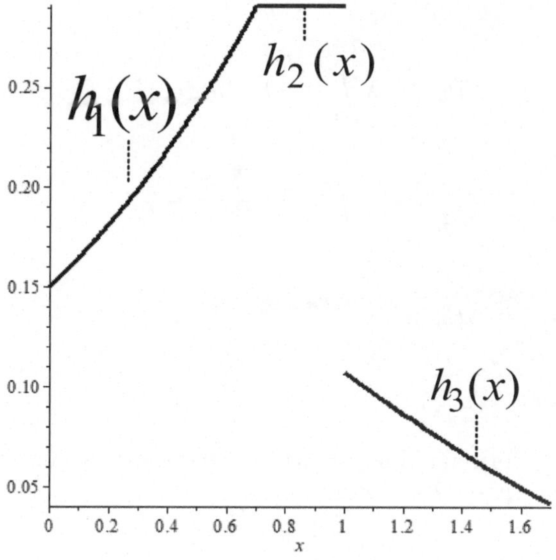

FIGURE 3.7
Plot of $h_i(x), i = 1, 2.3$ when $D = 1$, $\lambda = 0.95$, $K = 1.7$, in workload-barrier M/D/1 queue with workload bounded at level K, $1 < K < 2$.

3.4.8.1 LC Derivation of Pdf $h_K(x)$ in the Workload-barrier M/D/1 Queue

We now present an LC derivation of $h_K(x)$ in formulas(3.49), (3.50), and (3.51). Let $\lim_{t\to\infty} \mathcal{D}_t(x)/t$ be the workload sample-path downcrossing rate, and let $\lim_{t\to\infty} \mathcal{U}_t(x)/t$ be the workload sample-path upcrossing rate, of level x in the state space. In the first equation of each derivation immediately below, the left side $h_K(x) := \lim_{t\to\infty} \mathcal{D}_t(x)/t$ and the right side $:= \lim_{t\to\infty} \mathcal{U}_t(x)/t$. Rate balance for all x in the state space, "$\lim_{t\to\infty} \mathcal{D}_t(x)/t = \lim_{t\to\infty} \mathcal{U}_t(x)/t$", yields the first equation.

The "accept/block" regulation mandates that jumps can occur only if the arrival-point workload is in state-space interval $[0, K - D)$.

When "$h_K(x)$, $0 < x < K - D$", P(SP jump upcrosses level x) $= 1$. Thus,

$$
\begin{aligned}
h_K(x) &= \lambda P_{K,0} + \lambda \int_0^x h_K(y)dy,\ 0 < x < K - D,\\
h'_K(x) &= 0 + \lambda h_K(x),\\
\frac{d}{dx} \ln h(x) &= \lambda,\\
h_K(x) &= Ae^{\lambda x},\ 0 < x < K - D, \text{ where } A \text{ is a constant},\\
h_K(0^+) &= \lambda P_{K,0} \implies A = \lambda P_{K,0},\\
h_K(x) &= \lambda P_{K,0} e^{\lambda x},\ 0 < x < K - D;
\end{aligned}
$$

which agrees with formula (3.49).

Next, when "$h_K(x), K-D<x<D$",

$$h_K(x) = \lambda P_{K,0} + \lambda \int_0^{K-D} h_K(y)dy; K-D<x<D,$$

$$h_K(x) = \lambda P_{K,0} + \lambda \int_0^{K-D} \lambda P_{K,0} e^{\lambda y} dy$$

$$h_K(x) = \lambda P_{K,0}(1 + e^{\lambda(K-D)} - 1)$$

$$h_K(x) = \lambda P_{K,0} e^{\lambda(K-D)}, K-D<x<D;$$

which agrees with formula (3.50) (constant on state-space interval $(K-D, D)$).

Finally when "$K-D<x<D$",

$$h_{K(x)} = \lambda \int_{y=x-D}^{K-D} h_K(y)dy;\ K-D<x<D,$$

$$h_K(x) = \lambda \int_{y=x-D}^{K-D} \lambda P_{K,0} e^{\lambda y} dy$$

$$h_K(x) = \lambda P_{K,0} \left[e^{\lambda(K-D)} - e^{\lambda(x-D)}\right], D<x<K,$$

which agrees with formula (3.51).

Note that the LC solution technique for the stationary pdf of wait in this Section, can be used for more general arrangements of the non-overlapping state-space intervals, with respect to service size D and workload

barrier level K.

3.4.9 $\mathrm{M}_\lambda/\mathrm{D}_N/1$ with N Discrete Service Times – A Brief Discussion

We briefly mention the single-server queue, denoted by $\mathrm{M}_\lambda/\mathrm{D}_{\mathcal{N}}/1$ (aka M_λ/Discrete/1), due to its theoretical and computational approach for the waiting time distribution in M/D/1 queues when there are several arrival independent Poisson arrival streams with mutually distinct "prior" service times. (This is equivalent to having a single Poisson stream with fixed probabilities of for "each arriving D_i prior service.) Assume that the total arrival rate is Poisson at rate λ. Assume there are N mutually distinct types of customers in the arrival stream, where type i arrivals brings a prior service time $D_i > 0$

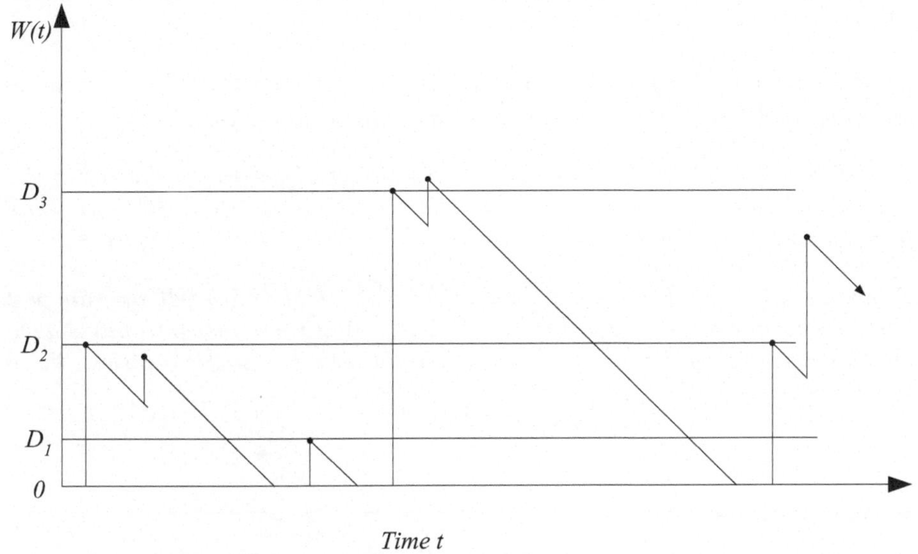

FIGURE 3.8
Virtual wait process $\{W(t)\}_{t\geq 0}$ in the M/D$_{\mathcal{N}}$/1 queue with $N = 3$. Arrivals enter the system at rate λ. There are N types of customers. An arrival is a type i with probability p_i, where$\sum_{i=1}^{\mathcal{N}} p_i = 1$.

with probability p_i $i = 1, ..., N$, and $\sum_{i=1}^{N} p_i = 1$. The D_is are mutually distinct positive constants. The pdfs and cdfs of the virtual wait, and actual wait, can be derived using LC. Also, a computational method can be applied to compute the pdfs and cdfs numerically (see [16] (1988), and Section 3.11, pp. 139–144 in [22] (2017) for more details). A sample path of the virtual wait is pictured in Figure 3.8. An immediate observation from viewing the sample path, is that the stationary pdf will have discontinuities at state-space levels D_1 D_2, and D_3.

3.5 Exercises

Exercise 3.5.1 *Give ttwo potential real-world applications of a standard M/D/1 queue.*

Exercise 3.5.2 *Give two potential examples of an M/D/1 queue where zero-wait customers get exceptional service.*

Exercise 3.5.3 *Reason from the sample path in Figure 3.4, that the size of* *the downward-jump discontinuity at $x = D$ in Figure 3.5 is equal in size to λP_0.*

Exercise 3.5.4 *Check that formula (3.39) follows from formula (3.38) algebraically, and that every term has the same dimension $[Time]^{-1}$.*

Exercise 3.5.5 *In an M/D/1 queue, formula (3.41) connects the cdf of the number of customers in the system and cdf of the virtual wait. State why this formula (3.41) can be derived by observing the sample path in Figure 3.4.*

Exercise 3.5.6 *In Figure 3.8, explain why the stationary pdf of wait will have discontinuities at levels D_1 D_2, and D_3. Will these discontinuities have downward or upward jumps? (No calculations are required to answer these questions).*

3.6 The Standard M/Erlang/1 Queue—$\mathbf{M_\lambda/Erl_{k.\mu}/1}$

We first derive equations for the stationary mixed pdf of wait in an $M_\lambda/Erl_{k,\mu}/1$ queue. The customer arrival rate is Poisson at rate λ. The service times are denoted by $S_k \stackrel{d}{=} Erl_{k.\mu}$, which is the "sum of k iid Exp_μ" r.v.s., where k is a positive integer. The expected service time is $E[S_k] = \sum_{i=1}^{k} E[Exp_\mu] = \frac{k}{\mu}$, since $E[Exp_\mu] = \frac{1}{\mu}$. To illustrate the derivation, we derive the pdf of wait when $k = 2$ (for general values of k, the derivation is similar (see [22]) (2017)).

Properties of the Erlang probability distribution are given in Section 2.1.14.1 in Chapter 2.

Let $\mathcal{G}_k(x) :=$ the event {number of "Poi_μ events" in interval $(0, x)$ is $\geq k$}, where "Poi_μ" denotes a Poisson process with rate μ (see Section 1.5.4 in Chapter 1). (For further discussion of applications of the Poisson process, see pp. 312–316 in [52] (2010).)

From the definition of $\mathcal{G}_k(x)$,

$$P(S_k \leq x) \equiv P(G_k(x)). \tag{3.57}$$

Denote the cdf of the service time S_k by: $B_k(x) = P(S_k \leq x) = P(G_k(x))$. The identity in formula (3.57) gives

$$B_k(x) = P(\mathcal{G}_k(x)) = \sum_{i=k}^{\infty} \frac{e^{-\mu x}(\mu x)^i}{i!},\ x > 0. \tag{3.58}$$

(See Section 1.5 in Chapter 1, for properties of Exp_μ, and Chapter 5 in [52] for more properties of Poi_μ.)

Taking $\frac{d}{dx}$ on both sides in (3.58) shows, on cancelling terms in the resulting "telescoping series" on the right side, that the pdf of the service time S_k, denoted by $b_k(x)$ ($:= \frac{d}{dx}B_k(x)$), is given by

$$b_k(x) = e^{-\mu x}\frac{(\mu x)^{k-1}\mu}{(k-1)!}, x > 0. \tag{3.59}$$

Formula (3.59) checks with formula (2.24) in Section 2.1.14.1. (Intuitively, formula (3.59) is equivalent to: $b(x)dx = P((k-1 \text{ Poi}_\mu$ events occur within the interval $(0, x)$ and a single k^{th} Poi_μ event occurs at instant x).

The expected value of $\text{Erl}_{k,\lambda}$ is the sum of the expected values of k Exp_λs, i.e.,

$$E\left[Erl_{k,\lambda}\right] = k \times E\left[Exp_\lambda\right] = \frac{k}{\lambda}. \tag{3.60}$$

From formula (3.59),

$$\sum_{i=0}^{\infty} e^{-\mu x}\frac{(\mu x)^i}{i!}$$

$$= e^{-\mu x} \times e^{\mu x} = 1.$$

The probability that $S > x$, is

$$\overline{B}(x) = 1 - B(x) = e^{-\mu x}\left(\sum_{i=0}^{k-1}\frac{(\mu x)^i}{i!}\right), \ x \geq 0. \tag{3.61}$$

The condition for existence of the stationary distribution of wait is:

$$E[\text{number of customers served in } S] < 1,$$
$$\text{or, equivalently } \lambda < \mu/k. \tag{3.62}$$

(see Chapter 3, equation M/Erl/1 stability condition.)

Intuitively, formula (3.59) is equivalent to: "$b(x)dx$ is equal to the probability $P(k-1$ Poisson events with rate λ occur during $(0, x)$ and the kth Poisson event occurs at time $x)dx$". The time interval between events is $\stackrel{d}{=}$ Exp_λ (see Figure 3.9).

3.6.1 Example of the $\text{M/Erl}_{2,\mu}/1$ Queue

Here, we give an example of the $\text{M/Erl}_{2,\mu}/1$ queue, with $k = 2$, in the notation of Section 3.6.

Example 3.6.1 *Next, we focus on the case $k = 2$, to illustrate deriving an analytic solution for the pdf of wait, and related results. (The case $k = 2$, corresponds to the $M/Erl_{2,\mu}/1$ queue.) Substituting $e^{-\mu x}\sum_{i=0}^{k-1}\frac{(\mu x)^i}{i!}$ for $\overline{B}(x)$*

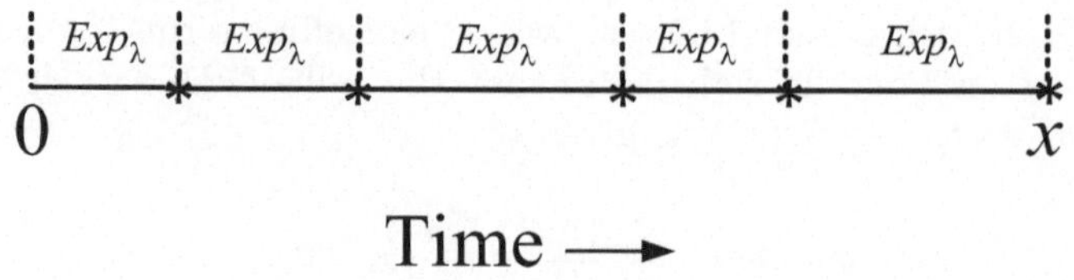

FIGURE 3.9
Illustration of $\text{Erl}_{k,\lambda}$ random variable, composed of the sum of $k = 5$ iid Exp_λ random variables where the fifth Exp_λ random variable ends at time point x.

in the integral equation in Figure 2.5 in Chapter 2, gives the integral equation for the pdf of wait: as

$$f(x) = \lambda P_0 e^{-\mu x} \sum_{i=0}^{k-1} \frac{(\mu x)^i}{i!} + \lambda \int_{y=0}^{x} e^{-\mu(x-y)} \sum_{i=0}^{k-1} \frac{(\mu(x-y))^i}{i!} f(y)dy, \; x > 0, \tag{3.63}$$

where $k = 2$, *and* $P_0 = 1 - \lambda E[S_2] = 1 - \lambda \times \frac{2}{\mu}$. *The normalizing condition is*

$$P_0 + \int_{x=0}^{\infty} f(x)dx = 1.$$

Equation (3.63) yields the equation for the pdf $f(x)$ *as:*

$$f(x) = \lambda P_0 e^{-\mu x}(1+\mu x) + \lambda \int_{y=0}^{x} e^{-\mu(x-y)}(1+\mu(x-y))f(y)dy, x > 0. \tag{3.64}$$

Differentiating both sides of (3.64) with respect to x *"twice in succession" results in the second order differential equation*

$$f''(x) + (2\mu - \lambda)f'(x) + (\mu^2 - 2\lambda\mu)f(x) = 0, x > 0. \tag{3.65}$$

The solution of equation (3.65) is:

$$f(x) = a_1 e^{r_1 x} + a_2 e^{r_2 x}, x > 0, \tag{3.66}$$

where a_1, a_2 *are constants to be determined, and* r_1 *and* r_2 *are given by*

$$r_1 = -\mu + \frac{\lambda}{2} - \frac{1}{2}\sqrt{\lambda^2 + 4\mu\lambda}, \quad r_2 = -\mu + \frac{\lambda}{2} + \frac{1}{2}\sqrt{\lambda^2 + 4\mu\lambda}. \tag{3.67}$$

In (3.67), r_1 *and* r_2 *are the solutions of the "characteristic equation" of the second-order differential equation in (3.65) which is given by:*

$$z^2 + (2\mu - \lambda)z + (\mu^2 - 2\lambda\mu) = 0.$$

In formula (3.67), $r_1 < 0$ and $r_2 < 0$. The constants a_1, a_2 and P_0 can be determined by combining the initial condition $f(0) = \lambda P_0$, and the normalizing condition $P_0 + \int_{y=0}^{\infty} f(y)dy = 1$, resulting in

$$a_1 = \tfrac{r_1 r_2}{r_1 - r_2}(1 - P_0 + \tfrac{\lambda P_0}{r_2}),$$

$$a_2 = \lambda P_0 - a_1, \tag{3.68}$$

$$P_0 = 1 - \tfrac{2\lambda}{\mu}.$$

In the equilibrium (stationary) $M/Erl_{2,\mu}/1$ queue with arrival rate λ, the traffic intensity is $\rho = \lambda E[S] = \lambda \times \frac{2}{\mu} < 1$ (see formula (3.60) in Section 2.1.14.1 in Chapter 2).

3.6.2 The Sojourn a_x above Level x in the $M_\lambda/Erl_{2.\mu}/1$ Queue

The sojourn of the virtual wait process above a state-space level x in the $M_\lambda/Erl_{k,\mu}/1$ queue is interesting, because the service time has k successive phases, each $\stackrel{d}{=} Exp_\mu$. Thus the length of a_x will depend on that stage of $Erl_{k,\mu}$ where the actual upcrossing of level x occurs. The excess over level x of the actual upcrossing will be $\stackrel{d}{=}$ sum of n Exp_μs, where $n = 1$, or $2, .., ,$or k.

To illustrate the analysis, consider a sample path of the virtual wait$\{W(t)\}_{t\geq 0}$ in the standard $M_\lambda/Erl_{2,\mu}/1$ queue, where $k = 2$. The service time $S_2 \stackrel{d}{=} Erl_{2,\mu}$ is the sum of 2 iid Exp_μ random variables; we call these "phase 1" and "phase 2" respectively. A quantity of interest in the $M_\lambda/Erl_{2.\mu}/1$ queue is $a_x :=$ the sojourn above an arbitrary level $x \geq 0$. There are two different types of sojourns above level x, depending on whether level x is upcrossed during phase 1, or upcrossed during phase 2 of the Erlang service that upcrosses level x to initiate a_x. Thus, $a_x \stackrel{d}{=} \mathcal{B}_1$, which is the busy period for the standard $M_\lambda/Erl_{2.\mu}/1$ queue; or $a_x \stackrel{d}{=} \mathcal{B}_2$, which is the busy period for the $M_\lambda/Erl_{2,\mu}/1$ queue where zero-wait customers get special (exceptional) service time $S_0 \stackrel{d}{=} Exp_\mu$, and the remaining service times within a_x are standard $Erl_{2,\mu}$ r.v.s. Hence the interesting situation arises, where a_xs' initial excess jump sizes above level x occur either during phase 1 of the $Erl_{2,\mu}$, or during phase 2 of the $Erl_{2,\mu}$ service time, in a typical sample path of the virtual wait.

If phase 1 of $Erl_{2.\mu}$ is the phase that overshoots x, then the excess initial jump above level $x \stackrel{d}{=} Erl_{2,\mu}$, due to the memoryless property of Exp_μ in phase 1. This initial jump size is the same as all other jump sizes during a_x, so that $a_x = B_1$. Then,

$$P_0 = 1 - \rho = 1 - \lambda\frac{2}{\mu},$$

where $\rho = \lambda\frac{2}{\mu}$, and the value for both $E\left[a_x\right]$ and $E\left[\mathcal{B}_1\right]$ is as follows:

$$E\left[a_x\right] = E\left[\mathcal{B}_1\right] = \frac{1 - P_0}{\lambda P_0} = \frac{2}{\mu - 2\lambda}. \tag{3.69}$$

If phase 2 of $\mathrm{Erl}_{2.\mu}$ is the phase that overshoots x, then the initial excess above x which begins the sojourn, is $\overset{d}{=} \mathrm{Exp}_\mu$ by the memoryless property of Exp_μ. Then the initial service time in $a_x \overset{d}{=} Exp_\mu$. This results in an $\mathrm{M}_\lambda/\mathrm{Erl}_{2,\mu}/1$ busy period above level x, with

$$P_0 = \frac{1 - \rho}{1 - \rho + \rho_0},$$

and thus the value for $E\left[a_x\right]$ and $E[\mathcal{B}_2]$ is as follows:

$$E[a_x] = E[\mathcal{B}_2] = \frac{1 - P_0}{\lambda P_0} = \frac{1}{\mu - 2\lambda}, \tag{3.70}$$

because $a_x \overset{d}{=} \mathcal{B}_2$, in which zero-wait customers receive special (exceptional) service Exp_μ different from the rest of the service times, during a_x, which are all r.v.s $\overset{d}{=} \mathrm{Erl}_{2,\mu}$.

The foregoing analysis implies that $E\left[a_x\right]$ is obtained by conditioning on the phase of $\mathrm{Erl}_{2.\mu}$ that first upcrosses level x. Hence,

$$E[a_x] = P_1^{\mathcal{U}}(x)\left(\frac{2}{\mu - 2\lambda}\right) + P_2^{\mathcal{U}}(x)\left(\frac{1}{\mu - 2\lambda}\right) = \frac{1 - F(x)}{fx)}, x \geq 0, \tag{3.71}$$

$$P_1^{\mathcal{U}}(x) + P_2^{\mathcal{U}}(x) = 1 \tag{3.72}$$

where $P_{i1}^{\mathcal{U}}(x) = P(\text{phase } n \text{ of the SP jump that overshoots } x|\text{SP upcrosses } x)$, $n = 1, 2$; and where $f(x)$ and $F(x)$ are known for every $x \geq 0$.

3.6.2.1 LC Explanation of Equation (3.71)

For each $x \geq 0$, the number of sample-path downcrossings of level x, $\{\mathcal{D}_t(x)\}_{t \geq 0}$, is a regenerative process. This is because of the memoryless property of the Exp_λ interarrival times, which takes effect just after every downcrossings of level x. Let $\mathcal{C}_x$:= time between successive downcrossings of level x.

By the basic LC theorem and the elementary renewal theorem,

$$E\left[\mathcal{C}_x\right] = \frac{1}{f(x)}.$$

By the renewal reward theorem,

$$\frac{E\left[a_x\right]}{E\left[\mathcal{C}_x\right]} = \lim_{t \to \infty} \frac{\text{time spent above level } x \text{ during } (0, t)}{t} = 1 - F(x),$$

where the ccdf $1 - F(x) = P(W_Q > x)$. Therefore

$$\frac{E[a_x]}{E[C_x]} = \frac{E[a_x]}{\frac{1}{f(x)}} = 1 - F(x),$$

$$E[a_x] = \frac{1 - F(x)}{f(x)}, x \geq 0, \tag{3.73}$$

which proves the validity of Equation (3.71).

Notice that

$$a_0 = \frac{1 - F(0)}{f(0)} = \frac{1 - P_0}{\lambda P_0} = E[\mathcal{B}].$$

3.6.2.2 Hazard Rate of a Random Variable

Here, we introduce the "hazard rate" (aka instantaneous failure rate, in Engineering), because of its inverse appearance in formula (3.71). The hazard rate of a random variable (e.g., the virtual waiting time) is defined as, and denoted by

$$\phi(x) = \frac{f(x)}{1 - F(x)}, x \geq 0,$$

(Note: dimension $[\phi(x)] = [Time]^{-1}$ – a rate), where $f(x)$ denotes the pdf, and $F(x)$ denotes the cdf, of the random variable of interest. Intuitively, the dimensionless quantity (after multiplying by dx)

$$\phi(x)dx = \frac{f(x)\,dx}{1 - F(x)},$$

is the probability that the r.v. "survives" for an additional small time dx, given that the random variable has already survived up to time x.

Remark 3.7 *From formula (3.73) in Section 3.6.2.2, the hazard rate at level x is the reciprocal of the expected sojourn time above level x, that is,*

$$\phi(x) = \frac{1}{E[a_x]}. \tag{3.74}$$

By definition, the hazard rate is

$$\phi(x)dx = P(W_Q \in (x, x + dx)|W_Q > x) = \frac{P(x < W_Q < x + dx)}{P(W_Q > x)},$$

where W_Q is the stationary queue waiting time (see, e.g., p. 299 in [52] (2010)). Formula (3.74) suggests an intuitive meaning based on $\phi(x) = 1/E[a_x]$, i.e., "$\phi(x)$ is large iff "$E[a_x]$ is small", and "$\phi(x)$ is small" iff "$E[a_x]$ is large".

This suggests it may be useful to study connections between hazard rates of state random variables, and their sample-path expected sojourn times with respect to state-space levels, in related stochastic models.

3.6.2.3 Solving for $P_1^{\mathcal{U}}(x)(x)$ and $P_2^{\mathcal{U}}(x)(x)$ for a_x in M/Erl$_{2,\mu}$ Queue

Inverting equation (3.71) givers

$$\frac{f(x)}{1-F(x)} = \frac{1}{P_1^{\mathcal{U}}(x)E[\mathcal{B}_1] + P_2^{\mathcal{U}}(x)E[\mathcal{B}_2]}, \; x \geq 0.$$

Thus,

$$\frac{d}{dx}\ln(1-F(x)) = \frac{-1}{P_1^{\mathcal{U}}(x)E[\mathcal{B}_1] + P_2^{\mathcal{U}}(x)E[\mathcal{B}_2]}, \tag{3.75}$$

$$\ln(1-F(x)) = -\int_{y=0}^{x} \frac{1}{P_1^{\mathcal{U}}(y)\times\frac{2}{\mu-2\lambda} + P_2^{\mathcal{U}}(x)(y)\times\frac{1}{\mu-2\lambda}}dy + (\text{constant}), \tag{3.76}$$

$$1-F(x) = Ae^{-\int_{y=0}^{x}\frac{1}{P_1^{\mathcal{U}}(y)\times\frac{2}{\mu-2\lambda}+P_2^{\mathcal{U}}(x)(y)\times\frac{1}{\mu-2\lambda}}dy}, \text{ where } A \text{ is constant}, \tag{3.77}$$

$$A = 1-F(0) = 1-P_0, \tag{3.78}$$

$$F(x) = 1-(1-P_0)\,e^{-\int_{y=0}^{x}\frac{1}{P_1^{\mathcal{U}}(y)\times\frac{2}{\mu-2\lambda}+P_2^{\mathcal{U}}(x)(y)\times\frac{1}{\mu-2\lambda}}dy}, \tag{3.79}$$

$$F(x) = 1-\rho e^{-\int_{y=0}^{x}\frac{1}{P_1^{\mathcal{U}}(y)\times\frac{2}{\mu-2\lambda}+P_2^{\mathcal{U}}(x)(y)\times\frac{1}{\mu-2\lambda}}dy}, \tag{3.80}$$

where $\rho = \lambda\times\frac{2}{\mu}$ in the M/Erl$_{2.\mu}$/1 queue.

In Example 3.6.1, Equation (3.64) for M$_\lambda$/Erl$_{2,\mu}$/1 yields

$$P_1^{\mathcal{U}}(x) = \frac{\lambda\left(P_0e^{-\mu x} + \int_{y=0}^{x} e^{-\mu(x-y)}f(y)dy\right)}{f(x)}$$

$$= \frac{P_0e^{-\mu x} + \int_{y=0}^{x} e^{-\mu(x-y)}f(y)dy}{P_0e^{-\mu x}(1+\mu x) + \int_{y=0}^{x} e^{-\mu(x-y)}(1+\mu(x-y))f(y)dy}, x \geq 0,$$

$$P_2^{\mathcal{U}}(x) = \frac{P_0e^{-\mu x}\mu x + \int_{y=0}^{x} e^{-\mu(x-y)}\mu(x-y)f(y)dy}{P_0e^{-\mu x}(1+\mu x) + \int_{y=0}^{x} e^{-\mu(x-y)}(1+\mu(x-y))f(y)dy}, x \geq 0. \tag{3.81}$$

where $\{P_0, f(y)\}_{y\geq 0}$ is specified in formula (3.66).

Example 3.6.2 *In the M/Erl$_{2,\mu}$/1 queue*

$$S \stackrel{d}{=} Erl_{2,\mu} \text{ and } E[\mathcal{B}] = E[S]\,P_0 = \frac{\frac{2}{\mu}}{1-\lambda\times\frac{2}{\mu}} = \frac{2}{\mu-2\lambda}.$$

Then

$$E\left[a_x\right] = P_1^{\mathcal{U}}(x) \times \frac{2}{\mu - 2\lambda} + P_2^{\mathcal{U}}(x) \times \frac{1}{\mu - 2\lambda},$$

where $P_1^{\mathcal{U}}(x) + P_2^{\mathcal{U}}(x) = 1$, *and* $P_1^{\mathcal{U}}(x) > 0$, $P_2^{\mathcal{U}}(x) > 0$. *Thus,*

$$E\left[a_x\right] < \frac{2}{\mu - 2\lambda} = E[\mathcal{B}_1].$$

For a further discussion of the $M/Erl_{k,\mu}/1$ queue using LC see [22] (2017).

3.7 Exercises

Exercise 3.7.1 *Check the validity of formula (3.57).*

Exercise 3.7.2 *Prove the validity of formula (3.69) starting with* $P_0 = 1 - \rho$.

Exercise 3.7.3 *Prove the validity of formula (3.70) starting with* $P_0 = P_0 = \frac{1-\rho}{1-\rho+\rho_0}$.

Exercise 3.7.4 *Obtain the solution algebraically of equations (3.71) and (3.72), when* $x = 0$. *Explain the rationale for this result.*

Exercise 3.7.5 *In formula (3.81) calculate the values of* $p_1(0)$ *and* $p_2(0)$, *and give an intuitive explanation for their values.*

4

M/M/c Queues

4.1 Introduction

In the standard M/M/c queue customers arrive to the system in a single Poisson stream with rate $\lambda > 0$, and form a single waiting line in front of c ($c \geq 2$) independent servers in a parallel arrangement (as opposed to series arrangement; see Figure 4.1). Customers begin service by one of the c servers in order of arrival due to the single line before service. If a customer arrives when some or all of the c servers are idle, the customer begins service immediately by one of the idle servers, with equal probability on the set of idle servers. In the standard M/M/c queue, each server serves a customer at the same exponential rate $\mu > 0$ independently of the other servers. In the standard M/M/c queue customers complete service and depart from the system in random order due to random service times and the parallel arrangement of the c servers. In the standard $M_\lambda/M_\mu/c$ queue we will derive the stationary probability distributions of key random variables. (Note: For a more general LC analysis of transient (time-dependent) probability distributions of key random variables in M/M/c queues, (see p. 188ff in Chapter 4 in [22] (2017), [12] (1988), and [14] (1988).)

One classical method of stationary analysis for M/M/c queues focuses first on deriving the stationary probability distribution of the "number of customers in the system" using "state transition diagrams" (see Figure 3.7 on p. 102 in [41] (1975)). This classical method then uses the probability distribution of the number in the system to obtain other related results. This method is straightforward.

A second classical method of stationary analysis for M/M/c queues focuses first on obtaining the stationary probability distribution of the customer virtual waiting time using"Lindley recursions", in [10] (1975), such as on pp. 3–7 in [22] (2017). The stochastic level crossing method (LC) evolved from the author's research work while using Lindley recursions in his PhD thesis (see [10] (1975)).

The analyses of the various stochastic models in this book shows that the"stochastic level crossing method" (LC), bypasses much algebra required in the Lindley recursion method, in order to create integral equations for the pdfs of the key random variables. In doing so, LC gives results exactly equivalent to the LIndley recursion method, and extra viewpoints for studying the

DOI: 10.1201/9780429297601-4

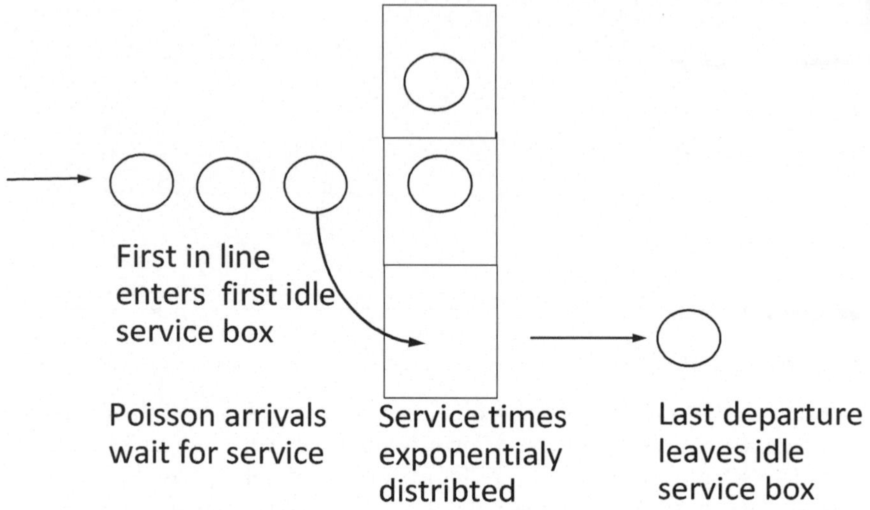

FIGURE 4.1
Shows dynamic of M/M/c queue with $c = 3$. First, three customers are waiting in line because all three servers are busy, then third server completes service. Customer at head of waiting line begins service immediately in third service box.

models (see pp. 7–17 in [22] (2017) and [10] (1975)). An advantage of LC is that it is more efficient in the analysis of many models, especially where the state variable is involved in dependencies, e.g., where "service time depends on waiting time" (see Section 2.3 in Chapter 2); or where the customer arrival rate depends on the current waiting time (see [15]) (1988). Another advantage of LC is that it connects immediately the "motion" of the sample path of a model's key r.v., to each analytic term of an integral equation for the stationary pdf of the key r.v.

(Note: Lindley recursions are currently very useful in analyzing stochastic models; e.g., in computational work and theoretical analyses. See, e.g., [58] (1990) and the references therein; and others.)

Exercises

Exercise 4.1.1 *(a) Give an example of a production process of a complex item where it may be an advantage to design the production process with parallel servers rather than with a single server. (b) Give an example where using parallel servers may be a disadvantage compared with single server.*

Exercise 4.1.2 *In an M/M/10 queue how many distinct ways can there be four idle and six busy servers? (See [34](1960) Chapter 2 for a classical review of combinatorial analysis.)*

Exercise 4.1.3 *How does LC circumvent a considerable amount of algebra compared with using Lindley recursions to derive stationary pdfs of waiting time? Hint: See pp. 3–7 and pp. 16–17 in [22](2017).*

4.1.1 LC Method of "Lines and Pages" for the Standard M/M/c Queues

In LC, the SP (leading point of sample path "thought of dynamically") in the M/M/c queue can make transitions between disjoint state-space sets (see Section 1.2 in Chapter 1). These disjoint state-space sets are called "lines"; or "Pages" (aka "Sheets"). This nomenclature comes about on applying geometric analogies with "real pages in a real regular book", or with "real sheets of paper (aka cards) in a box", respectively. (The Pages or sheets are like cards in a real deck of cards.)

For example, consider the SP of the sample path of the virtual wait in the M/M/c queue. See Figure 4.2. The SP moves in Time, making a trace on the "horizontal lines", making vertical upward jumps onto "Pages" (in state-space set $(0, \infty)$), and making "sloped traces" on the Pages. In the method of Lines and Pages, the waiting time of an arriving customer is "zero" if the SP is on a "horizontal line" which represents a state $(0, n)$, where n is the number of zero-wait customers in the system, $0 \leq n \leq c-1$, at the arrival instant. The waiting time of an arriving customer is "positive" (> 0) when the SP is on the "single "Page c" representing the composite state $(x, c), 0 < x < \infty$. Further details are given in Section 4.1.2.

4.1.2 Comparison of the Two Queueing Models: "M/M/c" Queue versus "M/M/1 Queue"

In the standard M/M/c queue ($c \geq 2$), a set of $c-1$ discrete parallel "Lines" which indicate the customer zero-wait states is closely linked by the "border state" $(0, c-1)$" to a state-space "Page c" (see Figure 4.2 in Section 4.2).

In the single-server queue $M_{\lambda}/M/_{\mu}1$ there is only "one arriving-customer zero-wait state", i.e., when the server is idle. This one zero-wait "horizontal Line" is placed at level zero just below the single sheet of customer positive-wait states $(0, \infty)$ (see, e.g., Figure 2.2 in Section 2.1.6 in Chapter 2).

In the multiple-server M/M/c variants there may be one, two, several, or many separate Pages depending on the context. Pages may not connected to each other by a continuous segment of the sample path. in order for the SP to move from a containing state-space Page to a different state-space Page, the SP must jump from the present Page to the receiving Page, at essential time points in the model dynamics. The jumps may be vertical, up or down, or

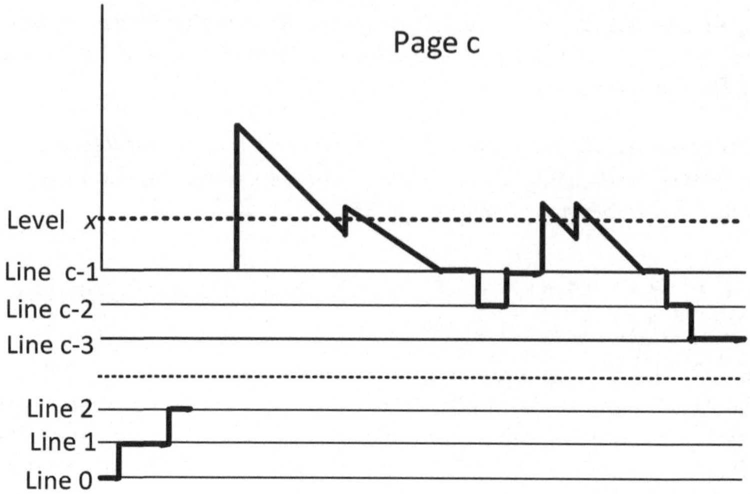

FIGURE 4.2
Sample path of virtual wait $\{W(t)\}_{t\geq 0}$ in standard M/M/c queue (heavy black line going from left to right in Time).There are c lines $0, ..., c-1$, for zero-wait states. and one Page c for positive-wait states. Line $c-1$ at level 0 of Page c is a "border state". The SP begins at level 0 when all servers are idle. The SP moves to Line 1 just after the first arrival. The SP moves through all the Lines $n = 0, ..., c-1$ randomly, in a birth-death stochastic process, one step up or one step down until reaching Line $c-1$. If there is an arrival while the SP is in Line $c-1$, the SP jumps upward onto Page c with jump size $\stackrel{d}{=} Exp_{c\mu}$.

parallel (from a level on the containing Page to the same level on the receiving Page). Examples of essential time points are: when a new customer arrives to the system (e.g., adding to its workload); or a customer in service departs from the system (e.g., lowering the number in the system and/or lowering the number of busy servers), etc.

The "concrete" sample path follows the motion of the SP, and traces a recording over Time of important events with respect to the queueing model. This "sample path is a template" which assists the researcher to develop precise analytic integral equations efficiently for the pdfs of the random variables of interest using the principle of rate balance across state-space levels; or to write solutions of the equations quickly.

(Note: Complex single-server M/G/1 queues can also be modelled using the method of Lines and Pages (aka Sheets) – see, e.g., [25] (1981); [17] (1997); [38] (1893).)

The method of Lines and Pages gives LC flexibility to analyze different types of stochastic models: e.g., queues; dams; inventories;

production-inventories; actuarial risk models; replacement models; pharmacokinetic models; natural-science models, etc.

Section 4.2, introduces the method of "Lines and Pages" by deriving some known results for the standard M/M/c queue where c $\geq$ 2. The remaining sections of this Chapter study variants of M/M/c queues.

4.2 Analysis of Standard M/M/c Queue

We use LC to analyze the standard M/M/c queue, which is one of the first models analyzed to validate the (then new) LC methodology (see pp. 37–39 in [10] (1975); see also [11] (1975)). It is instructive to derive some classical (known) results for M/M/c queues using LC, which were obtained originality by other analytic methods (e.g., birth and death processes, which are a special case of continuous-time Markov chains; or Lindley recursions), for comparison. Those comparison methods can lead to a broader view of the LC-derived results (which are often more efficiently obtained).

Assume that the number of servers in the standard M/M/c queue is $c \geq 2$. Let all arriving customers receive the same exponential service rate $\mu > 0$ when in service, independent of any other customers in service. The stability condition for the M/M/c queue is

$$\lambda < c\mu, \tag{4.1}$$

where λ is the customer arrival rate to the system, and $c\mu$ is the maximum customer departure rate (service-completion rate) from the system, which occurs when all c servers are busy simultaneously, each serving at rate μ) (see Section 1.7 on system stability in Chapter 1)

Definition 4.1 *The "inter-start-of-service departure time" is the duration from the "start of service instant" of an arriving customer until the "first departure instant" thereafter, of any customer completing service from any one of the c busy servers.*

Definition 4.1 implies that if a customer arrives when the state is $(c-1)$ (i.e., virtual wait $y = 0$), or when all c servers are busy (i.e., virtual wait $y > 0$), then that arrival's start of service will begin time y later with all c servers being occupied. The first departure from the thereafter will occur at time $\stackrel{d}{=} y+$ $\text{Exp}_{c\mu}$ later. The latter $\text{Exp}_{c\mu}$ occurs because of the memoryless property of all c Exp_μs, and $P(\min(c \text{ iid Exp}_\mu\text{s} > x) = e^{-c\mu x}$ (see Section 1.5.3 in Chapter 1).

The stability condition in formula (4.1) for the standard M/M/c queue is intuitive (see Section 1.7, and formula (1.25) in Section 1.5.3 in Chapter 1, which gives the exponential rate of service completions when all c independent servers are operating at the same exponential rate μ.

To briefly review, in LC, the "method of Lines and Pages", utilizes a total of $c-1$ horizontal Lines, each corresponding to the number of customers: $0, 1, 2, ..., c-1$, who wait zero time to begin service by "one of the idle servers". Also, there is one "state-space Page c" corresponding to all c servers being busy (representing $\geq c$ customers in the system). If a new customer arrives when the SP is on Page c, then the arrival must wait a positive time to begin service. A sample path of the virtual wait) $\{W(t)\}_{t \geq 0}$ is shown in Figure 4.2. The sample path has $c-1$ Lines for the zero-wait states $(0, n)$, $n = 0, ..., c-1$, with probabilities P_n, $n = 0, ..., c-1$, respectively. The "0" in the first position of state $(0, n)$ denotes a "zero wait" for a potential new arrival, and the "n" in the second position of state $(0, n)$ denotes the number of customers in the system that a new arrival "sees" just before joining the system. Thus state $(0, 0)$ denotes an empty system (all servers idle), and $P_0 := P$(System state is $(0, 0)$). The single Page c for the "composite state" $((0, \infty), n \geq c)$,indicates that a new arrival to the system would wait a positive time before beginning service.

Definition 4.2 *A "border state" in the standard M/M/c queue, is such that the next arrival to the system changes the system state from a zero-wait discrete state $(0, n)$, $n = 0, ..., c-1$, to a "positive-wait" continuous state on Page c $(0, \infty)$, with $n \geq c$), which is equivalent to "virtual wait > 0".*

State $(0, c-1)$, corresponding to Line $c-1$ is the "border". Geometrically in a sample-path diagram Line $c-1$ is placed (usually) at the bottom of Page c (see Figure 4.2).

(Note : Downcrossings of level $x > 0$ on Page c are all smooth crossings with slope $= -1$. The reason for slope -1 is that the start-of-service departure-time, viewed logically, is like the "look-ahead" service time of an arrival in a single-server $M_\lambda/M_{c\mu}/1$ queue (with service rate $c\mu$) when considering the virtual wait. If the moment of arrival had been one minute later (horizontal line of size +1 minute) and no new arrivals intervened during that minute, the arrival would wait one minute less (vertical jump of size -1 minute), resulting in a slope:

$$\text{slope} = \frac{\text{vertical jump of size } -1 \text{ minute}}{\text{horizontal line of size } +1 \text{ minute}} = -1, \tag{4.2}$$

as long as the system is non-empty.

Upcrossings of level $x > 0$ on Page c are all due to SP vertical jumps which may originate from: (1) level 0 on Page c, i.e., from Line state $(0, c-1)$; or, (2) any level y $(0 < y < x)$ on Page c.

(Note: The placement in a sample-path diagram of Line $c-1$ (e.g., below or above Page c, does not change the analysis because a jump into composite state $(0, \infty)$ on Page c starting with a new arrival when the SP is on Line $(c-1)$, is balanced at a later time by a smooth downcrossing into level 0 on Page c with slope -1, i.e., back into state $(0, c-1)$, as long as stability holds

($\lambda < c\mu$). Thus "local rate balance between Page c and state $(0, c-1)$" is preserved no matter where Line $(c-1)$ is placed in the diagram. In Figure 4.2, this local rate balance occurs at two time points.)

Denote the zero-wait probabilities as $P_n, n = 0, ..., c-1$, the stationary pdf of wait as $f(x), x > 0$, and the stationary cdf of wait by $F(x), x \geq 0$. Then

$$F(x) = \sum_{n=0}^{c-1} P_n + \int_0^x f(x)dx, x \geq 0, \tag{4.3}$$

$$F(0) = \sum_{n=0}^{c-1} P_n. \tag{4.4}$$

4.2.1 Stationary Pdf of Virtual Wait in Standard M/M/c Queue

Sections 4.2.1.1–4.2.3 compute the probabilities of the discrete zero-wait states $(0,0)$–$(0, c-1)$ using the "companion method" of birth and death processes. Section 4.2.4 computes the probabilities of the positive wait states. Section 4.2.5 recaps the linear algebraic equations for the state-space Lines, and the integral equation for Page c.

4.2.1.1 Probability That an Arrival Waits Zero Time before Beginning Service in an M/M/c Queue

We use a special case of formula (1.31) for "birth and death processes" given in Section 1.6 in Chapter 1, namely

$$P_n = \frac{\lambda_0 \lambda_1 \cdots \lambda_{n-1}}{\mu_1 \mu_{12} \cdots \mu_n} P_0, n = 1, ..., c-1. \tag{4.5}$$

In formula (4.5), the product in the numerator stops at $n = c-1$ (instead of continuing to "∞" as in (1.31)). The symbol $P_n :=$ probability that the SP is in state $(0, n), n = 0, 1, ..., c-1$, where n denotes the number of zero-wait customers seen by an arrival. In the numerator of formula (4.5), the symbol $\lambda_n :=$ customer arrival rate when the state is $(0, n), n = 0, 1, ..., c-1$, and the "0" in $(0, n)$ indicates that an arrival begins service immediately, with no waiting.

In the standard M/M/c queue under consideration, the arrival rate to the system is the same for all customers, independent of the state of the system. That is, λ is the arrival rate of new customers for every zero-wait target sate: $(0,0), ..., (0, c-2)$. Therefore

$$\lambda_n \equiv \lambda, \ n = 0, 1, ..., c-1. \tag{4.6}$$

(Note that the arrival rate is also he same λ for positive-wait states.)

In the denominator of formula (4.5), the product stops at $n = c - 1$. The symbol $\mu_n :=$ minimum of n iid exponential r.v.s with service rates μ, when the states are $(0, n), n = 1, ..., c - 1$ ("n" denotes "number in system" that waited 0 time for service, and are each being served independently). So,

$$\mu_n \cong n\mu, n = 1, ..., c - 1, \tag{4.7}$$

(See Section 1.5.3 in Chapter 1 for the pdf of the minimum of n iid Exp_μs.)

The symbol P_0 is the probability that the SP is in state $(0, 0)$, where all servers are idle and the system is empty.

4.2.2 Model Equations for Zero-wait States

Next, we derive model equations for the stationary pdf of wait. Further explanations are given in Section 4.2.5. Plugging the value λ_n from formula (4.6) and $n\mu$ from formula (4.7) into formula (4.5) gives

$$P_n = \frac{\left(\frac{\lambda}{\mu}\right)^n}{n!} P_0, n = 0, ..., c - 1. \tag{4.8}$$

Specifically,

$$P_{c-1} = \frac{\left(\frac{\lambda}{\mu}\right)^{c-1}}{(c-1)!} P_0\,. \tag{4.9}$$

4.2.3 Further Discussion on Formulas for Probabilities P_{c-1} and P_0 in Standard M/M/c Queue

As a check on formula (4.9) for P_{c-1}, the total SP entrance rate into state $(0, c - 1)$ must be taken into account, including the rate $f(0^+)$ from Page c into Line $(0, c - 1)$.

(Note: We can obtain formula (1.4) also by using "SP rate out" = "SP rate in" for each state: $(0, 1) ..., (0, c - 2)$ as follows in formula (4.10)):

$$\begin{gathered}
\textbf{Rate out} = \textbf{Rate in} \\
\lambda P_0 = \mu P_1 \\
(\lambda + \mu)P_1 = \lambda P_0 + 2\mu P_2 \\
(\lambda + 2\mu)P_2 = \lambda P_1 + 3\mu P_3 \\
\cdots \\
(\lambda + (c-2)\mu)P_{c-2} = \lambda P_{c-3} + (c-1)\mu P_{c-1}
\end{gathered} \tag{4.10}$$

Special attention is given separately to the "border state" $(0, c - 1)$ as follows:

$$\begin{gathered}
\textbf{SP Rate out} = \textbf{SP Rate in} \\
(\lambda + (c-1)\mu)P_{c-1} = \lambda P_{c-2} + f(0^+).
\end{gathered}$$

$$(\lambda + (c-1)\mu)P_{c-1} = \lambda P_{c-2} + \lambda P_{c-1}. \tag{4.11}$$

Due to "local rate balance" between Line $(0, c-1)$ and Page c, we have the equality,

$$f(0^+) = \lambda P_{c-1}, \tag{4.12}$$

In formula (4.12) $f(0^+)$ is the downcrossing rate at which the SP enters state $(0, c-1)$ from above, and $\lambda P_{c-1}$is the upcrossing rate out of state $(0, c-1)$ which causes the SP to jump vertically onto Page c. We obtain the last terms in Equation (4.11), based on "local rate balance". The SP exit rate from Page c into state $(0, c-1)$ is equal to the SP entrance rate into Page c from state $(0, c-1)$. (The only path for the SP to exit Page c is by an exit (downcrossing) into state $(0, c-1)$. Conversely, the only path for the SP to enter Page c is by an exit from state $(0, c-1)$ due to a customer arrival when the system state is $(0, c-1)$). Thus, equation (4.11) becomes

$$(c-1)\mu P_{c-1} = \lambda P_{c-2}$$

$$P_{c-1} = \frac{\lambda P_{c-2}}{(c-1)\mu} = \frac{\lambda \frac{\left(\frac{\lambda}{\mu}\right)^{c-2}}{(c-2)!} P_0}{(c-1)\mu} \tag{4.13}$$

$$P_{c-1} = \frac{\left(\frac{\lambda}{\mu}\right)^{c-1}}{(c-1)!} P_0. \tag{4.14}$$

Equation (4.14) connects the continuous part of the pdf of virtual waiting time on Page c with the probabilities of the "atoms" (i.e., states $(0, c-1)$, $(0, c-2)$, ..., $(0, 1)$). The formula for $f(x), 0 < x < \infty$ and value of $f(0^+)$ are given in Section 4.2.4, in formula (4.20).

(The connection between the continuous part of the pdf of wait and the discrete part of the pdf of wait, and other examples, led the author to coin the term "border state" for state $(0, c-1)$ in the present context.).

The normalizing condition for the waiting time is "All probabilities, for zero-wait states plus positive-wait states, sum to 1." The sum of the probabilities for the zero-wait states is $P_0 + \sum_{n=1}^{c-1}\left(\frac{\left(\frac{\lambda}{\mu}\right)^n}{n!} P_0\right)$. The sum of the probabilities for the positive-wait states is $\int_0^\infty f(x)dx$. Therefore

$$P_0 + \sum_{n=1}^{c-1} \frac{\left(\frac{\lambda}{\mu}\right)^n}{n!} P_0 + \int_0^\infty f(x)dx = 1, \tag{4.15}$$

where the pdf on Page c is

$$f(x) = \lambda P_{c-1} e^{-(c\mu-\lambda)x}, 0 < x < \infty. \tag{4.16}$$

Equation (4.16) for $f(x)$ is derived by rate balance across level x,in formula (4.19) in Section 4.2.4. Thus, collecting the terms in equation (4.15),

$$\int_0^\infty f(x)dx = \lambda P_{c-1}\frac{1}{c\mu - \lambda},$$

$$\text{Left side} = P_0 + \sum_{n=1}^{c-1}\frac{\left(\frac{\lambda}{\mu}\right)^n}{n!}P_0 + \lambda P_{c-1}\frac{1}{c\mu - \lambda},$$

$$P_0 + \sum_{n=1}^{c-1}\frac{\left(\frac{\lambda}{\mu}\right)^n}{n!}P_0 + \lambda\frac{\left(\frac{\lambda}{\mu}\right)^{c-1}}{(c-1)!}P_0\frac{1}{c\mu - \lambda} = 1,$$

where we applied formula (4.14) for the value of P_{c-1} in the last term on the right side. This gives the following formula for P_0:

$$P_0 = \frac{1}{1 + \sum_{n=1}^{c-1}\frac{\left(\frac{\lambda}{\mu}\right)^n}{(n-1)!} + \lambda\frac{\left(\frac{\lambda}{\mu}\right)^{c-1}}{(c-1)!}\frac{1}{c\mu-\lambda}},$$

$$P_0 = \frac{1}{\sum_{n=0}^{c-1}\frac{\left(\frac{\lambda}{\mu}\right)^n}{n!} + \frac{\left(\frac{\lambda}{\mu}\right)^c}{c!}\frac{c\mu}{c\mu-\lambda},} \tag{4.17}$$

where

$$f(x) = \lambda\frac{\left(\frac{\lambda}{\mu}\right)^{c-1}}{(c-1)!}P_0 e^{-(c\mu-\lambda)x},$$

and

$$\int_{y=0}^\infty f(x)dx = \lambda\frac{\left(\frac{\lambda}{\mu}\right)^{c-1}}{(c-1)!}\frac{1}{c\mu - \lambda}P_0.$$

4.2.4 Analysis of Positive-Wait States in Standard M/M/c Queue

Using LC, the integral equation for the stationary pdf of wait on Page c, $f(x), 0 < x < \infty$, is the rate balance equation:

$$f(x) = \lambda P_{c-1}e^{-c\mu x} + \lambda\int_{y=0}^{x} e^{-c\mu(x-y)}f(y)dy, x > 0. \tag{4.18}$$

In equation (4.18), the left side is the SP downcrossing rate of level x; the right side is the vertical upcrossing rate of level x. By analogy with Equation

(2.4) for the stationary pdf of wait in the $M_\lambda/M_\mu/1$ queue in Chapter 2, the solution of equation (4.18) is similar (assuming stability, $\lambda < c\mu$). Thus, taking $\frac{d}{dx}$ on both sides of equation (4.18) gives,

$$f'(x) = -c\mu f(x) + \lambda f(x) = f(x)\left(-c\mu + \lambda\right),$$

$$\frac{d \ln f(x)}{dx} = -c\mu + \lambda,$$

$$\ln fx) = (-c\mu + \lambda)\, x + A_1 \text{ (constant)},$$

$$f\,(x) = Ae^{-c\mu - \lambda)x} \quad (A \text{ constant}),$$

$$f(0) = A = \lambda P_{c-1},$$

$$f(x) = \lambda P_{c-1} e^{--(c\mu-\lambda)x}, 0 < x < \infty. \tag{4.19}$$

Importantly, equation (4.19) gives

$$f(0^+) = \lambda P_{c-1}. \tag{4.20}$$

(recall that $f(0) \cong f(0^+) =$ SP downcrossing rate into level 0 on Page c by the basic LC theorem; see Figure 4.2).

To explain further Equation (4.45), composite state $((0,\infty), c-1)$ is accessible only by means of a single vertical SP upward jump at an arrival instant when the system is in the "border state" $(0, c-1)$. The normalizing condition (to be solved together with equation (4.18), expresses that the sum of all probabilities with respect to the virtual wait, is equal to 1, namely:

$$\sum_{n=0}^{c-1} P_n + \int_{x=0}^{\infty} f(x)dx = 1. \tag{4.21}$$

Formula (4.21) leads to the value of P_0 as follows:

$$\sum\nolimits_{n=0}^{c-1} \frac{\left(\frac{\lambda}{\mu}\right)^n}{n!} P_0 + \lambda \frac{\left(\frac{\lambda}{\mu}\right)^{c-1}}{(c-1)!)(c\mu-\lambda)} P_0 = 1$$

$$P_0 = \left[\sum\nolimits_{n=0}^{c-1} \frac{\left(\frac{\lambda}{\mu}\right)^n}{n!} + \lambda \frac{\left(\frac{\lambda}{\mu}\right)^{c-1}}{(c-1)!)(c\mu-\lambda)}\right]^{-1}$$

$$P_0 = \left[\sum\nolimits_{n=0}^{c-1} \frac{c^n\left(\frac{\lambda}{c\mu}\right)^n}{n!} + \frac{c^c\left(\frac{\lambda}{c\mu}\right)^c}{c!\left(1-\frac{\lambda}{c\mu}\right)}\right]^{-1} \tag{4.22}$$

$$P_0 = \frac{1}{\sum_{n=0}^{c-1} \frac{\left(\frac{\lambda}{\mu}\right)^n}{n!} + \frac{\left(\frac{\lambda}{\mu}\right)^c}{c!} \frac{c\mu}{c\mu-\lambda}},$$

which is the same as formula (4.17) in Section 4.2.3. Also, every term in the formula is dimensionless. Note that formula (4.23) shows the normalization in terms of $\lim_{x\uparrow\infty} F(x) = F(\infty)$, that is

$$F(\infty) = \sum_{n=0}^{c-1} P_n + \int_{x=0}^{\infty} f(x)dx = 1, \tag{4.23}$$

where $F(x), 0 \leq x < \infty$, denotes the stationary cdf of the virtual wait. Formulas (4.21) and (4.23) show that the total probability "1" is shared by the probabilities of he $c-1$ discrete states denoted by $P_n, n = 0, ..., c-1$, and by the area under the curve $f(x), 0 < x < \infty$, on Page c, expressed by $\int_{x=0}^{\infty} f(x)dx$ (Figure 4.2).

4.2.5 Recap of Linear Algebraic Equations for the Lines and Integral Equation for Page c in M/M/c Queue

This recap clarifies further some details for obtaining the stationary pdf of the virtual wait and related results.

4.2.5.1 Linear Algebraic Equations for the Probabilities Representing the Lines

Equations (4.10) in Section 4.2.3 are rate-balance equations, which equate SP rates out of, and into, the discrete zero-wait states $(0, n)$, $n = 0, ..., c-1$. The term $f(0^+)$ (aka $f(0)$) in Equation (4.20) in Section 4.2.4 is the SP downcrossing rate into level 0 on Page c, i.e., the SP entrance rate into state $(0, c-1)$ from above.

4.2.5.2 Integral Equation for the Stationary Pdf of Virtual Wait on Page c

We now recap details to derive the customer positive-wait integral Equation (4.18) in Section 4.2.4. Consider composite state $((x, \infty), c)$ on Page c (Figure 4.2). The SP exit rate from $((x, \infty), c)$ on Page c (due only to exits which are downcrossings of level x – left side of equation) is equal to the SP entrance rate into $((x, \infty), c)$ (consisting of the upcrossing rate of vertical jumps that overshoot level x, in two possible ways – right side of equation):

(1) upward jumps beginning from Line $c-1$ at level 0 of Page c due to arrivals when the state is $(0, c-1)$;

(2) upward jumps beginning on Page c in the state-space interval $(0, x)$ that overshoot level x due to arrivals when the SP is in a composite state (y, x) where $0 < y < x$.

The downcrossing rate of level x is $f(x)$ by the basic LC theorem (stated in formula (1.1) in Section 1.4, Chapter 1).

Another way to express the same idea is that the SP entrance rate into composite state $((x, \infty), c)$ is from two different sources resulting in the two terms on the right side of Equation (4.18 in Section 4.2).

These two terms are:

Term 1, on right side of (4.18). Entrances are generated by upward jumps due to arrivals when the SP is in the border state $(0, c-1)$, beginning at level 0 of Page c and overshooting level x on Page c when a customer arrives. Since there is only "one Page", the only access to composite state $((x, \infty), c)$ in one step beginning from a zero-wait state $(0, (c-1))$, i.e., from Line $c-1$ when a customer arrives (Figure 4.2). The SP entrance rate from this source is "$\lambda P(S > x)P_{c-1}$", where P_{c-1} is the stationary probability of state $(0, c-1)$, and the jump size $S \stackrel{d}{=} \text{Exp}_{c\mu}$ since there would be c customers in service simultaneously each with rate μ, just after such an arrival begins service; also $S :=$ "minimum" of c iid Exp_μ random variables. Thus, $P(S > x)$ $= e^{-c\mu x}$ (see Section 1.5.3 in Chapter 1). Thus Term 1 is $\lambda P_{c-1}e^{-c\mu x}$ on the right side of (4.18).

Term 2, on right side of (4.18). Entrances into composite state $((x, \infty), c)$ may be generated also by vertical jumps due to arrivals when the SP is at any level on Page c, say level y, where $0 < y < x$. These SP jumps begin at level y and overshoot level x on Page c with probability $e^{-c\mu(x-y)}f(y)dy$. Just after such an arrival begins service (after waiting time y to begin service) all c servers will be busy and the inter-start-of-service–departure time $S \stackrel{d}{=}$ $\text{Exp}_{c\mu}$, independent of any arrivals to the system that arrived earlier or may arrive later. The SP will enter immediately into composite state $((x, \infty), c)$ with probability $e^{-c\mu(x-y)}$. This reasoning leads to Term 2 (the integral) on the right side of the integral equation (4.18).

4.2.6 More Explicit Solution of Equations in M/M/c Queue

We first solve integral Equation (4.18). Differentiating both sides with respect to x, accounting for differentiation under the integral sign, and solving the resulting first-order differential equation, gives

$$f(x) = Ae^{-(c\mu-\lambda)x}, x > 0, \tag{4.24}$$

where A is a constant. Letting $x \downarrow 0$ in (4.25) we get the initial condition

$$f(0) = A = \lambda P_{c-1}, \tag{4.25}$$

since $f(0) \equiv f(0^+)$ is the downcrossing rate into level 0 (rate of entrance into state $(0, c-1)$ from Page c), and λP_{c-1} is the upward jump rate out of level 0 (rate of egress out of $(0, c-1)$ into Page c). (Equivalently, $f(0)$ is the exit rate out of $((0, \infty), c)$ and λP_{c-1} is the entrance rate into $((0, \infty), c)$.) Thus $A = \lambda P_{c-1}$ and

$$f(x) = \lambda P_{c-1}e^{-(c\mu-\lambda)x}, x > 0. \tag{4.26}$$

Next, from (4.14) and (4.26) we obtain

$$P_n = \left(\frac{\lambda}{\mu}\right)^n \frac{1}{n!} P_0, n = 0, ..., c-1.$$

In particular

$$P_{c-1} = \left(\frac{\lambda}{\mu}\right)^{c-1} \frac{1}{(c-1)!} P_0. \tag{4.27}$$

Substituting formula (4.27) in Section 117 into (4.26) gives

$$f(x) = \lambda \left(\frac{\lambda}{\mu}\right)^{c-1} \frac{1}{(c-1)!} P_0 e^{-(c\mu-\lambda)x}, x > 0.$$

The normalizing condition (4.23) is

$$\left(\sum_{n=0}^{c-1} \left(\frac{\lambda}{\mu}\right)^n \frac{1}{n!}\right) P_0 + \lambda \left(\frac{\lambda}{\mu}\right)^{c-1} \frac{1}{(c-1)!} P_0 \int_{x=0}^{\infty} e^{-(c\mu-\lambda)x} dx = 1,$$

$$\left(\sum_{n=0}^{c-1} \left(\frac{\lambda}{\mu}\right)^n \frac{1}{n!}\right) P_0 + \lambda \left(\frac{\lambda}{\mu}\right)^{c-1} \frac{1}{(c-1)!(c\mu-\lambda)} P_0 = 1,$$

leading to

$$P_0 = \frac{1}{\sum_{n=0}^{c-1} \left(\frac{\lambda}{\mu}\right)^n \frac{1}{n!} + \left(\frac{\lambda}{\mu}\right)^c \frac{c\mu}{c!(c\mu-\lambda)}}. \tag{4.28}$$

Formula (4.28) agrees with formulas (4.22) in Section 4.2.4, and with formula 4.2.3 in Section 4.17, which are obtained from different perspectives.

The cdf of wait is :

$$F(x) = P_0 + \int_{y=0}^{x} \lambda P_{c-1} e^{-(c\mu-\lambda)y} dy$$

$$= P_0 \left(1 + \lambda \left(\frac{\lambda}{\mu}\right)^{c-1} \frac{1}{(c-1)!(c\mu-\lambda)} \left(1 - e^{-(c\mu-\lambda)x}\right)\right), x \geq 0. \tag{4.29}$$

Note that $F(0) = P_0$ and

$$F(\infty) = P_0\left(1 + \lambda\left(\frac{\lambda}{\mu}\right)^{c-1} \frac{1}{c)!(c\mu - \lambda)}\right)$$

$$= P_0 + P_0\lambda\left(\frac{\lambda}{\mu}\right)^{c-1} \frac{1}{(c-1)!(c\mu - \lambda)}$$

$$= P_0 + \int_{x=0}^{\infty} \lambda P_{c-1} e^{-(c\mu-\lambda)} dx$$

$$= P_0 + \int_{x=0}^{\infty} f(x)dx = 1, \tag{4.30}$$

where (4.30) follows from formula (4.26).

Moreover $F(x)$ is concave downward on $(0, \infty)$ since the second derivative is

$$F"(x) < 0, 0 < x < \dot{\infty}. \tag{4.31}$$

4.2.6.1 Boundedness of the Pdf of Wait

From the pdf (4.26), "$0 < f(x) < \lambda$", $x > 0$, since $0 < P_{c-1} < 1$ and $0 < e^{-(c\mu-\lambda)x} < 1$, $x > 0$, provided the queue is stable, i.e., $c\mu - \lambda > 0$.

Example 4.2.1 *This example compares the single-server $M_\lambda/M_\mu/1$ queue and the two-server $M_\lambda/M_\mu/2$ queue as follows (using superscript "(1)" for the M/M/1 and superscript "(2)" for the M/M/2, pdfs and probabilities, respectively). Assume $\lambda = 1$; $\mu = 2$.*

For the $M_\lambda/M_\mu/1$ queue, where $P_0^{(1)}$is the probability that the system is empty, and an arrival is a zero-wait customer.

$$P_{0Wt}^{(1)} = P_0^{(1)} = 1 - \frac{\lambda}{\mu} = \frac{1}{2} = \frac{5}{10};$$

$$f^{(1)}(x) = \lambda P_0^{(1)} e^{-(\mu-\lambda)x}, 0 < x < \infty.$$

$$f^{(1)}(x) = \frac{5}{10}e^{-x}, x > 0.$$

For the $\mathrm{M}_\lambda/\mathrm{M}_\mu/2$ queue, where $P_0^{(2)}$ is the probability of state $(0,0)$,

$$P_0^{(2)} = \frac{1}{\sum_{n=0}^{c-1}\left(\frac{\lambda}{\mu}\right)^n \frac{1}{n!} + \left(\frac{\lambda}{\mu}\right)^c \frac{c\mu}{(c-1)!(c\mu-\lambda)}}$$

$$= \frac{1}{\sum_{n=0}^{1}\left(\frac{\lambda}{\mu}\right)^n \frac{1}{n!} + \left(\frac{\lambda}{\mu}\right)^2 \frac{c\mu}{(c-1)!(c\mu-\lambda)}}$$

$$= \frac{1}{1+\frac{\lambda}{\mu}+\left(\frac{\lambda}{\mu}\right)^2 \frac{2\mu}{2(2\mu-\lambda)}}$$

$$= \frac{1}{1+\frac{1}{2}+\frac{1}{6}} = \frac{6}{10}.$$

The probability of a zero wait by an arrival, where $P_1^{(2)}$ is the probability of state $(0,1)$,is

$$\begin{aligned} P_{0Wt}^{(2)} &= P_0^{(2)} + P_1^{(2)} \\ &= P_0^{(2)} + \left(\frac{\lambda}{\mu}\right)^{c-1} \frac{1}{(c-1)!} P_0^{(2)} \\ &= \left(\frac{1}{1+\frac{1}{2}+\frac{1}{3}}\right) + \frac{1}{2} P_0^{(2)} \\ &= \frac{6}{10} + \frac{3}{10} = \frac{9}{10}. \end{aligned}$$

The continuous part of the pdf of the virtual wait when $c = 2$, is

$$\begin{aligned} f^{(2)}(x) &= \lambda P_{c-1}^{(2)} e^{-(c\mu-\lambda)x} = \lambda\left(\frac{\lambda}{\mu}\right)^{c-1} \frac{1}{(c-1)!} P_0^{(2)} e^{-(c\mu-\lambda)x} \\ &= \lambda\left(\frac{\lambda}{\mu}\right) P_0^{(2)} e^{-(2\mu-\lambda)x} = \frac{1}{2} P_0^{(2)} e^{-3x} \\ &= \frac{1}{2}\left(\frac{1}{1+\frac{1}{2}+\left(\frac{1}{2}\right)^2 \frac{2\cdot 2}{2(2\cdot 2-1)}}\right) e^{-(2\mu-\lambda)x}; \\ f^{(2)}(x) &= \frac{3}{10} e^{-3x}, x > 0. \end{aligned}$$

$$\text{where } P_0^{(2)} = \left[\sum_{n=0}^{c-1} \frac{\left(\frac{\lambda}{\mu}\right)^n}{n!} + \frac{\left(\frac{\lambda}{\mu}\right)^c c\mu}{(c)!)\,(c\mu-\lambda)}\right]^{-1} = \frac{6}{10};$$

$$\int_{x=0}^{\infty} \frac{3}{10} e^{-3x} dx = \frac{1}{10}.$$

When $P_{0Wt}^{(2)} := P_0^{(2)} + p_1^{(2)}$ and $P_{+Wt}^{(2)} := \int_{x=0}^{\infty} f^{(2)}(x)dx$, a straightforward computation validates the normalizing condition

$$P_{0Wt}^{(2)} + P_{+Wt}^{(2)} = P_0^{(2)} + p_1^{(2)} + \int_{x=0}^{\infty} \frac{3}{10} e^{-3x} dx = 1.$$

$$P_{0Wt}^{(2)} = P_0^{(2)} + p_1^{(2)} = \frac{9}{10}.$$

4.2.6.2 A Brief Comparison of Single-server $M_\lambda/M_\mu/1$ Queue and Multiple-server $M\lambda/M_\mu/2$ Queue

Figure 4.3 shows that there is a much larger probability of a customer waiting a positive time in the $M_\lambda/M_\mu/1$ queue, than in the $M\lambda/M_\mu/2$ queue ($\lambda = 1$, $\mu = 2$), in stationary sate (equilibrium). (These probabilities are the areas under the curves Figure 4.3.)

The sum of all probabilities of the virtual wait to sum to 1, is a necessary condition for the mixed pdf of the virtual wait. A theoretical conclusion is that in the M/M/1 queue "50% of customers begin service immediately" after

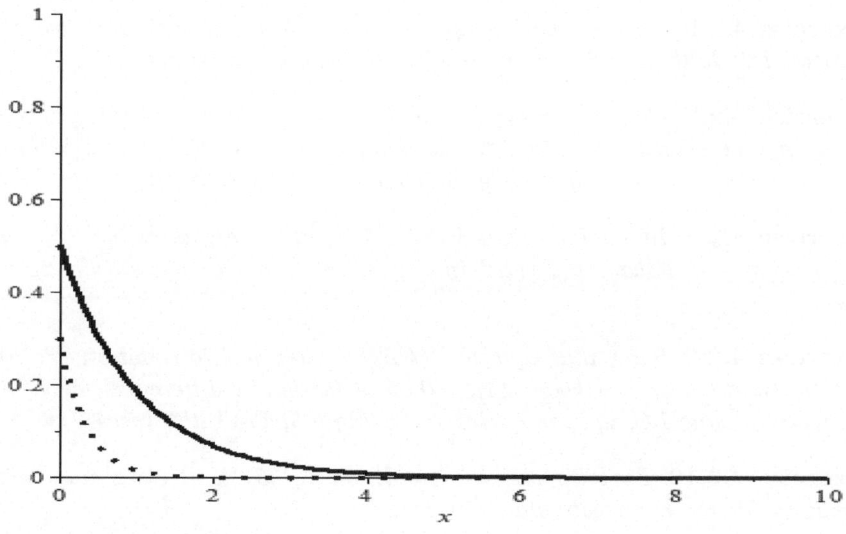

FIGURE 4.3
The "solid line" is the continuous part of wait $f^{(1)}(x)$ in the M/M/1 queue. The "dotted line" is the continuous part of pdf of wait $f^{(2)}(x)$ in M/M/2 queue. In both queueing models, $\lambda = 1$ and $\mu = 2$.

arriving (zero wait). In the M/M/2 queue "90% of customers begin service immediately" on arriving (zero wait). Theoretically the M/M/2 queue looks superior. In practice a manager can decide which queueing system is "better", based on a decision analysis accounting for costs of providing two servers, space available, etc.

Exercises

Exercise 4.2.1 *In formula (4.12) explain the role of the last term* $f(0^+)$ *regarding SP rate balance. What makes this instance an example of the basic LC theorem stated in Section 1.4 in Chapter 1?*

Exercise 4.2.2 *Explain why* $\lambda_n \equiv \lambda$ *in formula (4.6).*

Exercise 4.2.3 *Explain why* $\mu_n = n\mu$ *in formula (4.7). (Hint: Don't forget the memoryless property).*

Exercise 4.2.4 *In formula (4.8) explain why the* $n = 0, ..., c-2$.

Exercise 4.2.5 *Explain why the right side of Equation (4.18) is equal to the upcrossing rate of level* x

Exercise 4.2.6 *Derive equation (4.18) starting from the basic LC theorem in Section 1.4. Is there a faster, more efficient way to derive it?*

Exercise 4.2.7 *Compare integral Equation (4.18) for the* $M_\lambda/M_{c\mu}/c$ *queue with integral equation (2.4) for the* $M_\lambda/M_\mu/1$ *queue in Chapter 2. Why does this comparison give immediately the solution of Equation (2.4).*

Exercise 4.2.8 *In light of Exercise 4.2.7, is there an advantage for using connections, or discovering valid analogies, between different stochastic models?*

Exercise 4.2.9 *Show that equation (4.22) is identical to equation (3.39) on p. 103 in the classical book [41]. What is the main difference between the derivation using LC and the derivation using a birth-death process?*

Exercise 4.2.10 *Explain why the condition (4.25) is itself a rate-balance equation for the composite state* $((0,\infty),c)$.

Exercise 4.2.11 *Complete the mathematical steps leading to formula (4.26).*

Exercise 4.2.12 *In equation (4.26) if* $c = 1$ *then the stationary stationary pdf of workload is given by* $f(x) = \lambda P_0 e^{-(\mu-\lambda)x}, x > 0$ *(equation (2.14) for the M/M/1 queue in Section 2.1.10 in Chapter 2). Explain why the stationary* *pdf of workload in the M/M/c model* $f(x) = \lambda P_{c-1} e^{-(c\mu-\lambda)x}, x > 0$ *looks so similar, yet the physical models are quite different. In M/M/1 there is one waiting line and one server; customers begin and complete service in order of arrival. in M/M/c there is one waiting line and* $c \geq 2$ *independent servers; customers begin service in order of arrival but complete service in random order.*

Exercise 4.2.13 *Compare the mixed pdfs* $\{P_0, f(x)\}_{x>0}$ *of the virtual wait in (1) a stable single-server* $M_\lambda/M_\mu/1$ *queue and (2) a stable three-server* $M_\lambda/M_\mu/3$ *queue, with respect to:* P_0 *;* $f(x)$*;* E *[Busy cycle];* E *[Busy period];* E *[Idle period] (Hint: See Example 4.2.1 in this chapter.)*

4.2.7 Geometric Derivation of Stationary Pdf and Cdf of Virtual Wait in the Standard M/M/c Queue Using Only Geometric Properties of Sample Paths

This section derives the stationary pdf and cdf of the virtual wait in the standard $M_\lambda/M\mu/c$ queue ($c \geq 2$) using only geometric properties of the sample path and LC concepts. These derivations bypass deriving and solving a differential and the integral Equation (4.18) in Section 4.2.4. The strictly geometric approach utilizes the rate-balance principle in LC.

Consider level $x > 0$ on Page c (Figure 4.2). The rate-balance principle across level x is equivalent to:

$$\text{SP upcrossing rate of level } x$$

$$= \text{SP downcrossing rate of level } x = f(x),\ 0 < x < \infty.$$

(The second equality is due to the basic level crossing theorem in Section 1.4 in Chapter 1). Equivalently:

$$\lim_{t\to\infty} \frac{\mathcal{U}_t(x)}{t} = \lim_{t\to\infty} \frac{\mathcal{D}_t(x)}{t} \overset{a.s.}{=} f(x)\ ,\ 0 < x < \infty,$$

where $f(x)$ is the stationary pdf of the virtual wait on Page c. Thus, formula (4.25), follows.

4.2.8 Sojourns by the SP in the State Space in M/M/c Queue

Definition 4.3 *A sojourn by the SP in a state-space set is the duration of a single SP visit to that state-space set, i.e., the duration from an entrance into the state-space set until the first exit from that state-space set, thereafter.*

In the standard M/M/c queue, each sojourn "above level $x > 0$ on Page c" is initiated by a vertical upcrossing of level x. The size of the vertical excess above level x is $\stackrel{d}{=}$ the arriving customer's "inter-start-of-service departure time" (see Definition 4.1 near beginning of Section 4.2). The SP jumps which overshoot level $x > 0$, begin from two possible sources with regard to level x: (1) from a level in state-space interval $(0, x)$ on Page c, or (2) from the Line $(c - 1)$.

Let a_x denote a sojourn "above level x". Any "time-extension" of a_x will be due to new vertical upward jumps that begin while the SP is at levels above x, i.e., in state-space interval (x, ∞) on Page c. Each sojourn a_x on Page c ends at the instant when the SP next makes a continuous downcrossing of level x smoothly with slope -1.

Each vertical upward jump that begins on Page c, from (1) $(0, x)$ or from (2) LIne $c-1$, which overshoots level x is $\stackrel{d}{=} \text{Exp}_{C\mu}$ by the memoryless property of $\text{Exp}_{c\mu}$. Upward jumps originating from levels in (x, ∞)) also are $\stackrel{d}{=} \text{Exp}_{c\mu}$, because looking ahead from the arrival instant to the instant when the arriving customer begins service, the first departure by any customer from the system thereafter, occurs after a time $\stackrel{d}{=} \text{Exp}_{c\mu}$ (i.e., when the minimum of c iid remaining service times, ends (see Definition 4.1 near the beginning of Section 4.2). The memoryless property of $\text{Exp}_{c\mu}$ ensures that the excess over level x by vertical jumps beginning below level x and ending above level x are $\stackrel{d}{=}$ $\text{Exp}_{c\mu}$. (For the memoryless property of exponential variates see Section 1.5.1 in Chapter 1.)

4.2.9 More Explanation on SP Sojourn above Level x on Page c in Standard M/M/c Queue

Lemma 4.4 shows that a_x, for all $x \geq 0$, in the standard $\text{M}_\lambda/\text{M}_\mu/\text{c}$ queue, is distributed identically as the busy period in a stable standard $\text{M}_\chi/\text{M}_{c\mu}/1$ queue.

Lemma 4.4 *In the stable standard $M_\lambda/M_\mu/c$ $(c \geq 2)$ queue each sojourn a_x above level $x \geq 0$ is $\stackrel{d}{=}$ busy period in a stable, standard single-server $M_\lambda/M_{c\mu}/1$ queue with arrival rate λ and service rate $c\mu$.*

Proof *In the $M_\lambda/M_\mu/c$ $(c \geq 2)$ queue the sizes of "all SP vertical jumps ending on Page c" are $\stackrel{d}{=}$ $Exp_{c\mu}$ by the discussion above following Definition 4.3. The arrival rate of vertical jumps during the sojourn a_x is Poisson with rate λ, where $\lambda < c\mu$. The arrival rate and jump sizes are the same as in the busy period of a stable standard "$M_\lambda/M_{c\mu}/1$" queue with arrival rate λ and service $\stackrel{d}{=}$ $Exp_{c\mu}$.(See Sections 2.1.7.1 and 2.1.7.2 in Chapter 2.)*

Corollary 4.5 *The expected value of a sojourn above level $x \geq 0$ in $M_\lambda/_\mu/c$ $(c\geq 2)$ is equal to*

$$E\left[a_x\right] = \frac{1}{c\mu - \lambda}, x \geq 0. \tag{4.32}$$

Proof *In formula (4.32) the exponentially distributed "start-of-service departure rate "$c\mu$" on Page c in M/M/c, replaces the service rate "μ" in formula (2.36) for "E [busy period] $= \frac{1}{\mu-\lambda}$" in the standard $M_\lambda/M_\mu/1$ (see Section 2.1 in Chapter 2).*

Note that, by Lemma 4.4 and Corollary 4.5,

$$E\left[a_x\right] \text{ is independent of level } x \geq 0. \tag{4.33}$$

4.2.10 Busy Period and "Zero-wait Period" in Standard M/M/c Queue

4.2.10.1 Busy Period in Standard M/M/c Queue

Definition 4.6 *The "busy period" in the standard M/M/c queue is defined here to be the period when "all c servers are busy". It is the same as the sojourn above level* 0, *i.e., "a_0 ", and is denoted by "$\mathcal{B}_{c-1,c}$". Thus,*

$$\mathcal{B}_{c-1,c} \equiv a_0 \ .$$

Note that a_0 is equal to the duration from a customer's arrival instant when the system state is $(0, c-1)$, until the first instant thereafter when the SP dowhcrosses back into state $(0, c-1)$. (The arrival when the system is in state $(0, c-1)$ "increases" the number in the system up to c, so that all servers are busy. The first downcrossing thereafter back into state $(0, c-1)$ when the SP first hits level 0 from above on Page c, "decreases" the number in the system from c back down to $(c-1)$.

We now use the fact that all vertical upward jumps which end on Page c are $\stackrel{d}{\equiv} \text{Exp}_{c\mu}$. Those jumps, which upcross level x have an excess $\stackrel{d}{=} \text{Exp}_{c\mu}$ above level x by the "memoryless property". Thus each sojourn a_x abobe level x, has the property $a_x \stackrel{d}{\equiv} a_0 \equiv \mathcal{B}_{c-1,c}$, $x \geq 0$. Another useful observation is that "$\mathcal{B}_{c-1,c}$"= "$\mathcal{B}$" in a single-server "$M_\lambda/M_{c\mu}/1$" queue. Thus, the expected value of a_x for every $x \geq 0$ on Page c is:

$$E\left[a_x\right] \equiv E\left[\mathcal{B}_{c-1,c}\right] = \frac{1}{c\mu - \lambda}, 0 \leq x < \infty. \tag{4.34}$$

(See formula (2.37) for $E[\mathcal{B}]$ in the M/M/1 queue, in Section 2.1.17 in Chapter 2.)

4.2.10.2 Zero-wait Period in Standard M/M/c Queue

Just after each level 0 downcrossing on Page c, the SP motion is horizontal (left to right) on Line $c-1$, i.e., within state $(0, c-1)$. Then the SP may jump onto Page c with probability $\frac{\lambda}{\lambda+c\mu}$ (= probability of an arrival before a departure), or into state $(0, c-2)$ with probability $\frac{c\mu}{\lambda]+c\mu}$ (= probability of a departure before an arrival). In the latter case, the SP may "circulate randomly" among the zero-wait states $(0,c-2),...,(0,0)$;until returning (due to stability) into state $(0,c-1)$. While the SP is in state $(0,c-1)$, an arrival causes a vertical SP upward jump onto Page c. All arrivals to the system during this "zero-wait period" following the SP downcrossing of level 0, wait zero time and begin service immediately on arrival. So, we call this period the "zero-wait period" for "arrivals" (abbreviated "0-Wt").

The "zero-wait period" in the M/M/c queue generalizes the "idle period" in the single-server M/M/1 queue. (From an arrival's viewpoint, the "idle period" in M/M/1 queues may also be called the "zero-wait period".).

4.2.11 The Time Period between Successive Downcrossings of Level $x \geq 0$ in the Standard $M_\lambda/M_\mu/c$ Queue

Definition 4.7 *Let "d_x" denote the SP inter-downcrossing time of a fixed level $x \geq 0$. Then d_x is the time period between any two successive SP downcrossings of level x in the state space. (Let a downcrossing of level* 0 *be the same as an entrance into level* 0 *(i.e., into state* $(0,c-1)$ *in Line* $c-1$*, at the bottom of Page c). (For geometric appearance, see "Left limit, entrance from above", in Figure 1.2 in Section 1.4.5 in Chapter 7.)*

Thus, $d_x :=$ inter-downcrossing time of level x on Page c ($x \geq 0$) including downcrossings into Line $c-1$. By the memoryless property of inter-arrivals in a Poisson process, just after each downcrossing of level x ($\dot{x} \geq 0$) the SP continues its motion for a time $\overset{d}{=} \text{Exp}_\lambda$, until the next arrival to the system, due to the memoryless property of Exp_λ. Thus for each fixed level x, the SP returns from above back to level x in times which form a "level-x dependent renewal process", denoted by $\{d_{x,n}\}_{n=1,....}$ where $d_{x,n} \overset{d}{=} d_x$, $n = 1, 2, ...$ In general, if $x \neq y$ then $d_x \neq d_y$, where x and y are horizontal levels in Page c.

By the elementary renewal theorem in Section 1.8.3 in Chapter 1,

$$E[d_x] = \frac{1}{f(x)}, x \geq 0. \tag{4.35}$$

4.2.11.1 Expected Value of Zero-Wait Period in Standard M/M/c Queue

We now use $E[d_0]$ to compute $E[0\text{-}Wt]$. The direct computation of $E[0\text{-}Wt]$ in M/M/c queue seems to be complex. But its computation is greatly simplified by using the expression for the inter-downcrossing time d_0 of level 0 at the bottom of Page c. That is, the time between successive "level-0 downcrossings on Page c" form a renewal process $\{d_{0,n}\}_{n=1,2,...}$, where the expected inte-renewals are $\overset{d}{=} d_0$. Thus, by formula (4.26) in 4.2.6,

$$E[d_0] = \frac{1}{f(0)} = \frac{1}{\lambda P_{c-1}} = \frac{1}{\lambda \frac{\left(\frac{\lambda}{\mu}\right)^{c-1}}{(c-1)!} P_0}. \tag{4.36}$$

We now use formula (4.36) with the value of P_0 given in formula (4.22) in Section 4.2.3 to derive $E[0Wt]$. Note that

$$d_0 = 0Wt + a_0,$$

$$0Wt = d_0 - a_0,$$

as shown on examination of Figure 4.2 in Section 4.2.

Thus, the expected value of the "zero-wait time period $0Wt$", is given byλ

$$E[0Wt] = E[d_0] - E[a_0]$$

$$E[0Wt] = \frac{1}{\lambda\left(\frac{\lambda}{\mu}\right)^{c-1}\frac{1}{(c-1)!}P_0} - \frac{1}{c\mu - \lambda} \tag{4.37}$$

Lemma 4.8

$$E[d_x] > E[a_x], \ rx \geq 0. \tag{4.38}$$

Proof *(1) We give a geometrical proof. Examination of Figure 4.2 in Section 4.2, shows that each sojourn a_x above level x, can start only at an instant strictly between two successive SP downcrosings of level x, i.e., at an interior point of a d_x. Also, a_x ends at the next SP downcrossing of level x, which ends both d_x and the contained a_x. Therefore, $d_x > a_x$ for every fixed level $x \geq 0$. Formula (4.38) follows.*

(2) An alternative way to prove formula (4.38) is

$$f(x) = \lambda P_{c-1} e^{-(c\mu-\lambda)x}, x > 0,$$

$$E[d_x] = \tfrac{1}{f(x)} = \tfrac{e^{(c\mu-\lambda)x}}{\lambda P_{c-1}} > \tfrac{1}{\lambda P_{c-1}} = \tfrac{1}{f(0)} \tag{4.39}$$

$$E[d_x] > E\left[busy\ period\ in\ M_\lambda/M_{c\mu-\lambda}/1\right] = \tfrac{1}{c\mu-\lambda} = E[a_x].$$

Example 4.2.2 *Consider the standard M/M/2 queue in Example 4.2.1 in Section 4.2.6 above. Assume the arrival rate $\lambda = 1$ and the service rate $= 2$ for each server. Then*

$$E[d_0] = \frac{1}{1\left(\frac{1}{2}\right)^1\frac{1}{(1)!}P_0} = \frac{1}{\left(\frac{1}{2}\right)^1\frac{6}{10}} = \frac{10}{3},$$

and

$$E[a_0] = \frac{1}{3},$$

resulting in

$$E[0Wt] = \frac{10}{3} - \frac{1}{3} = 3.$$

In this example, the zero-wait period is nine times longer than the busy period (when all servers are occupied).

4.2.11.2 Further Consideration of the Inter-downcrossing Time d_x, $x \geq 0$

Note that the renewal process $\{d_{x,n}\}_{n=1,2,...}$ in Section 4.2.11, is also a "regenerative" process where the regenerative cycles are $d_{x,n}, n-1,2,....$ By the "renewal reward theorem" (Section 1.8.4 in Chapter 1) and formula (4.33) for the sojourn a_x, which occurs as the second part of every d_x:

$$\frac{E\left[a_x\right]}{E\left[d_x\right]}=\lim_{t\to\infty}\frac{\text{time that } W(\cdot)\in(x,\infty) \text{ during } (0,t)}{t}=1-F(x),$$

$$\frac{\frac{1}{\frac{1}{c\mu-\lambda}}}{\frac{1}{f(x)}}=1-F(x);$$

equivalently

$$\frac{f(x)}{1-F(x)}=c\mu-\lambda, x>0. \tag{4.40}$$

Formula (4.40) is the "hazard rate". It is equivalent to the differential equation

$$\frac{\frac{d}{dx}(1-F(x))}{1-F(x)}=-\left(c\mu-\lambda\right),$$

$$\frac{d}{dx}\ln(1-F(x))=-\left(c\mu-\lambda\right),$$

with solution

$$\ln(1-F(x))=-\left(c\mu-\lambda\right)x+\text{constant},$$

$$1-F(x)=e^{-(c\mu-\lambda)x+\text{constant}},$$

$$1-F(x)=Ae^{-(c\mu-\lambda)x},$$

where A is a constant. Letting $x\downarrow 0$, yields $A=1-F(0)$, and the cdf of wait is

$$F(x)=1-(1-F(0))e^{-(c\mu-\lambda)x}, x\geq 0, \tag{4.41}$$

where $F(0)=P(\text{zero wait}):=P_{0\text{-}W_t}$. Taking $dF(x)/dx$, $x>0$, in (4.41) gives another form of the pdf of wait,

$$f(x)=(1-P_{0\text{-}W_t})(c\mu-\lambda)e^{-(c\mu-\lambda)x}, x>0. \tag{4.42}$$

Next, the equation (4.27) in Section 4.2.6 leads to:

$$F(0)=\sum_{n=0}^{c-1}P_n=P_0\sum_{n=0}^{c-1}\left(\frac{\lambda}{\mu}\right)^n\frac{1}{n!}. \tag{4.43}$$

Note that

$$f(0) = \lambda P_{c-1}, \tag{4.44}$$

i.e., the SP entrance rate into state $(0, c-1)$ from above is $f(0)$ (SP down-crossing rate into level 0), which is equal to the SP exit rate λP_{c-1} from state $(0, c-1)$ at arrival instants. Letting $x \downarrow 0$ In (4.42) and using (4.44) yields

$$f(0) = (1 - P_{0\text{-}Wt})(c\mu - \lambda)$$

$$= \int_{x=0}^{\infty} f(x)dx \times (c\mu - \lambda)$$

$$= \int_{x=0}^{\infty} \lambda P_{c-1} e^{c\mu-\lambda} dx \times (c\mu - \lambda)$$

$$= \lambda P_{c-1} \times \frac{1}{c\mu - \lambda} \times (c\mu - \lambda) = \lambda P_{c-1}. \tag{4.45}$$

Formulas (4.45) and (4.27) in Section 4.2.6 yield:

$$F(0) = 1 - \frac{\lambda}{c\mu - \lambda} P_{c-1} = 1 - \frac{\lambda}{c\mu - \lambda} \left(\frac{\lambda}{\mu}\right)^{c-1} \frac{1}{(c-1)!} P_0. \tag{4.46}$$

Substituting the value of $F(0)$ from (4.43), into (4.46) and solving for P_0 gives (4.28) in Section 4.2.6..

The upshot of this explanatory exercise is that there are two different ways to determine P_0; and two different, equivalent formulas for $f(x)$, $x > 0$: (4.26) and (4.42).

Remark 4.9 *Another way to obtain the second equality in (4.45) is to note that the SP expected sojourn time above* 0 *is*

$$E[a_0] = E[\textit{busy period of } M_\lambda/M_{c\mu}/1]$$

$$E[\mathcal{B}_{c-1,c}] = \frac{1}{c\mu - \lambda}.$$

Therefore the long-run proportion of time that the SP spends (strictly) above level 0 *is:*

$$\lim_{t\to\infty} \frac{E[\mathcal{U}_t(0)]}{t} \times \frac{1}{c\mu - \lambda} = \lim_{t\to\infty} \frac{E[\mathcal{U}_t(0)]}{t} \times \frac{1}{c\mu - \lambda}$$

$$= \lambda P_{c-1} \times \frac{1}{c\mu - \lambda} = 1 - F(0).$$

4.2.12 The Pmf of Number of Customers in Standard M/M/c Queue

The stationary "probability mass function, pmf" differs from the "pdf" (see Definition 3.1 in Section 3.4). The pdf is the "probability density function" on zero-wait states and on positive-wait state-space intervals. The pmf ("probability mass function") gives the stationary probabilities of the number of units (customers) in the system (discrete points of the state space).

To derive the pmf for the number of units in the M/M/c system for numbers greater than $c-1$, i.e., $P_n, n = c, c+1, ...,$, we utilize the pdf of wait

$$f(x) = \lambda P_{c-1} e^{-(c\mu-\lambda)x}, x > 0,$$

given in (4.26) of Section 4.2.6.

This approach is in the reverse order of the "classical method" for deriving the pmf. (Note: the order of operations is important in Operations Research and in Applied probability, as well as in other professions.)

The classical technique first derives the pmf of the number-in-system using a birth-death analysis for $P_n, n = 0, ..., \infty$. It then obtains the pdf of wait by conditioning on the number in the system, just before an arrival. The classical method may not work when there are within-model dependences present, e.g., when service time depends on waiting time, or service time depends on the number of customers left behind in the system after the service.

4.2.12.1 An Aside on the LC Method

Before continuing to derive P_n, $n \geq c$, we give a brief aside on LC. The LC method applied here is based mainly on deriving the stationary pdf of waiting time (which simplifies using "Lindley recursions" for deriving the integral equations for the waiting times). LC is a "stand-alone method" developed by the author, for bypassing the often tedious algebraic steps when using Lindley recursions (see Chapter 1 in [22] (2017); [10] (1975); [11] (1975)). Also, LC can be part of a hybrid-type derivation to derive the stationary pdf of waiting time. LC is flexible because it can work easily in conjunction with many other types of analyses of stochastic models. For example, when computing the pmf of the number-in-system in the standard M/M/c queue, LC utilizes a birth and death analysis for the discrete states $(0,0)$,...,$(0, c-1)$, i.e., for $P_0, ..., P_{c-1}$, and a stand-alone LC analysis using the pdf $f(x), x > 0$, of wait on Page c, for the number of customers in the system: $c, c+1, ...$ LC provides a broad perspective of the M/M/c model, and emphasizes the waiting time r.v. as basic in the queueing model.

4.2.12.2 Three Important Limiting Probability Distributions in a Stable Queue with Poisson Arrivals

There are three limiting (stationary) probabilities which help to analyze stable queues (see a detailed discussion on pp. 500–502 in Sections 8.2.2–8.2.3 in [52] (2010)).

"P_n", the stationary probability that there are n customers in the system "at an arbitrary time point". That is, P_n is the long-run "proportion of time" when there are n units in the system;

"A_n", the "proportion of customer arrivals", which sees just ahead, n units in the system;

"D_n",the "proportion of customer departures", which leaves just behind, n units in the system.

It can be shown that always $A_n \equiv D_n$. In addition, if arrivals occur in a Poisson process, then "$P_n = A_n = D_n$ ".

Reasoning as for the M/M/1 queue in Section 2.1.15 in Chapter 2 gives

$$P_n = D_n = P(n - c \text{ arrivals during a waiting time}), n = c, c+1, ...$$

Substituting from (4.26) in Section 4.2.6, and (4.27) in Section 4.2.6 gives

$$P_n = \int_{x=0}^{\infty} \frac{e^{-\lambda x}(\lambda x)^{n-c}}{(n-c)!} f(x)dx$$

$$= \left(\frac{\lambda}{\mu}\right)^{n-c+1} \frac{1}{c^{n-c+1}} P_{c-1} \int_{x=0}^{\infty} c\mu e^{-c\mu x} \frac{(c\mu x)^{n-c}}{(n-c)!} dx$$

$$P_n = \left(\frac{\lambda}{\mu}\right)^n \frac{1}{c^{n-c}c!} P_0, n = c, c+1, ... \tag{4.47}$$

In summary, we obtain the well-known formulas (e.g., p. 67 in [36] (2008)),

$$P_0 = \frac{1}{\sum_{n=0}^{c-1} \left(\frac{\lambda}{\mu}\right)^n \frac{1}{n!} + \left(\frac{\lambda}{\mu}\right)^c \frac{c\mu}{c!(c\mu - \lambda)}}, \tag{4.48}$$

$$P_n = \frac{\frac{\lambda}{\mu}^n}{n!} P_0, n = 0, ..., c-1;$$

$$P_n = \left(\frac{\lambda}{\mu}\right)^n \frac{1}{c^{n-c}} \frac{1}{c!} P_0, n = c, c+1, ...$$

The probability that all servers are occupied is

$$\sum_{n=c}^{\infty} P_n = P(\text{wait is} > 0) = \int_{x=0}^{\infty} f(x)dx$$

$$= \lambda P_{c-1} \int_{x=0}^{\infty} e^{-(c\mu-\lambda)x} dx = \frac{\lambda}{c\mu - \lambda} P_{c-1}$$

$$\sum_{n=c}^{\infty} P_n = \frac{\lambda \left(\frac{\lambda}{\mu}\right)^{c-1} \frac{1}{c!}}{c\mu - \lambda} P_0. \tag{4.49}$$

The probability that there is at least one idle server is

$$\sum_{n=0}^{c-1} P_n = P(\text{wait is} = 0) = 1 - \sum_{n=c}^{\infty} P_n = 1 - \frac{\lambda \left(\frac{\lambda}{\mu}\right)^{c-1} \frac{1}{c!}}{c\mu - \lambda} P_0 M, \tag{4.50}$$

where P_0 is given in formula (4.48).

4.2.13 More Details on the Inter-downcrossing Time d_x and the Sojourns a_x and b_x in the Standard M/M/c Queue

Consider d_x, a_x, b_x $(x \geq 0)$, respectively: $d_x$is the time between successive SP downcrossings of level $x \geq 0$; a_x is the sojourn above level $x \geq 0$, initiated by an SP upcrossing of level x; b_x is the sojourn at or below $x \geq 0$, initiated by a downcrossing of level x. Formula (4.35) in Section 4.2.11 shows that

$$E\left[d_x\right] = \frac{1}{f(x)} = \frac{1}{\lambda P_{c-1} e^{-(c\mu-\lambda)x}} = \frac{e^{(c\mu-\lambda)x}}{\lambda P_{c-1}}, x \geq 0. \tag{4.51}$$

Formula (4.33) in Section 4.2.9 shows that:

$$E\left[a_x\right] = \frac{1}{c\mu - \lambda}, x \geq 0, \tag{4.52}$$

independent of x,

The long-run proportion of time that the SP is above level x is:

$$\lim_{t\to\infty} \frac{\text{time that the SP is above level } x \text{ during } (0,t)}{t} = 1 - F(x), , x \geq 0.$$

By the "renewal reward theorem" (see Section 1.8.4 in Chapter 1), and formulas (4.51) and (4.52)),

$$\frac{E\left[a_x\right]}{E\left[d_x\right]} = \frac{E\left[a_x\right]}{1/f(x)} = 1 - F(x),$$

implying that

$$\frac{1-F(x)}{f(x)} = \frac{1}{c\mu - \lambda}, x > 0.$$

The last equality above confirms that formula (4.40) holds, when solving for $f(x)$ geometrically in Section 4.2.7. Also, we validate (4.40) in Section 4.2.7 as follows:

$$\frac{1-F(x)}{f(x)} = \frac{\int_{y=x}^{\infty} f(y)dy}{\lambda P_{c-1} e^{-(c\mu-\lambda)x}}$$

$$= \frac{\int_{y=x}^{\infty} \lambda P_{c-1} e^{-(c\mu-\lambda)y} dy}{\lambda P_{c-1} e^{-(c\mu-\lambda)x}} = \int_{y=x}^{\infty} e^{-(c\mu-\lambda)(y-x)} dy,$$

$$\int_{z=0}^{\infty} e^{-(c\mu-\lambda)z} dz = \frac{1}{c\mu - \lambda}, \tag{4.53}$$

where we used the transformation: "$z = y - x$" in equation in (4.53). Also

$$F(x) = \lim_{t\to\infty} \frac{\text{time that the SP spends at or below } x \text{ during } (0,t)}{t}, x \geq 0.$$

Each instant that the SP downcrosses $x \geq 0$ is a regenerative point, due to the memoryless property of the customers' inter-arrival times. From the renewal reward theorem (see Section 1.8.4 in Chapter 1) the expected value $E[b_x]$ of a sojourn below level x is given by

$$\frac{E[b_x]}{E[d_x]} = \frac{E[b_x]}{\frac{1}{f(x)}} = F(x),$$

implying:

$$E[b_x] = \frac{F(x)}{f(x)} = \frac{1-(1-F(0))e^{-(c\mu-\lambda)x}}{\lambda P_{c-1} e^{-(c\mu-\lambda)x}}$$

$$= \frac{e^{(c\mu-\lambda)x}}{\lambda P_{c-1}} - \frac{1-F(0)}{\lambda P_{c-1}}$$

$$= \frac{1}{\lambda \left(\frac{\lambda}{\mu}\right)^{c-1} \frac{1}{(c-1)!} P_0} \left(e^{(c\mu-\lambda)x} - (1 - P_0 \sum_{n=0}^{c-1} \left(\frac{\lambda}{\mu}\right)^n \frac{1}{n!}) \right)$$

$$E[b_x] = \frac{e^{(c\mu-\lambda)x} - 1}{\lambda \left(\frac{\lambda}{\mu}\right)^{c-1} \frac{1}{(c-1)!} P_0} + \frac{\sum_{n=0}^{c-1} \left(\frac{\lambda}{\mu}\right)^n \frac{1}{n!}}{\lambda \left(\frac{\lambda}{\mu}\right)^{c-1} \frac{1}{(c-1)!}}, \tag{4.54}$$

where $P_0 = \dfrac{1}{\sum_{n=0}^{c-1} \left(\frac{\lambda}{\mu}\right)^n \frac{1}{n!} + \left(\frac{\lambda}{\mu}\right)^c \frac{c\mu}{c!(c\mu-\lambda)}}$.

Remark 4.10 *From (4.33), when the SP upcrosses x, it next downcrosses x after a time a_x where "$E[a_x]$" is constant independent of x. In contrast to this constant expected sojourn above x, formula (4.54) implies that $E[b_x]$ grows exponentially with increasing values of x, due to the term $e^{(c\mu-\lambda)x}-1$.*

4.3 The $M_\lambda/M_{\mu-\frac{\lambda}{c}}/c/c$ Queue and Its Relation to the $M_\lambda/M_\mu/c$ Queue

4.3.1 Description of the $M_\lambda/M_{\mu-\frac{\lambda}{c}}/c/c$ Queue

The M/M/c/c queue is a special case of the M/M/c/k queue, where k is the "upper limit of the number of customers allowed in the system at any time" (see, e.g., Section 2.5, p. 76ff in [36] (2008)). In the M/M/c/c queue, if there is at least one idle server then an arriving customer enters the system and starts service immediately using one of the idle servers. If all c servers are busy, arriving customers balk (are "blocked and cleared immediately") without waiting in line, and thus not receiving any service. In the M/M/c/c queue there is no waiting line before start of service. See Figure 4.4.

FIGURE 4.4
Sketch of layout of M/M/c/c queue ($c = 3$). Arriving customers start service immediately on arrival if there is at least one idle server. Arriving customers balk and are cleared immediately with no waiting if they find all c servers busy

4.3.2 Relation of the $M_\lambda/M_{\mu-\frac{\lambda}{c}}$/c/c Queue to the M_λ/M_μ/c Queue

In order to relate the M/M/c/c queue to the M/M/c queue, consider the case where arrivals are Poisson rate λ, and all service times are $\stackrel{d}{\equiv}$ "$\text{Exp}_{\mu-\frac{\lambda}{c}}$", where

$$\mu - \tfrac{\lambda}{c} > 0, \qquad \text{or } c\mu - \lambda > 0, \tag{4.55}$$

which is necessary for existence of the stationary pdf, as in the M_λ/M_μ/c queue analyzed in Section 4.2.

A sample path of the virtual wait in the $M_\lambda/M_{\mu-\frac{\lambda}{c}}$/c/c queue is depicted in Figure 4.5, in Section 4.3.3. The upward-jump structure of the sample path for positive values in the M/M/c/c queue is much simpler than the corresponding upward-jump structure of the standard M/M/c queue (see Figure 4.2). The simpler M/M/c/c jump structure makes it easier to derive the stationary pdf of the virtual wait in M/M/c/c. Notably, the stationary pdf of wait in a "parameter-modified" M/M/c/c queue (i.e., the service times are $\stackrel{d}{\equiv} Exp_{\mu-\frac{\lambda}{c}}$), can be derived efficiently without deriving and solving an integral equation (which is necessary in the standard M/M/c queue). Yet, the derived pdf of within $M_\lambda/M_{\mu-\frac{\lambda}{c}}$/c/c has the same functional formats as in M_λ/M_μ/c. (Note: The M/M/c/c is stable for "all positive values of $\lambda > 0$" because there is no waiting line and the number in the system never exceeds c.)

The M/M/c/c queue can be analyzed using a birth-death analysis (see Section 1.6 in Chapter 1). For consistency, we employ here, an LC approach. Consider an M/M/c/c queue where the service time for each customer that enters the system has exponential rate $\mu - \frac{\lambda}{c} > 0$. (We choose $\lambda < c\mu$ because the related M_λ/M_μ/c model is a stable standard M_λ/M_μ/c queue.)

In M/M/c/c all "actual waits are 0"; there is no waiting line before service. In a queue where blocking arrivals from service is possible, we shall define the virtual wait as the time that a potential arrival would wait to begin service, provided it was not blocked and cleared. The virtual wait is not 0 for every arrival. In M/M/c/c, customers that arrive when all servers are occupied (the virtual wait is positive), are blocked and cleared from the system. In both M/M/c/c and M/M/c queues, the virtual wait is positive if and only if all c servers are occupied. In M/M/c all arrivals eventually get service. If the SP is on Page c in the M/M/c queue, then the number of customers in the system is $\geq c$. If the SP is on Page c in the M/M/c/c queue, then exactly c customers are in the system.

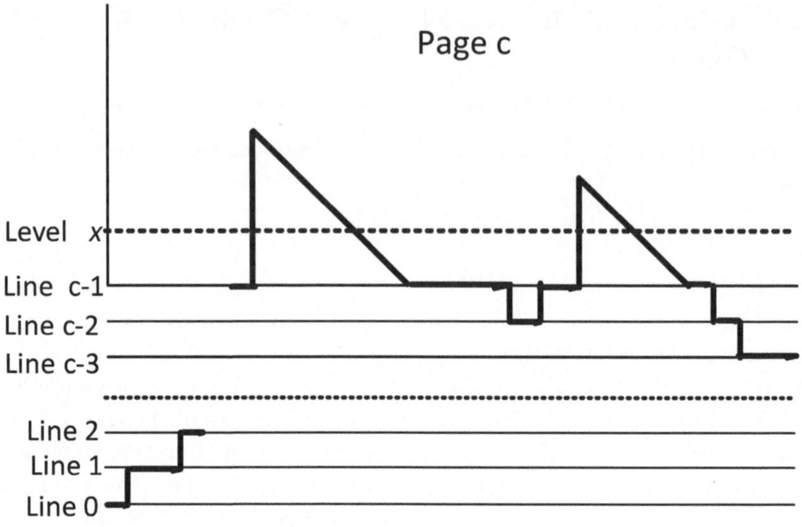

FIGURE 4.5
Sample path of the virtual wait $\{W(t)\}_{t\geq 0}$ in the M/M/c/c queue. All vertical jumps ending in $(0,\infty)$ begin at level 0 on Page c (from state $(0,c-1)$) with size $\stackrel{d}{=} \text{Exp}_{c(\mu-\frac{\lambda}{c}))} \equiv \text{Exp}_{c\mu-\lambda}$, which is the minimum of c iid exponential r.v.s each $\stackrel{d}{=} \text{Exp}_{\mu-\frac{\lambda}{c}}$. The SP then moves downward with slope -1 until it hits level 0–stae $(0,c-1)$. The SP descent on Page c is not interrupted because any new arrivals will balk (get blocked cleared immediately from the system) until the SP moves back into state $(0,c-1)$.

4.3.3 Sample Path of Virtual Wait $\{W(t)\}_{t\geq 0}$ in M/M/c/c Queue

Consider a sample path of $\{W(t)\}_{t\geq 0}$ in the standard $\text{M}_\lambda/\text{M}_{\mu-\frac{\lambda}{c}}/\text{c}/\text{c}$ queue (Figure 4.5). Without loss of generality, assume the system starts empty. The SP is on Line 0 at time $t=0$ (all servers idle). As the system evolves over time, the SP moves randomly among the c Lines until the SP enters Line $c-1$, where $c-1$ servers are occupied, similarly as in a standard M/M/c queue In Figure 4.2 (where Line $c-1$ is also situated at level 0 of Page c). This arrangement makes it easy to depict SP exchanges between Line $c-1$ and the virtual-wait positive states on Page c.

Let us further emphasize the difference between the two figures. In $\text{M}_\lambda/\text{M}_{\mu-\frac{\lambda}{c}}/\text{c}/\text{c}$, suppose a customer arrives when $c-1$ servers are busy. The arrival joins the system and begins service by the single idle server. All c servers are busy just after that arrival begins service. All c servers have iid service times $\stackrel{d}{=} \text{Exp}_{\mu-\frac{\lambda}{c}}$ once the new arrival begins service, due to the memoryless property of exponential service times (see Section 1.5.3 in Chapter 1). The SP jumps to a level $y \in (0, \infty)$ on the Page c, where $y \stackrel{d}{=} \mathrm{Exp}_{c(\mu-\frac{\lambda}{c})} = \mathrm{Exp}_{c\mu-\mu}$, which is distributed as the minimum of c iid exponential r.v.s each distributed with rate $\mu - \frac{\lambda}{c}$. When all c servers are busy, the remaining time until the next exit from the system is also equal to y because the slope of the sample path is -1, and any new arrivals during that period on Page c are blocked and cleared from the system. After a duration y the SP hits level 0 on Page c from above and enters horizontal Line $c-1$, which means $c-1$ servers are busy and a single server is idle.

Once the SP hits level 0, it may start moving randomly among Lines $0, ..., c-2$ with probability $\frac{\mu-\frac{\lambda}{c}}{\mu-\frac{\lambda}{c}+\lambda}$, or return to Page c via a vertical upward jump with probability $\frac{\lambda}{\mu-\frac{\lambda}{c}+\lambda}$. (See $P(\min(\mathrm{Exp}_{\mu_i}\ i = 1, ..., N)$ in Section 1.5.3 in Chapter 1.)

In M/M/c/c all SP vertical jumps begin from level 0. By contrast, in the standard M/M/c queue SP vertical upward jumps can begin from any positive level on Page c, or from level 0 (state $(0, c-1)$. Hence the jump structure in M/M/c/c is much simpler than that of the standard M/M/c queue, where SP vertical jumps can begin from any level in $[0, \infty)$.

4.3.4 Stationary Pdf of Virtual Wait in the M/M/c/c Queue

Denote the stationary pdf of the virtual wait by in the M/M/c/c queue by $f(x)$, $x > 0$. To derive the pdf of $f(x)$, fix level $x > 0$. The SP downcrossing rate of x is equal to $f(x)$ by the basic LC theorem (Section 1.4 in Chapter 1). Since all SP jumps ending on Page c begin from state $(0, c-1)$ at customer arrival instants, and all vertical jump sizes are $\stackrel{d}{=} \mathrm{Exp}_{c\mu-\lambda} \stackrel{d}{=} \min(c$ iid $\mathrm{Exp}_{\mu-\frac{\lambda}{\lambda}}$s), the upcrossing rate of x is $\lambda P_{c-1} e^{-(c\mu-\lambda)x}$. Applying rate balance across level x, i.e., balancing SP rates out of and into set $((x, \infty), c)$, yields

$$f(x) = \lambda P_{c-1} e^{-(c\mu-\lambda)x}, x > 0. \tag{4.56}$$

Remark 4.11 *Formula (4.56) has precisely the same functional form as the stationary pdf of wait in the standard $M_\lambda/M_\mu/c$ queue given by formula (4.26) in Section 4.2.6, except that the term P_{c-1} has different values in the standard M/M/c queue and in the M/M/c/c queue. For the $M_\lambda/M_{\mu-\frac{\lambda}{c}}/c/c$ queue, formula (4.56) is derived directly from observing a typical sample path of the virtual wait. There is no need to derive and solve an integral equation to get the formula for the pdf of wait, as is necessary in the $M_\lambda/M_\mu/c$ queue. In the $M_\lambda/M_{\mu-\frac{\lambda}{c}}/c/c$ queue, the formula for pdf $f(x)$ (4.56) is inherently in itself the "model equation" This is the main relationship between the $M_\lambda/M_\mu/c$ and the $M_\lambda/M_{\mu-\frac{\lambda}{c}}/c/c$ queues that is discussed here. The result for $M_\lambda/M_{\mu-\frac{\lambda}{c}}/c/c$ gives an analyst the ability to write the "functional form" of the pdf of wait in the $M_\lambda/M_\mu/c$ queue immediately (functional form—because the factor "P_{c-1}" differs in value in the two models).*

4.3.5 Blocking Time and the Erlang-B Loss Formula in the M/M/c/c Queue

Definition 4.12 *The blocking time T_{block} is the duration between the instant when all c servers become busy (when system begins blocking and clearing new arrivals) and the first instant thereafter when the SP enters state $(0, c-1)$ (when one server becomes idle allowing a new arrival to enter the system).*

The pdf of the virtual wait in the $M_\lambda/M_{\mu-\frac{\lambda}{c}}/c/c$ queue is the same as the pdf of S (inter start-of-service departure time) when an arrival "sees" state $(0, c-1)$. Also, from Definition 4.12, $S \overset{d}{=} T_{block}$.

Then

$$E\left[T_{block}\right] = E\left[\mathrm{Exp}_{c\mu-\lambda}\right] = \frac{1}{c\mu-\lambda}.$$

Let P_{prop} denote the "proportion of time" the system is blocked. Then

$$P_{prop} = \int_{x=0}^{\infty} f(x)dx = \lambda P_{c-1}\int_{x=0}^{\infty} e^{-(c\mu-\lambda)x}dx$$

$$= \lambda\left(\frac{\lambda}{\mu-\frac{\lambda}{c}}\right)^{c-1}\frac{1}{(c-1)!}P_0\int_{x=0}^{\infty} e^{-(c\mu-\lambda)x}dx$$

$$= \left(\frac{\lambda}{\mu-\frac{\lambda}{c}}\right)^{c}\frac{1}{c!}P_0$$

$$P_{prop} = \frac{\left(\frac{\lambda}{\mu-\frac{\lambda}{c}}\right)^{c}\frac{1}{c!}}{\sum_{n=0}^{c}\left(\frac{\lambda}{\mu-\frac{\lambda}{c}}\right)^{n}\frac{1}{n!}} \tag{4.57}$$

(Note: P_{prop} in formula (4.57) is the probability that a right-truncated Poisson r.v. (truncated at c), has value c. It is the classical "Erlang-B loss formula" for the $M_\lambda/M_{\mu-\frac{\lambda}{\mu}}/c/c$ queue (see, e.g., p. 82, Section 2.6 in [36] (2008)).)

4.3.6 Non-blocking States in the $M_\lambda/M_{\mu-\frac{\lambda}{c}}/c/c$ Queue

The rate-balance equations for the non-blocking states, i.e., states $(0,0), ..., (0, c-1)$ are the same as in formula (4.14) in Section 4.2.3 for $M_\lambda/M_\mu/c$, with "$\mu - \frac{\lambda}{c}$" substituted for "μ". In the resulting $M_\lambda/M_{\mu-\frac{\lambda}{c}}/c/c$ queue:

$$P_n = \left(\frac{\lambda}{\mu-\frac{\lambda}{c}}\right)^{n}\frac{1}{n!}P_0, n = 0, ..., c-1,$$

so that

$$P_{c-1}=\left(\frac{\lambda}{\mu-\frac{\lambda}{c}}\right)^{c-1}\frac{1}{(c-1)!}P_0.$$

The normalizing condition is

$$\left(\sum_{n=0}^{c-1}\left(\frac{\lambda}{\mu-\frac{\lambda}{c}}\right)^n\frac{1}{n!}\right)P_0+\int_{x=0}^{\infty}f(x)dx=1.$$

Applying the normalizing condition (4.56) gives

$$\left(\sum_{n=0}^{c-1}\left(\frac{\lambda}{\mu-\frac{\lambda}{c}}\right)^n\frac{1}{n!}\right)P_0$$

$$+\lambda\left(\frac{\lambda}{\mu-\frac{\lambda}{c}}\right)^{c-1}\frac{1}{(c-1)!}P_0\int_{x=0}^{\infty}e^{-(c\mu-\lambda)x}dx=1,$$

and gives

$$P_0=\frac{1}{\sum_{n=0}^{c-1}\left(\frac{\lambda}{\mu-\frac{\lambda}{c}}\right)^n\frac{1}{n!}+\lambda\left(\frac{\lambda}{\mu-\frac{\lambda}{c}}\right)^{c-1}\frac{1}{(c-1)!}\frac{1}{c\mu-\lambda}}$$

$$P_0=\frac{1}{\sum_{n=0}^{c}\left(\frac{\lambda}{\mu-\frac{\lambda}{c}}\right)^n\frac{1}{n!}}.$$

The blocking time is a "$[c-1,c]$ busy period" denoted by $\mathcal{B}_{c-1,c}$, implying that $\mathrm{T}_{block}\overset{d}{=}\mathcal{B}_{c-1,c}$. From Definition 4.6 in Section 4.2.10,

$$\mathrm{T}_{block}=E\left(\mathcal{B}_{c-1,c}\right)=\frac{1}{c\mu-\lambda}. \tag{4.58}$$

4.3.7 Further Discussion of Cdf of Wait in M/M/c/c Queue

We can derive formula (4.56) for the pdf $f(x)$ in the M/M/c/c queue in Section 4.3.4, i.e., "$f(x)=\lambda P_{c-1}e^{-(c\mu-\lambda)x}, x>0$", geometrically, as in Section 4.2.7.

Let $F(x), x\geq 0$, denote the "cdf of the sationary virtual wait. In Section 4.2.10, it is shown using RRT (renewal reward theorem) that

$$\frac{E\left[a_x\right]}{E\left[d_x\right]}=\frac{E\left[a_x\right]}{\frac{1}{f(x)}}=1-F(x),$$

$$\frac{1-F(x)}{f(x)}=E\left[a_x\right]=E\left[a_0\right]=E\left[\mathcal{B}_{c-1,c}\right]=\frac{1}{c\mu-\lambda}.$$

Using the calculus formula "$\frac{d}{dx}\ln(u(x)) = \frac{u'(x)}{u(x)}$", and the observation,

$$F(0) + \int_{x=0}^{\infty} f(x)dx = 1,$$

$$F(0) + \int_{x=0}^{\infty} \lambda P_{c-1} e^{-(c\mu-\lambda)x} dx = 1,$$

$$1 - F(0) = \frac{\lambda P_{c-1}}{c\mu - \lambda},$$

gives

$$\frac{d}{dx}\ln(1 - F(x)) = \frac{-f(x)}{1 - F(x)} = \frac{-1}{E\left[\mathcal{B}_{c-1,c}\right]} = -(c\mu - \lambda),$$

$$\frac{-\lambda P_{c-1} e^{-(c\mu-\lambda)x}}{1 - F(x)} = -(c\mu - \lambda) \tag{4.59}$$

$$(4.60)$$

$$F(x) = 1 - (1 - F(0))e^{-(c\mu-\lambda)x}, x \geq 0,$$

$$f(x) = (c\mu - \lambda)(1 - F(0))e^{-(c\mu-\lambda)x}. \tag{4.61}$$

Comparing formula (4.56) and formula (4.61) shows that

$$\lambda P_{c-1} = (c\mu - \lambda)(1 - F(0)) = (c\mu - \lambda)P_{prop}, \tag{4.62}$$

where P_{c-1} is the probability of $c - 1$ units in the system, and P_{prop} is the proportion of time when arrivals are blocked (see the immediately next paragraph).

In M/M/c/c an arrival enters the system only if the virtual wait is 0. Thus $F(0) = P$(an arrival enters the system). Hence $1 - F(0) = P$(an arrival is blocked and cleared) = "P_{prop}". Equation (4.62) is precisely the balance equation that would appear in a birth-death analysis of the number of units in the system.

4.4 M/M/2 Queue with Zero-Wait Customers Receiving Special Service

In this section we formulate and solve an M/M/2 queue with zero-wait customers receiving exceptional (special) service. The solution illustrates relevant LC ideas about the method of Lines and Pages, and related insights. A more general solution procedure of a two-server M/M/2 queue where service time depends on waiting time in a general manner is detailed in Chapter 4 of [10] (1975). We denote the queue by M/M/2/(μ_0, μ_1), $(0, (0, \infty))$

4.4.1 Dynamics of the Queue $M/M/2/(\mu_0, \mu_1), (0, (0, \infty))$, M/M/2 Queue with Zero-wait Customers Receiving Special Service

This section gives the system dynamics of M/M/c ($c \geq 2$) queues where zero-wait customers get special service. We derive the stationaary probabilities that customers get zero waiting times, and the stationary pdfs of positive-wait customers. The model is denoted by $M/M/2/(\mu_0, \mu_1), (0, (0, \infty))$. In this notation, in "$(0, (0, \infty))$", the left-most "0" indicates that 0-wait arrivals get service rate $\mu_0 > 0$. In "$(0, \infty)$" the state-space interval "$(0, \infty)$" indicates that arrivals who wait before service for an amount of rime in the interval $(0, \infty)$ get service rate $\mu_1 > 0$. Diagrammatically, $\mu_0 \leftrightarrow$ zero-wait; $\mu_1 \leftrightarrow$ positive-wait.

There are three zero-wait states for arrivals denoted by $(0, (0.0))$, $(0, (1, 0)$, $(0, (0, 1))$. In this notation the pairs in parentheses (m_0, m_1) represent the following: $m_0 :=$ number of busy servers containing zero-wait customers and $m_1 :=$ number of busy servers containing positive-wait customers. Briefly, the set of possible zero-wait states is:

$$\{(0, m_0 m_1)\} = \{(0, 00), (0, 10), (0, 01)\}.$$

Denote the stationary probabilities of the zero-wait states by P_{00}, P_{10}, P_{01} respectively.

For example, state $(0, 10)$ indicates that an arrival would wait 0 and would "see" a rate-μ_0 customer being served by the "other (busy) server". The arrival would be assigned rate μ_0 since it waits 0. There would then be two rate-μ_0 customers in service. The inter start-of-service depart time S would be $\stackrel{d}{=}$ $\text{Exp}_{2\mu_0}$, because $S := \min\{Exp_{\mu_0}, Exp_{\mu_0,}\}$ (see Section 1.5.3 in Chapter 1).

There are only two zer-wait states such that $m_0 + m_1 = 1$ (both border states). Denote the pdfs of the pos-wait states $(x, 10)$, $(x, 01)$, by $f_{10}(x)$, $f_{01}(x)$, $x > 0$, respectively. A would-be arrival that finds the state $(x, 10), x > 0$, would wait x before service, and be assigned service rate μ_1 (wait > 0). Just after that arrival's start of service, it would have a rate-μ_0 customer as the neighbor in the other server. The inter start-of-service departure time would be $S \stackrel{d}{=} \text{Exp}_{\mu_0+\mu_1}$. The rate-$\mu_1$ customer would finish service first with probability $\frac{\mu_1}{\mu_0+\mu_1}$. (The probability that the minimum of two exponential r.v.s, Exp_{μ_0} and Exp_{μ_1}, is $\text{Exp}_{\mu_0+\mu_1}$.). Similarly the probability is $\frac{\mu_0}{\mu_0+\mu_1}$ of leaving the rate-μ_1 customer in service ($m_0 m_1 = 01$).

If an arrival "sees" a pos-wait state $(x, 01)$, $x > 0$, then the start-of-service-departure time S would be $\stackrel{d}{=} \text{Exp}_{2\mu_1}$, because the arrival and its neighbor in service would both have received service rate μ_1. The first customer to complete service would have rate μ_1 with certainty (probability $= 1$). The customer remaining in service just after that service completion would have service rate μ_1 by the memoryless property of Exp_{μ_1} (leaving the system in stare $m_0 m_1 = 01$).

A sample-path diagram of the virtual wait process $\{W(t)\}_{t\geq 0}$, has "three Lines" and "two Pages" (see Figure 4.6).

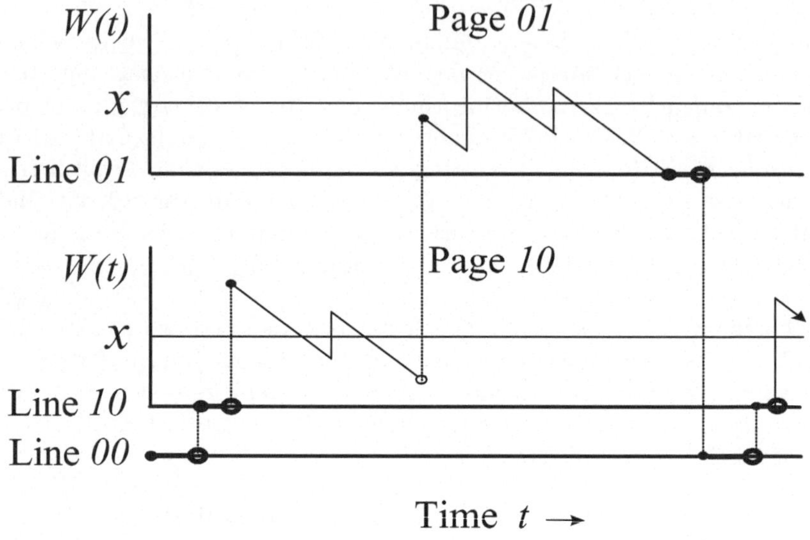

FIGURE 4.6
Sample path of virtual wait in M/M/2/(μ_0, μ_1), $(0, (0, \infty))$. Lines for states $(0, 10)$, $(0, 01)$ are at level 0 of the corresponding pages. Line for state $(0, 00)$ is isolated below. The SP can enter state $(0, 01)$ only by downcrossing level 0 on page 01.

The total (marginal) probability of a zer-wait is denoted by

$$P_{0wait} = P_{00} + P_{10} + P_{01}.$$

The total pdf of wait at level $x > 0$ is pdf on Page 10 plus pdf on Page 2, at level x:

$$f(x) = f_{10}(x) + f_{01}(x), x > 0.$$

4.4.2 Model Equations in M/M/2 Queue with Zero-Wait Customers Receiving Special Service

4.4.2.1 Equations for Zer-Wait States

Applying "SP exit rate = SP entrance rate" for the zero-wait states $(0, 00)$, $(0, 10)$, $(0, 01)$ gives, respectively,

$$\lambda P_{00} = \mu_0 P_{10} + \mu_1 P_{01},$$

$$(\lambda + \mu_0)P_{10} = \lambda P_{00} + f_{10}(0), \tag{4.63}$$

$$(\lambda + \mu_1)P_{01} = f_{01}(0).$$

In (4.63), the terms $f_{10}(0)$, $f_{01}(0)$ (same as $f_{10}(0^+)$, $f_{01}(0^+)$) are the rates at which the SP hits level 0 from above on Page 10, and Page 01, respectively. Immediately following such zero-level hits, the SP moves (to right) on Lines 10 and 01 respectively.

4.4.2.2 Equations for Positive-Wait States in M/M/2 Queue with Zero-Wait Customers Receiving Special Service

Applying "SP exit rate = SP entrance rate" for $((x,\infty),10)$ (on Page 10) yields the integral equation

$$f_{10}(x) + \lambda \int_{y=x}^{\infty} \frac{\mu_0}{\mu_0+\mu_1} f_{10}(y)dy$$

$$= \lambda P_{10} e^{-2\mu_0 x} + \lambda \frac{\mu_1}{\mu_0+\mu_1} P_{01} e^{-(\mu_0+\mu_1)x}$$

$$+ \lambda \frac{\mu_1}{\mu_0+\mu_1} \int_{y=0}^{x} e^{-(\mu_0+\mu_1)(x-y)} f_{10}(y)dy, x > 0. \tag{4.64}$$

When formulating Equation (4.64), note that the SP cannot jump directly from a pos-wait state on Page 01 into set $((x,\infty),10)$. An arrival that "sees" state $(y, 01)$, $y > 0$, will be assigned rate μ_1 and start service "after a wait y"; its neighbor in the other server will also have service rate μ_1 (because $m_0 m_1 = 01$). The random variable "inter-start-of-service departure time S" will be distributed as $\text{Exp}_{2\mu_1}$, and the remaining customer in service just after the first departure thereafter, will have rate μ_1. At the arrival instant, the SP will start a jump at level y on Page 01, which ends at level $y + S \stackrel{d}{=} y + \text{Exp}_{2\mu_1}$, also on Page 01. The configuration remains 01 just after the arrival. The only exit route from Page 01 is via a downcrossing of level 0 (continuous hit of level 0 from above—see Figure 4.6).

Now, we balance the SP exit and entrance rates for $((x,\infty),01)$ (Page 01), giving integral equation

$$f_{01}(x) = \lambda \frac{\mu_0}{\mu_0+\mu_1} P_{01} e^{-(\mu_0+\mu_1)x}$$

$$+ \lambda \int_{y=0}^{x} e^{-2\mu_1(x-y)} f_{01}(y)dy$$

$$+ \lambda \frac{\mu_0}{\mu_0+\mu_1} \int_{y=0}^{x} e^{-(\mu_0+\mu_1)(x-y)} f_{10}y)dy$$

$$+ \lambda \frac{\mu_0}{\mu_0+\mu_1} \int_{y=x}^{\infty} f_{10}(y)dy. \tag{4.65}$$

When formulating (4.65), note that the SP can exit $((x,\infty),01)$ only by downcrossing level x. Also, the SP cannot enter $((x,\infty),01)$ from state $(0,10)$ at arrivals, since all jumps that start from Line 10 (corresponding to state $(0,10)$) must end on Page 10, at an ordinate $\stackrel{d}{=} \text{Exp}_{2\mu_0}$.

The equation for the total pdf is

$$f(x) = f_{10}(x) + f_{01}(x),$$

as viewed from the "cover"("Cover" is the projection of all the Pages onto a single blank sheet (see Chapter 4 in [22] (2017) for more details). The path on the cover is the result of projecting sample-path segments on Pages 10 and 01 onto a single sheet. An integral equation for $f(x)$ is obtained by balancing the SP total downcrossing and total upcrossing rates of level $x > 0$. This is equivalent to equating the exit and entrance rates for the state-space set

$$((x,\infty),10) \cup ((x,\infty),01).$$

The resulting equation is

$$\begin{aligned} f(x) &= \lambda P_{10} e^{-2\mu_0 x} + \lambda P_{01} e^{-(\mu_0+\mu_1)x} \\ &+\lambda \int_{y=0}^{x} e^{-(\mu_0+\mu_1)(x-y)} f_{10}(y)dy \\ &+\lambda \int_{y=0}^{x} e^{-2\mu_1(x-y)} f_{01}(y)dy, \ x > 0. \end{aligned} \tag{4.66}$$

Equation (4.66) can also be derived by summing the corresponding sides of (4.64) and (4.65). It is intuitive and instructive to interpret Equation (4.66) as total SP rate-balance across level $x > 0$.

The normalizing condition is

$$P_{00} + P_{10} + P_{01} + \int_{x=0}^{\infty} f_{10}(x)dx + \int_{x=0}^{\infty} f_{01}(x)dx = 1,$$

or

$$P_{0\text{-}wait} + \int_{x=0}^{\infty} f(x)dx = 1. \tag{4.67}$$

4.4.3 Solution of Equations in M/M/2 Queue with Zero-wait Customers Receiving Special Service

Equation (4.64) is an integral equation in $f_{10}(x)$, which is not confounded by the presence of $f_{01}(x)$; so we utilize it to obtain the functional form of $f_{10}(x)$. Applying differential operator $\langle D\rangle\langle D+\mu_0+\mu_1\rangle$ to both sides of (4.64) leads to the second order differential equation

$$f_{10}''(x) + (\mu_0 + \mu_1 - \lambda)f_{10}'(x) - \lambda\mu_0 f_{10}(x)$$

$$= 2\lambda\mu_0(\mu_0 - \mu_1)P_{10}e^{-2\mu_0 x}, \ x > 0, \tag{4.68}$$

with solution

$$f_{10}(x) = C_{10}e^{ax} + C_{10}^1 e^{bx} + \lambda K_{10}P_{10}e^{-2\mu_0 x}, \ x > 0,$$

where a and b are the roots of the auxiliary quadratic equation, namely

$$a = \frac{1}{2}\left(\lambda - \mu_0 - \mu_1 - \sqrt{\lambda^2 + 2\lambda\mu_0 - 2\lambda\mu_1 + \mu_0^2 + 2\mu_0\mu_1 + \mu_1^2}\right) < 0,$$

$$b = \frac{1}{2}\left(\lambda - \mu_0 - \mu_1 + \sqrt{\lambda^2 + 2\lambda\mu_0 - 2\lambda\mu_1 + \mu_0^2 + 2\mu_0\mu_1 + \mu_1^2}\right) > 0,$$

$$K_{10} = \frac{2(\mu_0 - \mu_1)}{\lambda + 2\mu_0 - 2\mu_1},$$

and "C_{10}, C_{10}^1 are constants of integration". A necessary condition for system stability is $\lim_{x\to\infty} f_{10}(x) = 0$, which implies $C_{10}^1 = 0$ (since $b > 0$). Thus the functional form of $f_{10}(x)$ is

$$f_{10}(x) = C_{10}e^{ax} + \lambda K_{10}P_{10}e^{-2\mu_0 x}, x > 0, \tag{4.69}$$

where C_{10} is a constant to be determined.

The term K_{10} will be undefined if $\lambda + 2\mu_0 - 2\mu_1 = 0$. If $\lambda + 2\mu_0 - 2\mu_1 \neq 0$ and $\mu_0 - \mu_1 \neq 0$, then K_{10} may be positive or negative. If $\mu_0 - \mu_1 = 0$ the model reduces to a standard M/M/c queue with $c = 2$ (see Section 4.2 in Chapter 4, on the standard M/M/c queue, with c = 2.); the computed distribution of wait should then match that of a standard M/M/2 queue. (We will utilize this property later as a necessary check on the correctness of the present solution.)

We obtain the functional form of $f_{01}(x)$ by substituting the expression for $f_{10}(x)$ (4.69) into (4.66). Since

$$f_{01}(x) = f(x) - f_{10}(x),$$

this substitution gives the integral equation

$$f_{01}(x) = \lambda(1 - K_{10})P_{10}e^{-2\mu_0 x} + \lambda P_{01}e^{-(\mu_0+\mu_1)x} - C_{10}e^{ax}$$

$$+ \lambda\int_{y=0}^{x} e^{-(\mu_0+\mu_1)(x-y)}(C_{10}e^{ay} + \lambda K_{10}P_{10}e^{-2\mu_0 y})dy$$

$$+ \lambda\int_{y=0}^{x} e^{-2\mu_1(x-y)} f_{01}(y)dy. \tag{4.70}$$

The first integral term in (4.70) is

$$\lambda \int_{y=0}^{x} e^{-(\mu_0+\mu_1)(x-y)}(C_{10}e^{ay} + \lambda K_{10}P_{10}e^{-2\mu_0 y})dy$$

$$= \frac{\lambda C_{10}}{\mu_0+\mu_1+a}e^{ax} - \frac{\lambda^2 K_{10}P_{10}e}{\mu_0-\mu_1}^{-2\mu_0 x}$$

$$- \left(\frac{\lambda C_{10}}{\mu_0+\mu_1+a} - \frac{\lambda^2 K_{10}P_{10}}{\mu_0-\mu_1}\right) e_1^{-(\mu_0+\mu_1)x}.$$

Thus (4.70) is equivalent to the integral equation

$$\begin{aligned} f_{01}(x) =& H_{01}C_{10}e^{ax} + \lambda B_{01}P_{10}e^{-2\mu_0 x} \\ &+ D_{01}e^{-(\mu_0+\mu_1)x} \\ &+ \lambda \int_{y=0}^{x} e^{-2\mu_1(x-y)} f_{01}(y)dy, \end{aligned} \tag{4.71}$$

where

$$H_{01} = \frac{\lambda}{\mu_0+\mu_1+a} - 1,$$

$$B_{01} = 1 - K_{10} - \frac{\lambda K_{10}}{\mu_0-\mu_1},$$

$$D_{01} = \lambda P_{01} - \frac{\lambda C_{10}}{\mu_0+\mu_1+a} + \frac{\lambda^2 K_{10}P_{10}}{\mu_0-\mu_1}.$$

Applying the differential operator $\langle D + 2\mu_1 \rangle$ to both sides of (4.71) yields the differential equation for $f_{01}(x)$,

$$\begin{aligned} f'_{01}(x) + (2\mu_1 - \lambda)f_{01}(x) = (2\mu_1 + a)H_{01}C_{10}e^{ax} \\ + 2\lambda(\mu_1 - \mu_0)B_{01}P_{10}e^{-2\mu_0 x} \\ + (\mu_1 - \mu_0)D_{01}e^{-(\mu_1+\mu_0)x}, \end{aligned} \tag{4.72}$$

whose solution is

$$f_{01}(x) = \frac{2\lambda(\mu_1 - \mu_0)}{2\mu_1 - \lambda - 2\mu_0} B_{01} P_{10} e^{-2\mu_0 x}$$

$$+ \frac{\mu_1 - \mu_0}{\mu_1 - \lambda - \mu_0} D_{01} e^{-(\mu_1 + \mu_0)x}$$

$$+ \frac{2\mu_1 + a}{2\mu_1 - \lambda + a} H_{01} C_{10} e^{ax}$$

$$+ C_{01} e^{-(2\mu_1 - \lambda)x}, \tag{4.73}$$

where C_{01} is a constant of integration to be determined (see Section 4.4.5).

4.4.4 Stability Condition in M/M/2 Queue with Zero-Wait Customers Receiving Special Service

Consider the functional forms of $f_{10}(x)$ and $f_{01}(x)$ in (4.69) and (4.73). In the exponents, all the coefficients of x are negative except possibly the coefficient $-(2\mu_1 - \lambda)$ in $e^{-(2\mu_1 - \lambda)x}$ of (4.73). A necessary condition for stability is that

$$f_{10}(\infty) = f_{01}(\infty) = f(\infty) = 0;$$

implying $-(2\mu_1 - \lambda) < 0$, equivalent to $\lambda < 2\mu_1$. That is, the arrival rate must be less than the system departure rate when both servers are occupied by pos-wait customers, regardless how large x is. Thus, for stability, if the waiting time is large and customers are arriving, then the mean inter-arrival time should exceed the mean inter-departure time. This ensures that the waiting time will return to level zero in a finite time.

4.4.5 Determination of Constants in M/M/2 Queue with Zero-Wait Customers Receiving Special Service

A complete solution for the distribution of wait requires the values of five unknown constants:

$$P_{00},\ P_{10},\ P_{01},\ C_{10},\ C_{01},$$

which we obtain from five independent equations.

In (4.69) letting $x \downarrow 0$ to obtain $f_{10}(0)$, and referring to (4.63) gives

$$C_{10} + \lambda K_{10} P_{10} = (\lambda + \mu_0) P_{10} - \lambda P_{00}. \tag{4.74}$$

In (4.73) letting $x \downarrow 0$ to obtain $f_{01}(0)$ gives

$$f_{01}(0) = \frac{2\lambda(\mu_1 - \mu_0)}{2(\mu_1 - \mu_0) - \lambda} B_{01} P_{10}$$

$$+ \frac{\mu_1 - \mu_0}{\mu_1 - \mu_0 - \lambda} D_{01}$$

$$+ \frac{2\mu_1 + a}{2\mu_1 + a - \lambda} H_{01} C_{10} + C_{01}. \tag{4.75}$$

Substituting $f_{01}(0)$ from (4.75) into (4.63) gives

$$C_{01} = (\lambda + \mu_1) P_{01} - \frac{2\lambda(\mu_1 - \mu_0)}{2(\mu_1 - \mu_0) - \lambda} B_{01} P_{10}$$

$$- \frac{\mu_1 - \mu_0}{\mu_1 - \mu_0 - \lambda} D_{01}$$

$$- \frac{2\mu_1 + a}{2\mu_1 + a - \lambda} H_{01} C_{10}. \tag{4.76}$$

We get another independent equation by substituting the functional form

$$f_{10}(x) = C_{10} e^{ax} + \lambda K_{10} P_{10} e^{-2\mu_0 x}$$

into the integral Equation (4.64) and equating the coefficients of corresponding exponential terms on both sides after evaluating the integral (different exponentials are linearly independent – see, e.g., Section 3.3, pp. 99ff and p. 205 in [9]). The coefficient of $e^{-(\mu_0+\mu_1)x}$ on the right side of (4.64) must be 0. This yields the linear equation

$$\frac{\lambda\mu_1}{\mu_0 + \mu_1} P_{01} - \frac{1}{\mu_0 + \mu_1 + a} - \frac{\lambda K_{10}}{\mu_1 - \mu_0} P_{10} = 0. \tag{4.77}$$

The normalizing condition is

$$1 = P_{00} + P_{10} + P_{01} + \frac{C_{10}}{(-a)} + \frac{\lambda K_{10} P_{10}}{2\mu_0}$$

$$+ \frac{\lambda(\mu_1 - \mu_0)}{\mu_0(2(\mu_1 - \mu_0) - \lambda)} B_{01} P_{10} + \frac{\mu_1 - \mu_0}{(\mu_1 + \mu_0)(\mu_1 - \mu_0 - \lambda)} D_{01}$$

$$+ \frac{2\mu_1 + a}{(-a)(2\mu_1 + a - \lambda)} H_{01} C_{10} + \frac{1}{2\mu_1 - \lambda} C_{01}. \tag{4.78}$$

We now have a set of five equations to solve for the five constants: from (4.63)

$$\lambda P_{00} = \mu_0 P_{10} + \mu_1 P_{01},$$

and (4.74), (4.76), (4.77), (4.78).

In the derivation of the functional forms of $f_{10}(x)$, $f_{01}(x)$, the expressions

$$\mu_1 - \mu_0, \quad 2\mu_1 - \lambda - 2\mu_0, \quad \mu_1 - \lambda - \mu_0, \quad 2\mu_1 - \lambda + a,$$

appear in various denominators. "If any of these four expressions were equal to 0", the functional forms would have to be modified. The five equations used to solve for the constants in the present model would have to be modified accordingly. In this book we emphasize the system-point level-crossing approach to derive model equations, and various techniques to solve them. There are many techniques to solve systems of integral equations, outside the scope of the present book. We give "numerical solutions of the equations in several examples", below.

For further details, about the functional forms of the pdfs see Chapter 4 in [22] (2017).

4.4.5.1 Solution of Integral Equations in Numerical Example of M/M/2 Queue with Zero-Wait Customers Receiving Special Service

We first give a mild necessary numerical check on the five equations by letting $\mu_1 - \mu_0 = 0$. In this case $M/M/2/(\mu_0, \mu_1), (0, (0, \infty))$ reduces to a standard M/M/2 queue. We arbitrarily take

$$\lambda = 1, \ \mu_0 = 1.5, \ \mu_1 = 1.5.$$

Then $a = -2.581139$. Then, the solution for the constants is:

$$C_{10} = 0.0, \ P_{01} = .133333, \ P_{10} = .20, \ C_{01} = .333333, \ P_{00} = .50.$$

We compare this solution with that of the standard M/M/2 queue with $\lambda = 1$, $\mu = 1.5$. In M/M/2, the probability of an empty system is $P_0 = 0.5$. The probability of having 1 customer in the system is indeed $P_1 = 0.33333$. The values match P_{00} and $P_{10} + P_{01}$ in $M/M/2/(\mu_0, \mu_1), (0, (0, \infty))$ model, as expected.

Also, in $M/M/2/(\mu_0, \mu_1), (0, (0, \infty))$, we see from (4.73) that

$$f_{01}(x) = C_{01} e^{-(2\mu_1 - \lambda)x}$$

$$= \lambda P_1 e^{-(2\mu_1 - \lambda)x}$$

$$= 1 \cdot (0.33333) e^{-2x}, x > 0,$$

since $\mu_1 - \mu_0 = 0$ and $C_{10} = 0$.

Let $\lambda = 1$, $\mu_0 = 1.1$, $\mu_1 = 2.21$. These values preclude that any of the four denominators in formulas (4.75)–(4.78) is 0. We get $a = -2.715136$. We solve the equations and obtain

$$P_{00} = .417715,\ P_{10} = 0.339103,\ P_{01} = 0.0202270,$$

$$C_{01} = 0.022818,\ C_{10} = -0.322655.$$

The functions $f_{10}(x), x > 0$, and $f_{01}(x), x > 0$, are linear combinations of exponentials,

$$f_{10}(x) = -0.322655e^{-2.715136x} + 0.617056e^{-2.2x},$$

$$\begin{aligned} f_{01}(x) &= 0.505784e^{-2.2x} + 0.067831e^{-3.31x} \\ &\quad - 0.531504e^{-2.715136x} + 0.022818e^{-3.42x}. \end{aligned}$$

We substitute the values of P_{00}, P_{10}, P_{01}, $f_{10}(x)$, $f_{01}(x)$ into the normalizer (4.67), and obtain 1; it checks.

The partial pdfs of wait $f_{10}(x)$, $f_{01}(x)$ and total pdf of wait $f(x)$ are depicted in Figures 4.7, 4.8, and 4.9 respectively.

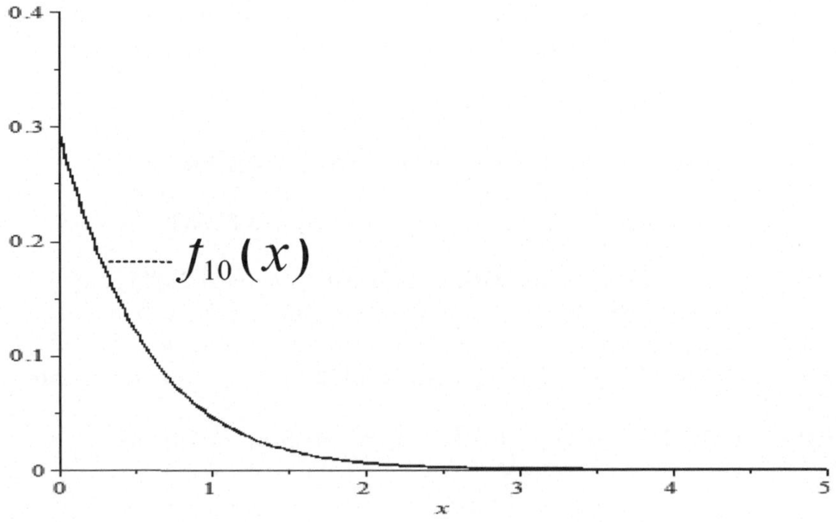

FIGURE 4.7
Partial pdf of wait $f_{10}(x)$ in M/M/2/(μ_0, μ_1), $(0, (0, \infty))$. $\lambda = 1$, $\mu_0 = 1.1$, $\mu_1 = 2.21$.

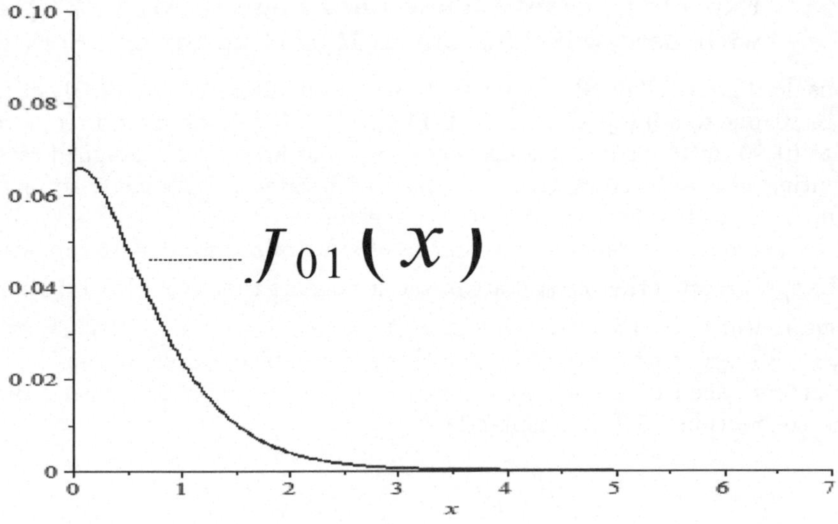

FIGURE 4.8
Partial pdf of wait $f_{01}(x)$ in M/M/2/(μ_0, μ_1) , $(0, (0, \infty))$. $\lambda = 1$, $\mu_0 = 1.1$, $\mu_1 = 2.21$.

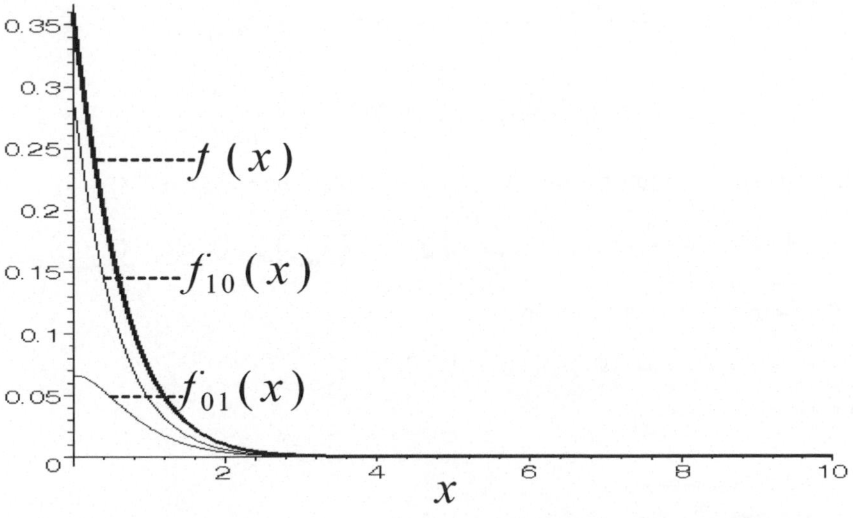

FIGURE 4.9
Total pdf of wait $f(x) = f_{10}(x) + f_{01}(x)$ in M/M/2/(μ_0, μ_1) , $(0, (0, \infty))$. $\lambda = 1$, $\mu_0 = 1.1$, $\mu_1 = 2.21$.

4.4.6 Expected Sojourn Time on a Page in M/M/2 Queue with Zero-wait Customers Receiving Special Service

Consider Page 01. The SP can enter Page 01 from discrete state $(0, 01)$ or from Page 10, due to a jump at an arrival (Figure 4.6). It cannot enter directly from state $(0, 10)$ at an arrival instant, since zero-wait arrivals are assigned rate μ_0 resulting in both servers being occupied with rate-μ_0 customers; so any SP jump to a positive level must end on Page 10.

In a sojourn on Page 01, the first inter start-of-service depart timewill be $\stackrel{d}{=} \text{Exp}_{\mu_0+\mu_1}$; any other inter start-of-service depart times that follow while on Page 01 will be $\stackrel{d}{=} \text{Exp}_{2\mu_1}$. While the SP is on Page 01, each departure will leave a rate-μ_1 customer in the neighboring occupied server. Given that the SP enters Page 01, its source state was $(0, 01)$ with probability (using Bayes' rule–see Section 1.5.2 in Chapter 1)

$$q = \frac{P_{01}}{P_{01} + \int_{y=0}^{\infty} f_{10}(y)dy}.$$

Its source was composite state $((0, \infty), 10)$ with probability

$$1 - q = \frac{\int_{y=0}^{\infty} f_{10}(y)dy}{P_{01} + \int_{y=0}^{\infty} f_{10}(y)dy}.$$

Let H denote the level above level 0 (ordinate) at which the SP enters Page 01 (see Figure 4.10). A sojourn on Page 01 starts at level H, where

$$E\,[H]\,|\text{source is sate } (0, 01)) = \frac{1}{\mu_0 + \mu_1},$$

and

$$E[H]|\text{source is level } y \text{ on Page } 10) = y + \frac{1}{\mu_0 + \mu_1}, y > 0,$$

since the size of a jump from either source onto Page 01 is $\stackrel{d}{=} \text{Exp}_{\mu_0+\mu_1}$. Thus

$$E\,[H] = \frac{1}{\mu_0 + \mu_1} \cdot q + \left(\int_{y=0}^{\infty} \left(y + \frac{1}{\mu_0 + \mu_1} \right) f_{10}(y)dy \right) \cdot (1 - q)\,.$$

From (4.69) $f_{10}(y)$ is given by

$$f_{10}(y) = C_{10}e^{ay} + \lambda K_{10} P_{10} e^{-2\mu_0 y}, y > 0,$$

and thus

$$\begin{aligned} E\,[H] &= \frac{1}{\mu_0 + \mu_1} \cdot q \\ &\quad + \left(\int_{y=0}^{\infty} \left(y + \frac{1}{\mu_0 + \mu_1} \right) \left(C_{10}e^{ay} + \lambda K_{10} P_{10} e^{-2\mu_0 y} \right) dy \right) \cdot (1 - q) \\ &= \frac{1}{\mu_0 + \mu_1} \cdot q + \left(\frac{1}{4} \Big(4C_{10}\mu_0^2\mu_1 + 4C_{10}\mu_0^3 \right. \\ &\quad + 3\lambda K_{10} P_{10} a^2 \mu_0 + \lambda K_{10} P_{10} a^2 \mu_1 \\ &\quad \left. - 4C_{10} a \mu_0^2 \Big) / \left(a^2 \mu_0^2 (\mu_0 + \mu_1) \right) \right) \cdot (1 - q). \end{aligned} \tag{4.79}$$

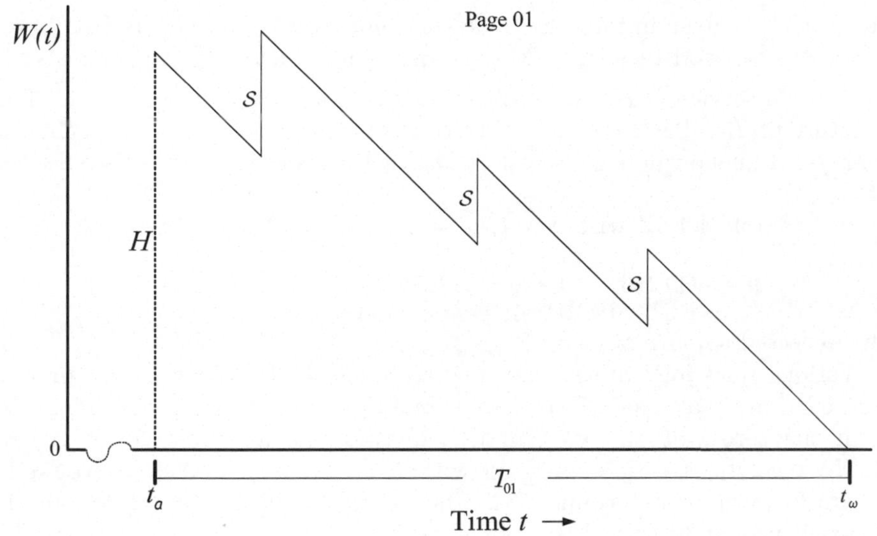

FIGURE 4.10
T_{01} := sojourn on page 01. t_α := start of T_{01}, t_ω := end of T_{01}. $\mathcal{S} \overset{d}{=} \text{Exp}_{2\mu_1}$. (See Figure 4.6.)

Let T_{01} denote a sojourn time on Page 01, i.e., the time from SP entrance until the first exit from Page 01 thereafter. The only possible exit is due to a downcrossing of level 0 (Figure 4.6). Thus

$$T_{01} = H + \sum_{i=1}^{N_H} \mathcal{B}_i,$$

where N_H is the number of arrivals during time H and $\mathcal{B}_i$ represents a busy period of an M/M/1 queue with service rate $2\mu_1$, since both servers are busy with rate-μ_1 customers. The expected busy period is obtained from formula (2.37) in Chapter 2 by substituting $2\mu_1$ for μ. Thus

$$E\left[\mathcal{B}_i\right] = \frac{1}{2\mu_1 - \lambda}, i = 1, ..., N_H.$$

The r.v.s N_H and $\mathcal{B}_i$, $i = 1, ..., N_H$ are independent and identically distributed (iid), since the $\mathcal{B}_i$s are iid, each distributed as an $M_\lambda/M_{2\mu_1}/1$ busy period. The expected sojourn time on Page 01 is

$$\begin{aligned} E\left[T_{01}\right] &= E\left[H\right] + E\left[\sum_{i=1}^{N_H} \mathcal{B}_i\right] = E\left[H\right] + E\left[N_H\right] E\left[\mathcal{B}_i\right] \\ &= E\left[H\right] + \lambda E\left[H\right] \frac{1}{2\mu_1 - \lambda} = \frac{E\left[H\right]}{1 - \lambda/\left(2\mu_1\right)}, \end{aligned} \tag{4.80}$$

where $E[H]$ is given in formula (4.79). It is noteworthy that T_{01} is distributed as the busy period in an $M_\lambda/M_{2\mu_1}/1$ queue in which zero-wait arrivals obtain special service $\stackrel{d}{=} H$, and positive-wait arrivals get service rate $2\mu_1$. This structure of T_{01} illustrates an interesting application, and the versatility, of the M/G/1 queue where zero-wait arrivals get exceptional service (see Section 2.5).

In Example 4.4.5.1 with $\lambda = 1$, $\mu_0 = 1.1$, $\mu_1 = 2.21$, we obtain

$$q = .111216, \qquad 1 - q = .888784, \qquad E[H] = 0.151416.$$

The expected sojourn time on Page 01 is $E[T_{01}] = 0.195689$.

Various questions arise regarding Example 4.4.6. What is the proportion of time that the SP spends circulating on Page 01, on Page 10, or in the zero-wait states? Can this question be answered for a general $M/M/c/(\mu_0, \mu_1), (0, (0, \infty))$ queue with $c > 2$? If yes, then it would be straightforward to determine P_{00}. This would facilitate solving for all the zero-wait probabilities and the partial pdfs of wait.

5

G/M/1 Queues

5.1 Standard G/M/1 Queue

In the standard G/M/1 queue, assume that arrivals occur according to a renewal process (for definition of a renewal process, see Section 1.8.1 in Chapter 1). For the common inter-arrival time denote the cdf, complementary cdf (ccdf), and pdf, respectively, by $A(x)$, $x > 0$, $\overline{A}(x) = 1 - A(x)$, $x \geq 0$ and $a(x) = dA(x)/dx$ wherever the derivative exists. It can be shown that the stationary pdf of the virtual wait has an "exponential form" (see Section 5.1.5). We derive the stationary pdf and cdf of the virtual wait, the stationary pdf and cdf of the actual (arrival-point) wait just before arrival instants, the expected busy period, expected idle period, and related results.

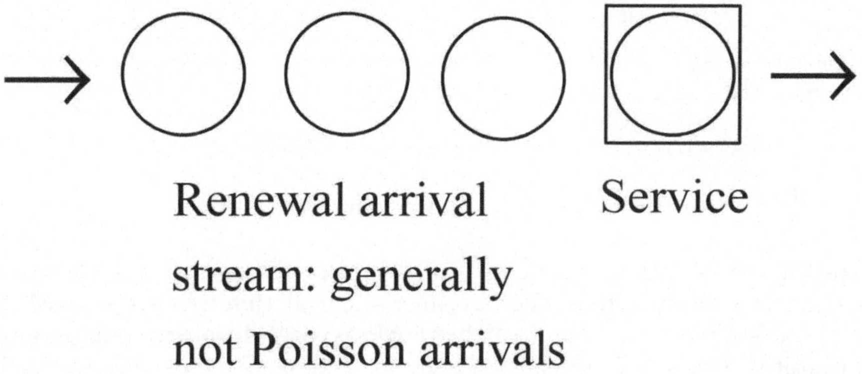

FIGURE 5.1
In the standard G/M/1 queue the customer arrival stream is a renewal process. The inter-renewal times are generally not exponentially distributed (i.e., not like in an M/M/1 queue).

5.1.1 Virtual Wait and Extended Age Processes in the G/M/1 Queue

Let $\{W(t)\}_{t \geq 0}$ denote the virtual wait process having state space $\boldsymbol{S} = [0, \infty)$ (e.g., similar to Figure 2.2 in Chapter 2).

DOI: 10.1201/9780429297601-5

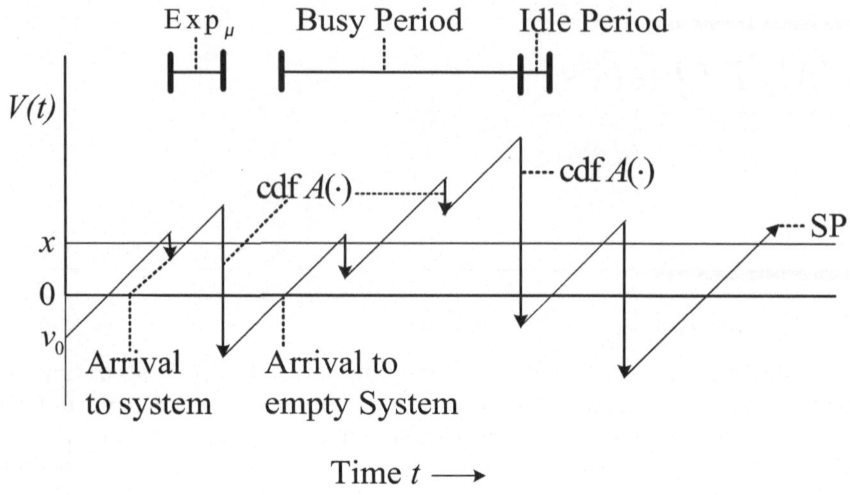

FIGURE 5.2
Sample path of "extended age process" $\{V(t)\}_{t\geq 0}$ in G/M/1 queue. Inter-arrival times have cdf $A(\cdot)$ (cdf of vertical downward jumps sizes). "Service" times are $\stackrel{d}{=}$ Exp$_\mu$. Slope between downward jumps $\equiv dV(t)/dt = +1$.

Consider the "extended age process" denoted by $\{V(t)\}_{t\geq 0}$ having state space $\boldsymbol{S}_{ext} = (-\infty, \infty)$, defined as follows. For $t > 0$,

$$V(t) = \begin{cases} \text{“age in system of customer who is in service” at time } t, \text{ if } V(t) \geq 0; \\ \text{or} \\ -\ \text{“(time from instant } t \text{ until new arrival instant)” if } V(t) < 0, \end{cases} \tag{5.1}$$

where "age in system of customer who is in service" means "time spent in the system" measured from the customer's arrival time. A sample path of $\{V(t)\}_{t\geq 0}$ is depicted in Figure 5.2. All SP vertical downward jumps start from positive levels (i.e., when $V(\cdot) > 0$) at customers' "departure instants" from the system. All jumps are in a vertical downward direction. By contrast, jumps in the "virtual wait" $\{W(t)\}_{t\geq 0}$ occur at arrival instants, and are in "vertical upward" direction.

5.1.2 Duality between Extended Age Process and Virtual Wait Process in Standard G/M/1 Queue

We can establish duality properties between $\{V(t)\}_{t\geq 0}\{W(t)\}_{t\geq 0}$ by examining both sample paths in Figure 5.3. The sample path of virtual wait $\{W(t)\}_{t\geq 0}$ is ≥ 0 during busy periods and idle periods, i.e., for all $t \geq 0$. The sample path of extended age $\{V(t)\}_{t\geq 0}$ is ≥ 0 during busy periods only. There is a one-to-one correspondence between the "sample-path peaks" (relative maxima) of $\{V(t)\}_{t\geq 0}$ and the "sample-path peaks" of $\{W(t)\}_{t\geq 0}$. There is also a one-to-one correspondence between their respective "sample-path troughs" (relative minima or infima), when both processes are restricted to a common busy period. Within the common busy periods, corresponding peaks have "equal ordinates" and occur "in the same time order" in both processes; similarly for corresponding troughs (see Figure 5.3).

5.1.2.1 Properties of the Busy Period in G/M/1 Queue

The extended age process $\{V(t)\}_{t\geq 0}$ has slope "+1" between SP vertical downward jumps. On the other hand, the virtual wait process $\{W(t)\}_{t\geq 0}$ has slope "-1" between SP vertical upward jumps within every busy period; the sample-path slope is 0 within every idle period. The busy periods occur during identical time intervals in process $\{V(t)\}_{t\geq 0}$ and process $\{W(t)\}_{t\geq 0}$.

These properties guarantee that the long-run proportion of time that the SP spends within any positive-valued state-space interval, is the same in both of these processes (see Proposition 5.1, and [15] (1988)).

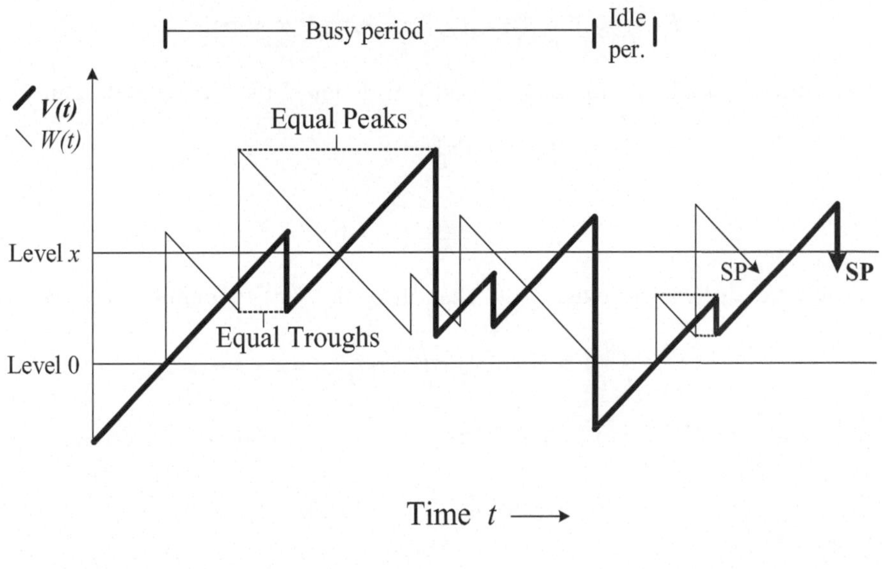

FIGURE 5.3
Sample path of the extended age process $\{V(t)\}_{t\geq 0}$ (slope $= +1$) superimposed on a sample path of the virtual wait process $\{W(t)\}_{t\geq 0}$ (slope $= -1$), in standard G/M/1 queue. The figure illustrates duality properties. Corresponding peaks and corresponding troughs have identical ordinates "> 0" in the same time sequence. Busy periods, idle periods, and busy cycles are equal (on identical time intervals) in the two processes.

The sojourn time of the extended age process $\{V(t)\}_{t\geq 0}$ below level 0 is identical to an idle period in the virtual wait process $\{W(t)\}_{t\geq 0}$ (see Figure 5.3). Thus the proportion of time that the sample path of $\{V(t)\}_{t\geq 0}$ spends below level 0 is equal to the proportion of time that sample paths of $\{W(t)\}_{t\geq 0}$ spend at level 0. The lengths of busy cycles (cycle $=\mathcal{B}+\mathcal{I}$) are identical in both processes (see Figure 5.3).

The stability condition for $\{V(t)\}_{t\geq 0}$ is:

$$\frac{\frac{1}{\mu}}{E\,[\text{inter-arrival time}]} < 1$$

(e.g., see pp. 296–297 in [22] (2017); and pp. 261–262 in [36] (2008)). Stability holds if E[number of arrivals in a service time] < 1. (See Proposition 5.6 in Section 5.1.9.)

5.1.2.2 Relation between Probability Distributions of the Extended Age and Virtual Wait in G/M/1 Queues

Denote the stationary cdf of the extended age by

$$F(x) = \lim_{t\to\infty} P(V(t) \leq x), -\infty < x < \infty,$$

having pdf denoted by $f(x), x \geq 0$, and pdf denoted by $h(x), x < 0$. Thus

$$\begin{aligned} f(x) &= \tfrac{dF(x)}{dx}, x \geq 0; \\ h(x) &= \tfrac{dF(x)}{dx}, x < 0, \end{aligned} \tag{5.2}$$

wherever the derivatives exist. Note that the pdf "$h(x)$" is defined on $(-\infty, 0)$. Thus

$$\int_{y=-\infty}^{0} h(y)dy + \int_{y=0}^{x} f(y)dy = 1.$$

The probability of an empty system is

$$P_0 = F(0) = \int_{y=-\infty}^{0} h(y)dy. \tag{5.3}$$

Then,

$$\begin{aligned} F(x) &= P_0 + \textstyle\int_{y=0}^{x} f(y)dy, x \geq 0, \\ F(x) &= \textstyle\int_{y=-\infty}^{x} h(y)dy, x \leq 0, \\ F(0) &= P_0 = \textstyle\int_{y=-\infty}^{0} h(y)dy, \\ F(\infty) &= P_0 + \textstyle\int_{y=0}^{\infty} f(y)dy = 1. \end{aligned} \tag{5.4}$$

Proposition 5.1 *The stationary cdfs of the extended age* $\{V(t)\}_{t\geq 0}$ *and of the virtual wait*$\{W(t)\}_{t\geq 0}$ *as* $t \to \infty$, *are identical. That is,*

$$F(x) \equiv \lim_{t\to\infty} P(V(t) \leq x) = \lim_{t\to\infty} P(W(t) \leq x), x \geq 0.$$

Proof *We use sample paths and their definitions in this proof. There is a one-to-one correspondence between sample paths of* $\{V(t)\}_{t\geq 0}$ *and* $\{W(t)\}_{t\geq 0}$ *because of the duality properties discussed above in Section 5.1.2 (see Figure 5.3). Namely, the proportion of time spent in each state-space interval entirely above level* 0 *is the same in the two sample paths for "every* $\omega \in \Omega$*", where* Ω *is the sample space of the "underlying random experiment", and "*ω*" is a possible outcome. The proportion of time spent "at level* 0*" in* $\{W(t)\}_{t\geq 0}$ *is the same as the proportion of time spent "below level* 0*" in* $\{V(t)\}_{t\geq 0}$ *(see pp. 12–17 for "random experiment" and pp. 17–18 for "definitions of* ω *and* Ω*", in [30] (1967)). For* $\{V(t)\}_{t\geq 0}$, *a sojourn time below level* 0 *is the same as an idle period at level* 0 *in* $\{W(t)\}_{t\geq 0}$. *Thus* $F(0) = P_0 = \lim_{t\to\infty} P_0(t)$ *is the same for both processes (where* $P_0(t)$ *is the probability of a zero wait at time* t *(see [15] (1988)).*

The reason for employing $\{V(t)\}_{t\leq 0}$ to obtain the limiting pdf of $\{W(t)\}_{t\geq 0}$ in G/M$_\mu$/1, is to apply LC directly, because SP vertical downward jumps occur at end-of-service instants with "Poisson rate μ". Thus, for vertical downward jumps in $\{V(t)\}_{t\leq 0}$ PASTA applies ("Poisson Arrivals See Time Averages", in Section 2.1.2 in Chapter 2; also see [59] (1982)). PASTA does not apply for vertical upward jumps in the virtual wait process $\{W(t)\}_{t\geq 0}$. PASTA applies for G/M/1 queues only if inter-arrival times (i.e., service times in this case) are exponentially distributed.

Remark 5.2 *"Transient" probability distributions of the extended age* $\{V(t)\}_{t\leq 0}$ *and the virtual wait* $\{W(t)\}_{t\leq 0}$ *are not equal to each other . Proposition 5.1 holds only for stationary probability distributions.*

5.1.3 Equation for Stationary Pdf of Extended Age in G/M/1 Queue

By Proposition 5.1, the stationary pdfs of $\{V(t)\}_{t\geq 0}$ and $\{W(t)\}_{t\geq 0}$ are identical. Thus, for the G/M/1 queue, we obtain the stationary pdf of $\{W(t)\}_{t\geq 0}$ by first deriving the stationary pdf $f(x)$ of $\{V(t)\}_{t\geq 0}$.

Consider a sample path of $\{V(t)\}_{t\geq 0}$ (Figure 5.2). Fix level $x > 0$ in the state space. Similarly as for the basic level crossing limit theorems for SP downcrossing rates, stated in Section 1.4 in Chapter 1, the limit theorem for the SP "upcrossing" rate of level $x > 0$ is given by

$$\lim_{t\to\infty} \frac{\mathcal{U}_t(x)}{t} = f(x), x > 0. \tag{5.5}$$

The SP downcrossing rate of level $x > 0$ is given by

$$\lim_{t\to\infty} \frac{\mathcal{D}_t(x)}{t} = \mu \int_{y=x}^{\infty} \overline{A}(y-x)f(y)dy, x > 0. \quad (5.6)$$

Rate balance across level x equates the rates in (5.5) and (5.6): $\lim_{t\to\infty} \frac{\mathcal{U}_t(x)}{t} = \lim_{t\to\infty} \frac{\mathcal{D}_t(x)}{t}$,which,gives immediately the "integral equation for $f(x), x > 0$":

$$f(x) = \mu \int_{y=x}^{\infty} \overline{A}(y-x)f(y)dy. \quad (5.7)$$

(Formulas (5.5) and (5.6) are proved similarly as for the limiting down- and upcrossing rates in M/G/1, e.g., stated in Theorem 1.3 in Section 1.4 in Chapter 1).

5.1.3.1 LC Interpretation of Formula for Downcrossing Rate of Level x in Extended Age Process in G/M/1 Queue

The right-most term in formula (5.5) is the SP rate of vertical downward jumps starting where service times end at level $y > \dot{x}$. These downward jumps occur at Poisson rate $\mu f(y)dy$, due to service times $\stackrel{d}{=}$ Exp$_\mu$. Each "downward jump size" is $\stackrel{d}{=}$ "inter-arrival time", giving:

$$\overline{A}(y-x) = P(\text{downward jump size} > y-x).$$

Summing over all $y > x$ gives the integral on the right side of (5.7).

5.1.4 Alternative Equation for Stationary PDF of Extended Age in G/M/1 Queue

An alternative form of integral Equation (5.7) is

$$f(x) = \mu(1 - F(x)) - \mu \int_{y=x}^{\infty} A(y-x)f(y)dy, x > 0. \quad (5.8)$$

5.1.4.1 LC Interpretation of Alternative Equation for Stationary Pdf of Extended Age in G/M/1 Queue

We can obtain equation (5.8) directly by setting $\overline{A}(y-x) = 1 - A(y-x)$ in (5.7). The LC interpretation of (5.8) is that the left side is the SP upcrossing rate of level x. On the right side, the first term is the rate of "all service completions" which generate SP downward jumps that begin above level x. Such downward jumps end either above level x, or below level x. The second term subtracts off the rate of service completions which generate SP downward jumps that begin above level x and end "above level x". Thus, the right side is the SP downcrossing rate of level x. Rate balance across level x gives Equation (5.8).

(The LC technique for obtaining alternative Equation (5.8) for $f(x)$ in the G/M/1 queue is similar to the LC technique for deriving alternative integral equations for the M/G/1 queue, and for other stochastic models such as the M/G/r($\cdot$) dam in Chapter 7. All these equations can be reached directly using LC interpretations of SP motion in state spaces.)

5.1.5 The Exponential Form of Stationary Pdf of Extended Age Process $\{V(t)\}_{t\geq 0}$ in G/M/1 Queue

An interesting property of the standard G/M/1 queue is that the continuous part of the stationary pdf of the virtual wait $\{W(t)\}_{t\geq 0}$ has an exponential form despite the queue having renewal customer inter-arrivals. So, $M_\lambda/M_\mu/1$ queues are not the only single-server queues to have an exponential stationary form for the continuous part of the stationary pdf for the virtual wait. We demonstrate the exponential form for G/M/1, first using LC for the stationary pdf $f(x)$, $x > 0$, of the extended age process $\{V(t)\}_{t\geq 0}$. Then, the stationary pdf of virtual wait $\{W(t)\}_{t\geq 0}$ must have the same stationary pdf by Proposition 5.1. The exponential form is defined for the continuous part, on the state-space interval $(0, \infty)$, with an atom for P_0 at level 0. The LC derivation focuses on the meanings of each term, which differs from other derivations (e.g., pp. 261–263 in [36] (2008); pp. 251–253 in [41] (1975); pp. 400–401 in [57] (2003)).

In a sample path of the extended age $\{V(t)\}_{t\geq 0}$, the memoryless property of the exponential service-time distribution implies an SP sojourn above an arbitrary level $x \geq 0$ is $\stackrel{d}{\equiv} \mathcal{B}$ "independent of level x", where $\mathcal{B} :=$ busy period (see Figures 5.2 and 5.3).

Let $u_x :=$ the "inter-upcrossing time of level x" (see Section 5.1.14). Combine formula (5.5), i.e., "$1 - F(x)$ is the long-run proportion of time spent by the extended age process $\{V(t)\}_{t\geq 0}$ above level x", with the elementary renewal theorem (see Section 1.8.3 in Chapter 1). This implies that

$$E[u_x] = \frac{1}{f(x)}, \ 0 < x < \infty.$$

The sequence of inter-upcrossing times, i.e., the u_xs, forms a regenerative process (Section 1.8.2 in Chapter 1). By the renewal reward theorem (Section 1.8.4 in Chapter 1)

$$\frac{E[\mathcal{B}]}{E[u_x]} = \frac{E[\mathcal{B}]}{\frac{1}{f(x)}} = 1 - F(x),$$

$$\frac{f(x)}{1 - F(x)} = \frac{1}{E[\mathcal{B}]}. \tag{5.9}$$

Equation (5.9) is equivalent to the differential equation

$$\frac{\frac{d}{dx}(1-F(x))}{1-F(x)} = -\frac{1}{E[\mathcal{B}]},$$

or

$$\frac{d\ln(1-F(x))}{dx} = -\frac{1}{E[\mathcal{B}]},$$

with solution

$$1-F(x) = e^{-\frac{x}{E[\mathcal{B}]}+\text{constant}},$$

$$1-F(x) = Re^{-\frac{x}{E[\mathcal{B}]}},\ R \text{ constant},$$

$$R = 1-F(0) = 1-P_0,$$

$$F(x) = 1-(1-P_0)e^{-\frac{1}{E[\mathcal{B}]}x}, x \geq 0.$$

Taking $dF(x)/dx$ gives

$$f(x) = \frac{1-P_0}{E[\mathcal{B}]}e^{-\frac{1}{E[\mathcal{B}]}x}, x > 0, \tag{5.10}$$

where $f(x) = F'(x)$, $F(0) = P_0$, and $f(0) = \frac{1-P_0}{E[\mathcal{B}]}$.

From formula (5.10), $f(x)$ has the exponential form

$$f(x) = Ke^{-\gamma x}, x \geq 0, \tag{5.11}$$

where

$$K = \frac{1-P_0}{E[\mathcal{B}]}; \qquad \gamma = \frac{1}{E[\mathcal{B}]}. \tag{5.12}$$

From (5.12),

$$E[\mathcal{B}] = \frac{1}{\gamma}, \tag{5.13}$$

$$P_0 = 1-KE[\mathcal{B}] = 1-\frac{K}{\gamma}. \tag{5.14}$$

Substituting from (5.11) into (5.7) in Section 5.1.3 and cancelling K on both sides gives an equation for γ:

$$\begin{aligned} e^{-\gamma x} &= \mu \int_{y=x}^{\infty} \overline{A}(y-x)e^{-\gamma y}dy, \\ \frac{1}{\mu} &= \int_{y=x}^{\infty} \overline{A}(y-x)e^{-\gamma(y-x)}dy. \end{aligned} \tag{5.15}$$

In (5.15), divide both sides of the first equation by "$\mu e^{-\gamma x}$", let $z = y - x$ in the second equation, and adjust the limits of integration to "$\int_{z=0}^{\infty}$"; this yields the equation

$$\int_{z=0}^{\infty} \overline{A}(z)e^{-\gamma z}dz = \frac{1}{\mu}. \tag{5.16}$$

Formula (5.16) is an "implicit equation" for γ, since μ and $\overline{A}(z)$ are assumed to be known. The left side is the Laplace transform of $\overline{A}(z)$ evaluated with parameter γ. Once (5.16) is solved for γ, we also have $E[\mathcal{B}] = \frac{1}{\gamma}$ by formula (5.13).

Since $\overline{A}(z) = P(\text{inter-arrival time} > z) = 1 - A(z)$, equation (5.16) gives

$$\int_{z=0}^{\infty} e^{-\gamma z}dz - \int_{z=0}^{\infty} A(z)e^{-\gamma z}dz = \frac{1}{\mu},$$

$$\frac{1}{\gamma} - A^*(\gamma) = \frac{1}{\mu},$$

$$A^*(\gamma) = \frac{1}{\gamma} - \frac{1}{\mu}, \tag{5.17}$$

where $A^*(z)$ denotes the LT (Laplace transform) of $A(z)$. (See Appendix C, "Transforms and Generaating Functions", in [36] (2008) or a Calculus text for definitions of LT and LST). Thus γ is the solution of Equation (5.17). Some forms of $A(\gamma)$ allow an analytical solution for γ directly. Generally, γ is computed by numerical methods using computational software. (Note that dimension is $[Time]$ for all three terms in formula (5.17)).

5.1.5.1 More on the Value of P_0 in Standard G/M/1 Queue

This section considers the sample path of the extended age below level 0 and above level 0. For $\{V(t)\}_{t\geq 0}$ on the state-space interval $(-\infty, 0)$, fix a level x, $-\infty < x < 0$ (Figure 5.3). The SP upcrossing rate of level x is equal to pdf "$h(x)$" (proved as for the downcrossing rate for the basic LC limit theorem in the M/G/1 queue, in Section 1.4 in Chapter 1).

The SP downcrossing rate of level x is due to the downward vertical jump rate originating at peaks all of which are above level 0, i.e.,

$$\mu \int_{y=0}^{\infty} \overline{A}(y-x)f(y)dy = \mu \int_{y=0}^{\infty} \overline{A}(y-x)Ke^{-\gamma y}dy,$$

where $f(y) = Ke^{-\gamma y}, \dot{y} \geq 0$, by formula (5.11) in Section 5.1.5. Note that $\overline{A}(y-x) = P(\text{downward jump} > y - x) = P(\text{inter-arrival time} > y - x)$.All SP downward vertical jumps originate at end-of-service instants (peaks) when the SP is in state-space interval $(0,\infty)$. Rate balance across level x gives:

$$\text{upcrossing rate} = \text{downcrossing rate}$$
$$h(x) = \mu\int_{y=0}^{\infty}\overline{A}(y-x)Ke^{-\gamma y}dy, x < 0, \tag{5.18}$$

where the pdf $h(x)$, $-\infty < x < 0$, is the upcrossing rate of x with slope "+1"; the right side is the vertical downcrossing rate of level x. Using (5.18) and (5.3) in Section 5.1.2 leads to

$$P_0 = \int_{x=-\infty}^{0} h(x)dx = K\int_{x=-\infty}^{0}\mu\int_{y=0}^{\infty}\overline{A}(y-x)e^{-\gamma y}dydx.$$

The transformation "$z = -x$" in the double integral immediately above gives

$$P_0 = K\int_{z=0}^{\infty}\mu\int_{y=0}^{\infty}\overline{A}(y+z)e^{-\gamma y}dydz.$$

Thus

$$P_0 = \frac{K}{C_\gamma}, \text{ or } K = P_0 C_\gamma, \tag{5.19}$$

where we define (for brevity)

$$C_\gamma :== \frac{1}{\int_{z=0}^{\infty}\mu\int_{y=0}^{\infty}\overline{A}(y+z)e^{-\gamma y}dydz}, \tag{5.20}$$

$$= K\times\frac{1}{\int_{z=0}^{\infty}\mu\int_{y=0}^{\infty}\overline{A}(y+z)Ke^{-\gamma y}dydz}, \tag{5.21}$$

and $C_\gamma > 0$. The term C_γ has dimenstion $\left[\frac{1}{Time}\right]$ (same as $[K]$), which agrees with P_0 being a dimensionless number in (5.19).

To evaluate P_0, use the normalizing condition and (5.19). Thus

$$P_0 + K\int_{y=0}^{\infty}e^{-\gamma x}dx = 1, \tag{5.22}$$

$$P_0 + C_\gamma P_0\int_{y=0}^{\infty}e^{-\gamma x}dx = 1. \tag{5.23}$$

Equations (5.22) and (5.23) give, respectively,

$$P_0 = 1 - \frac{K}{\gamma}, \tag{5.24}$$

and

$$P_0 = \frac{\gamma}{\gamma + C_\gamma}. \tag{5.25}$$

From (5.25)

$$K = P_0 C_\gamma = \frac{\gamma}{\gamma + C_\gamma} \times C_\gamma = \gamma \times \frac{C_\gamma}{\gamma + C_\gamma} < \gamma. \tag{5.26}$$

Due to exponentially distributed service times, the instants of SP upcrossings of level 0 into state space $(0, \infty)$, are regenerative instants of $\{V(t)\}_{t \geq 0}$ which initiate busy cycles (see Figure 5.2). The renewal reward theorem leads to the formula for the expected idle period, $E[\mathcal{I}]$, discussed in the next section.

5.1.6 Expected Idle Period in G/M/1 Queue

Let $\boldsymbol{C}$:= "Busy cycle". By the renewal reward theorem,

$$P_0 = \frac{E[\mathcal{I}]}{E[\boldsymbol{C}]} = \frac{E[\mathcal{I}]}{E[\mathcal{B}] + E[\mathcal{I}]}. \tag{5.27}$$

By formula (5.25),

$$P_0 = \frac{\gamma}{\gamma + C_\gamma} = \frac{\frac{1}{C_\gamma}}{\frac{1}{\gamma} + \frac{1}{C_\gamma}}. \tag{5.28}$$

The last equality in formula (5.28) is due to division of numerator and denominator by $\gamma \times C_\gamma$. From (5.12) in Section 5.1.5

$$E[\mathcal{B}] = \frac{1}{\gamma}.$$

Comparing the two last ratios in formula (5.27) and in formula (5.28), simple algebra shows that

$$E[\mathcal{I}] = \frac{1}{C_\gamma} = \int_{z=0}^{\infty} \int_{y=0}^{\infty} \mu \overline{A}(y+z) e^{-\gamma y} dy dz. \tag{5.29}$$

5.1.7 PDF of Idle Period $\mathcal{I}$ in G/M/1 Queue

Consider a sample path of the extended age process $\{V(t)\}_{t \geq 0}$ (Figure 5.2 in Section 5.1.1). Let I denote an idle period, and let $\mathcal{C}$ denote a busy cycle; then $\mathcal{C} = \mathcal{B} + \mathcal{I}$. Note that "$\mathcal{I} \stackrel{d}{=}$ excess of the SP vertical downward jumps below level 0, which end the busy periods. That is, while the SP is below level 0, it moves upward at slope +1 and begins a busy period when it upcrosses level 0. Denote the cdf and pdf of $\mathcal{I}$ by $F_{\mathcal{I}}(z)$, and $f_{\mathcal{I}}(z), z > 0$, respectively. The downcrossing rate of level "$-z$" $(z > 0)$ can be expressed in two different ways: first, by

$$\frac{\lim_{t\to\infty} \mathcal{D}_t(-z)}{t} = (1 - F_{\mathcal{I}}(z)) \times f(0)$$

$$= P(\mathcal{I} > z | SP \text{ downcrosses level } 0) \times \frac{\lim_{t\to\infty} \mathcal{D}_t(0)}{t},$$

since $f(0)$ is both the upcrossing rate and the downcrossing rate of level 0; second, by

$$\frac{\lim_{t\to\infty} \mathcal{D}_t(-z)}{t} = \mu \int_{y=0}^{\infty} \overline{A}(y+z) f(y) dy,$$

since all vertical downward jumps start above level 0. Equating the two different formulas for $\frac{\lim_{t\to\infty} \mathcal{D}_t(-z)}{t}$ and recalling that $f(x) = Ke^{-\gamma x}, x > 0$, gives

$$(1 - F_{\mathcal{I}}(z)) \times f(0) = \mu \int_{y=0}^{\infty} \overline{A}(y+z) f(y) dy, z > 0,$$

$$(1 - F_{\mathcal{I}}(z)) \times K = \mu \int_{y=0}^{\infty} \overline{A}(y+z) K e^{-\gamma y} dy, z > 0,$$

$$F_{\mathcal{I}}(z) = 1 - \mu \int_{y=0}^{\infty} \overline{A}(y+z) e^{-\gamma y} dy, z > 0, \tag{5.30}$$

after cancelling K. Using $1 - F_{\mathcal{I}}(z)$, $z > 0$, in (5.30), another way to obtain $E[\mathcal{I}]$ is

$$E[\mathcal{I}] = \int_{z=0}^{\infty} (1 - F_{\mathcal{I}}(z))\, dz = \mu \int_{z=0}^{\infty} \int_{y=0}^{\infty} \overline{A}(y+z) e^{-\gamma y} dy dz, \tag{5.31}$$

which agrees with formula (5.29) in Section 5.31. (Note: it is well known that $E[\mathcal{I}] = \int_{z=0}^{\infty} (1 - F_{\mathcal{I}}(z))\, dz = \int_{z=0}^{\infty} x f_{\mathcal{I}}(z) dz$, integrating by parts.)

Remark 5.3 *As a check on the above results, suppose that $G/M_\mu/1$ were an $M_\lambda/M_\mu/1$ queue. Then, in (5.11) $E[\mathcal{B}] = \frac{1}{\mu-\lambda}$, $\gamma = \mu - \lambda$, $C_\gamma = \lambda$, implying (using also formula (5.20)),*

$$K = \frac{\gamma \times C_\gamma}{\gamma + C_\gamma} = \lambda \left(1 - \frac{\lambda}{\mu}\right) = \lambda P_0,$$

which necessarily checks with "$f(x) = \lambda P_0 e^{-(\mu-\lambda)x}$, $x > 0$" in the $M_\lambda M_\mu/1$ queue (see formula (2.14) in Chapter 2).

5.1.8 PDF of Actual Wait in G/M/1 Queue

In a stable G/M/1 queue, generally the limiting pdf of the actual wait (arrival-point wait) is $\neq$ the limiting pdf of the virtual wait. These two pdfs are equal only when the arrival stream is Poisson as in the M/M/1 queue (where "PASTA holds"). We can utilize the extended age process $\{V(t)\}_{t\geq 0}$ in Sections 5.1.1—5.1.5 to determine the pdf of the actual wait, as follows.

For distinction we use the subscript "ι" (Greek "iota") to signify "actual wait". Let the limiting cdf of actual wait be $F_\iota(x) := P(\text{actual wait} \leq x)$, $x \geq 0$, and let the pdf be $f_\iota(x) = dF_\iota(x)/dx$, $x > 0$. Also let $P_{0,\iota} := P(\text{actual wait} = 0)$. Recall that in G/M/1 γ $(= 1/E[\mathcal{B}])$ is the solution of Equation (5.16) or Equation (5.17).

Proposition 5.4 *The stationary cdf $F_\iota(x)$ and stationary mixed pdf $\{P_{0,\iota}, f_\iota(x)\}_{0<x<\infty}$ of actual wait are given by:*

$$F_\iota(x) = 1 - \left(1 - \frac{\gamma}{\mu}\right) e^{-\gamma x}, \tag{5.32}$$

$$f_\iota(x) = K_\iota e^{-\gamma x}, x > 0, \tag{5.33}$$

$$P_{0,\iota} = 1 - \frac{K_\iota}{\gamma}, \tag{5.34}$$

where

$$K_\iota = \gamma \left(1 - \frac{\gamma}{\mu}\right). \tag{5.35}$$

Proof *The long-run "proportion of actual waits" greater than x is:*

$$1 - F_\iota(x) = \frac{\mu(1 - F(x)) - f(x)}{\mu\,(1 - F(0))}, \; x \geq 0, \tag{5.36}$$

where $F\,(x)$, $f\,(x)$ denote the stationary cdf and pdf respectively of the extended age process $\{V(t)\}_{t\geq 0}$ (same stationary cdf and pdf as for virtual wait "due to duality" between the two processes $\{V(t)\}_{t\geq 0}$ and $\{W(t)\}_{t\geq 0}$–see Figure 5.3 in Section 5.1.1).

Consider a further explanation of formula (5.36). In the numerator the term $\mu(1-F(x))$ is the long-run departure rate of customers that have been in the system greater than x time units (wait + service $> x$). That is because each sample-path peak corresponds to a customer's departure from the system. Each departure (sample-path peak) is followed immediately by an SP downward vertical jump $\stackrel{d}{=}$ an interarrival time, which "ends at the next actual wait greater than 0 only if the end point of the downward jump is at a sample-path trough "above 0". If the endpoint of the downward jump is below level 0 (starting a new idle period) then the next actual wait time is equal to 0 (Figure 5.2). The *term "$-f(x)$" subtracts off the rate at which SP jumps downcross level x, by the basic LC theorem for downcrossings of a level. Given that all jumps start above x, some end above x and same end below x. Also, $f(x)$ is equal to the level-x downcrossing rate (= the upcrossing rate of x by rate balance across level x—see derivation of Equation (5.7)). Therefore, the numerator in (5.36) is the rate at which "next actual waits before service, at sample-path troughs," are strictly greater than x. The denominator $\mu(1-F(0))$ in (5.36) is the total departure rate, which is equal to the long-run rate of vertical downward jumps, and must be the same as the long-run arrival rate.*

From (5.10) and (5.11) we obtain (5.32) and (5.33).

To obtain probability $P_{0,\iota}$ and the value of K_ι, set $x=0$ in (5.36). This yields, since $P_{0,\iota} = F_\iota(0)$,

$$1-P_{0,\iota} = 1-\frac{f(0)}{\mu(1-F(0))} = 1-\frac{K}{\mu\left(1-\left(1-\frac{K}{\gamma}\right)\right)}$$

$$= 1-\frac{K}{\mu\left(\frac{K}{\gamma}\right)} = 1-\frac{\gamma}{\mu}$$

$$P_{0,\iota} = \frac{\gamma}{\mu}. \tag{5.37}$$

We obtain the value of K_ι using the normalizing condition for $\{P_{0,\iota}, f_\iota(x)\}_{x>0}$,

$$P_{0,\iota} + \int_{y=0}^{\infty} K_\iota e^{-\gamma y} dy = 1,$$

$$\frac{\gamma}{\mu} + \frac{K_\iota}{\gamma} = 1,$$

yielding formulas (5.35) and (5.34).

Proposition. 5.4 demonstrates that the form of the pdf of actual wait $f_\iota(x)$, is the same as the form of the pdf of the virtual wait $f(x)$, $x>0$. The term $e^{-\gamma x}$ is common to both pdfs. The form of $P_{0,\iota}$ is similar to the form of P_0 formula (5.14). However, "$K_\iota \neq K$, except when the arrival stream is Poisson".

Remark 5.5 *$P_{0,\iota}$ and $f_\iota(x)$ in (5.34) and (5.33) agree with formulas obtained by different techniques (e.g., pp. 259–263 in [36]); the constant γ above, is equal to $\mu(1-r_0)$ in that book, where r_0 is the solution for z in the equation:*

$$z = A^*(\mu(1-z)),\ z\in(0,1)\prime,$$

5.1.9 Stability Condition for G/M/1 Queue

Stability occurs iff the solution of Equation (5.16) for γ is positive and finite, i.e., iff the "stationary" pdf $f(x) = Ke^{-\gamma x}$ of $\{W(t)\}_{t\geq 0}$ and pdf $f_\iota(x) = K_\iota e^{-\gamma x}$ at arrival points, exist. These pdfs exist provided γ is positive and finite, in which case K and K_ι are also positive and finite by (5.12) in Section 5.1.5 and (5.35) in Section 5.1.8, respectively.

Denote the $E\left[\textit{inter-arrival time}\right]$ by $\frac{1}{a}$. The $E\left[\textit{service time}\right]$ is equal to $1/\mu$.

Proposition 5.6 *The G/M/1 queue is stable if and only if $a < \mu$. Equivalently, iff $\frac{1}{\mu} < \frac{1}{a}$ iff $E\left[\textit{service time}\right] < E\left[\textit{inter-arrival time}\right]$.*

Proof *The G/M/1 queue is stable iff the expected busy period is finite, i.e., $0 < E\left[\mathcal{B}\right] < \infty$; since $E\left[\mathcal{B}\right] = 1/\gamma$ (see (5.13) in Section 5.1.5). Thus*

$$0 < E\left[\mathcal{B}\right] < \infty \iff 0 < \frac{1}{\gamma} < \infty \iff 0 < \gamma < \infty,$$

and $0 < e^{-\gamma y} < 1$ for all $y > 0$. If in addition γ satisfies Equation (5.16) in Section 5.1.5 then

$$\begin{aligned} \tfrac{1}{\mu} = \int_{y=0}^{\infty} \overline{A}(y)e^{-\gamma y}dy < \int_{y=0}^{\infty} \overline{A}(y)dy = \tfrac{1}{\boldsymbol{a}} \\ \Longrightarrow \boldsymbol{a} < \mu. \end{aligned} \tag{5.38}$$

Hence $a < \mu$ is a "necessary" condition for stability.

From (5.38)

$$\frac{1}{\mu} < \frac{1}{\boldsymbol{a}} \equiv \int_{y=0}^{\infty} \overline{A}(y)dy. \tag{5.39}$$

In order to prove that $a < \mu$ is a "sufficient" condition for stability, we construct a function of γ, $\phi(\gamma) := \int_{y=0}^{\infty} \overline{A}(y)e^{-\gamma y}dy$, $0 < \gamma < \infty$, with $\phi(\gamma) > 0$, $\lim_{\gamma\downarrow 0} \phi(\gamma) = \frac{1}{\boldsymbol{a}}$, $\lim_{\gamma\to\infty} \phi(\gamma) = 0$, $\phi'(\gamma) = -\gamma\phi(\gamma) < 0$, $\phi''(\gamma) = \gamma^2\phi(\gamma) > 0$. Thus $\phi(\gamma)$ is continuous, convex and strictly monotone decreasing on $(0,\infty)$. Consequently $\phi(\gamma)$ assumes each value in its range $\left(0, \frac{1}{\boldsymbol{a}}\right)$. For a given μ such that $\frac{1}{\mu} \in (0, \frac{1}{\boldsymbol{a}})$, there is a unique value $\gamma \in (0,\infty)$ such that $\phi(\gamma) = \frac{1}{\mu}$. Hence for each such μ there exists exactly one finite root $\gamma > 0$ of equation (5.16) such that $\frac{1}{\mu} < \frac{1}{\boldsymbol{a}}$, which implies $a < \mu$ is a "sufficient" condition for stability.

In conclusion, $a < \mu$ is a necessary and sufficient condition for stability.

5.1.10 Stationary Probability Distribution of System Time in G/M/1 Queue

Let $W_{q,\iota}$ denote the actual wait before service, S denote the service time, and σ denote the the system time of a customer, in a stable G/M/1 queue. Then

$\sigma = W_{q,\iota} + S$. The cdf of $W_{q,\iota}$ is $P(W_{q,\iota} \leq x) = F_\iota(x)$, $x \geq 0$, with pdf $f_\iota(x)$, $x > 0$. Also $P(W_{q,\iota} = 0) = F_\iota(0) = P_{0,\iota}$ (see Proposition 5.4 in Section 5.1.9). Let $F_\sigma(x)$, $x \geq 0$, and $f_\sigma(x) = \frac{d}{dx}F_\sigma(x)$, $x > 0$ (wherever the derivative exists), denote the stationary cdf and pdf of σ, respectively. For the standard G/M/1 queue, S and $W_{q,\iota}$ are independent (e.g., S does not depend on $W_{q,\iota}$)

From Proposition 5.4, the cdf of σ is the convolution (i.e., cdf of "sum" of r.v.s.)

$$\begin{aligned} F_\sigma(x) &= P_{0,\iota} \times P(S \leq x) + \int_{y=0}^{x} P(S \leq x-y) f_\iota(y)dy \\ &= \tfrac{\gamma}{\mu}(1-e^{-\mu x}) + \int_{y=0}^{x} \left(1-e^{-\mu(x-y)}\right)\gamma\left(1-\tfrac{\gamma}{\mu}\right)e^{-\gamma y}dy. \end{aligned} \tag{5.40}$$

The last integral in (5.40) is equal to

$$\frac{1}{\mu}\left[\mu e^{(\mu+\gamma)x} - \gamma e^{(\mu+\gamma)x} + \gamma e^{\gamma x} - \mu e^{\mu x}\right]e^{-(\mu+\gamma)x}. \tag{5.41}$$

Summing (5.41) and $\frac{\gamma}{\mu}(1-e^{-\mu x})$ simplifies (5.40) to

$$F_\sigma(x) = 1 - e^{-\gamma x}, x \geq 0, \tag{5.42}$$

with pdf $f_\sigma(x) = \frac{dF_\sigma(x)}{dx}$, namely

$$f_\sigma(x) = \gamma e^{-\gamma x}, x > 0. \tag{5.43}$$

Remark 5.7 *The expressions for $F_\sigma(x)$ and $f_\sigma(x)$ in (5.42) and (5.43) for the G/M/1 queue are analogous to the same formulas for the standard $M_\lambda/M_\mu/1$ queue given in formula (2.19), where $\gamma = \mu - \lambda$. In the term $e^{-\gamma x}$ the coefficient of x is $-\gamma = -\frac{1}{E[\mathcal{B}]}$ in both G/M/1 and M/M/1 (where $\mathcal{B}$:= busy period).*

5.1.11 Arrival-point Pmf of Number in System in Standard G/M/1 Queue

This section derives the stationary arrival-point pmf of the number of units in the system. Let N_ι denote the number in the system just before an arrival instant, and let $P_{n,\iota} := P(N_\iota = n)$, $n = 0, 1, ...$ From formulas (5.35) and (5.34),

$$P_{0,\iota} = \frac{\gamma}{\mu}. \tag{5.44}$$

Let "$\boldsymbol{d}_n$" be the stationary "probability of n remaining in the system just after a departure". Then $P_{n,\iota} = \boldsymbol{d}_n$, $n = 0, 1, ...$ (see pp. 500–502 in [52]) (2010). Let $A^{(n)}(y)$ be the "cdf of the n-fold convolution" of the interarrival time evaluated at $y > 0$. (For "convolution" see pp. 56–58 in [52] (2010), or p. 178 in [37] (1995)).

Proposition 5.8 *In the standard G/M/1 queue, for $n = 1, 2, \ldots,$*

$$P_{n,\iota} = \boldsymbol{d}_n = \int_{y=0}^{\infty} \left[A^{(n)}(y) - A^{(n+1)}(y)\right] \times f_\sigma(y)dy$$

$$= \gamma \int_{y=0}^{\infty} \left[A^{(n)}(y) - A^{(n+1)}(y)\right] e^{-\gamma y}dy. \tag{5.45}$$

Proof *Let $N_{\boldsymbol{A}}(t)$ denote the number of arrivals in $(0,t)$ and let S_n denote the time of the nth arrival. A very useful basic renewal equivalence relation is (see, p. 423, Section 7.2 in [52] (2010); and pp. 167–168 in [40] (1975):*

$$N_{\boldsymbol{A}}(t) \geq n \iff \mathcal{S}_n \leq t. \tag{5.46}$$

From (5.46),

$$P(N_{\boldsymbol{A}}(t) = n) = P(N_{\boldsymbol{A}}(t) \geq n) - P(N_{\boldsymbol{A}}(t) \geq n+1)$$

$$= P(\mathcal{S}_n \leq t) - P(\mathcal{S}_{n+1} \leq t) = A^{(n)}(t) - A^{(n+1)}(t), t > 0$$

Also, $\boldsymbol{d}_n(\boldsymbol{\sigma}) := P(n$ arrivals during a system time $\sigma)$.

Note that the arrival stream is statistically independent of the system time σ of a customer. That is, the parameters of the arrival stream "A" are independent of the state of of the system. In contrast, "A" does statistically depend on the state of the system in the publication [15] (1988).

$$\begin{aligned}
\boldsymbol{d}_n(\boldsymbol{\sigma}) &= \int_{y=0}^{\infty} P(N_{\boldsymbol{A}}(y) = n|\sigma = y)f_\sigma(y)dy \\
&= \int_{y=0}^{\infty} P(N_{\boldsymbol{A}}(y) = n)f_\sigma(y)dy \\
&= \int_{y=0}^{\infty} \left[A^{(n)}(y) - A^{(n+1)}(y)\right] f_\sigma(y)dy,
\end{aligned}$$

and $P_{n,\iota} = \boldsymbol{d}_n(\boldsymbol{\sigma})$. Substituting from formula (5.43) gives (5.45).

5.1.11.1 Compact Expression for Pmf of Number in the System in G/M/1 Queue

Proposition 5.8 leads to a compact expression for $P_{n,\iota}$, $n = 0, 1, \ldots$. Integration by parts immediately below gives

$$\int_{y=0}^{\infty} A^{(n)}(y)e^{-\gamma y}dy = \frac{1}{\gamma}\int_{y=0}^{\infty} a^{(n)}(y)e^{-\gamma y}dy = \frac{A^{n*}(\gamma)}{\gamma},$$

where $a^{(n)}(y)$ is the pdf of the n-fold convolution of inter-arrival times. Thus (5.45) becomes

$$P_{n,\iota} = A^{n*}(\gamma) - A^{(n+1)*}(\gamma), n = 1, 2, \ldots, \tag{5.47}$$

where "$A^{n*}(\gamma)$" := Laplace transform of A^n. (For Laplace transforms, see Section C1, p. 455ff in [36](2008), or a calculus text.). From Laplace-transform theory and (5.17)

$$A^{n*}(\gamma) = (A^*(\gamma))^n = \left(1 - \frac{\gamma}{\mu}\right)^n, n = 1, 2, \dots .$$

Substituting into (5.47) yields

$$P_{n,\iota} = \left(1 - \frac{\gamma}{\mu}\right)^n - \left(1 - \frac{\gamma}{\mu}\right)^{n+1}$$

$$= \frac{\gamma}{\mu}\left(1 - \frac{\gamma}{\mu}\right)^n$$

$$= P_{0,\iota}\left(1 - P_{0,\iota}\right)^n, n = 0, 1, ..., \tag{5.48}$$

using (5.44). Formula (5.48) is analogous to the result for P_n in terms of P_0 in the M/M/1 queue, given in formulas (2.1) and (2.2) in Chapter 2.

As a caveat to Proposition 5.8, the probabilities of n in the system at an arbitrary time point are not equal to $P_{n,\iota}, n = 0, 1, 2, ...$ Equality holds only if the arrival stream is Poisson (see Proposition 8.2, p. 502 in [52] (2010); or in [59]) (1982).

5.1.12 A Necessary Condition That the Above-obtained Results for G/M/1 Must Satisfy Results for the Mλ/Mμ/1 Queue

We specialize the foregoing stationary results for the G/M$_\mu$/1 queue to an M$_\lambda$/M$_\mu$/1 queue, by replacing the general renewal arrivals by Poisson arrivals with rate λ. If we get the same results as for M$_\lambda$/M$_\mu$/1 then we have a necessary condition that the above-obtained results for G/M$_\mu$/1 queue must satisfy.

5.1.12.1 Virtual Wait in G/M/1 when "G" ≡ "M"

In equation (5.16) in Section 5.1.5 for γ, "with $\overline{A}(x) := e^{-\lambda x}$, $x \geq 0$", the solution becomes

$$\gamma = \frac{1}{E\left[\mathcal{B}\right]} = \mu - \lambda.$$

For all $x \geq 0$, the sojourn above level x $a_x \stackrel{d}{=}$ busy period $\mathcal{B}$ independent of x, due to exponential service times and the memoryless property of $\text{Exp}_{(.)}$. By formula (5.12)

$$P_0 = 1 - KE(\mathcal{B}) = 1 - K, \tag{5.49}$$

$$E\left[a_x\right] = E\left[\mathcal{B}\right] = \frac{1}{\gamma}, x \geq 0. \tag{5.50}$$

$E[\mathcal{B}] = \frac{1}{\mu-\lambda}$ where $\mathcal{B}$ is the busy period in $M_\lambda/M_\mu/1$. From formula (5.29) we get $C_\gamma = \lambda$. The formulas in Section 5.1.5 yield

$$P_0 = \tfrac{\gamma}{\gamma+C_\gamma} = \tfrac{\mu-\lambda}{\mu-\lambda+\lambda} = 1 - \tfrac{\lambda}{\mu},$$

$$K = \tfrac{\gamma \cdot C_\gamma}{\gamma+C_\gamma} = \lambda(1 - \tfrac{\lambda}{\mu}) = \lambda P_0,$$

$$f(x) = Ke^{-\gamma x} = \lambda P_0 e^{-(\mu-\lambda)x}, x > 0, \tag{5.51}$$

$$E[\mathcal{B}] = \tfrac{1}{\gamma} = \tfrac{1}{\mu-\lambda},$$

$$E[\mathcal{I}] = \tfrac{1}{C_\gamma} = \tfrac{1}{\lambda}.$$

All values in (5.51) check with the stationary virtual-wait quantities for the standard M/M/1 queue.

For $x < 0$, the stationary pdf of the extended age process $\{W(t)\}_{t\geq 0}$ is denoted by "$h(x), -\infty < x < 0$" where

$$h(x) = \mu \int_{y=x}^{\infty} e^{-\lambda(y-x)} K e^{-(\mu-\lambda)y} dy = Ke^{\lambda x}, x < 0,$$

so that $P_0 = \int_{x=-\infty}^{0} h(x)dx = 1 - \frac{\lambda}{\mu}$, which agrees with (5.51).

5.1.13 Actual Wait and Pmf of Actual Number in System When "G" in G/M/1 Is a Poisson Arrival Stream

We obtain a necessary condition for the correctness of relevant results in Sections 5.1.8–5.1.12.1. For the actual wait in G/M/1 with "G" replaced by Poisson arrivals,

$$\gamma = \mu - \lambda,$$

$$P_{0,\iota} = \gamma/\mu = 1 - \lambda/\mu,$$

and

$$K_\iota = \gamma\left(1 - \frac{\gamma}{\mu}\right) = (\mu - \lambda)\,(1 - \frac{\mu - \lambda}{\mu})$$

$$K_\iota = \lambda \times \frac{\lambda - \mu}{\mu} = \lambda P_{0,\iota},$$

giving

$$f_\iota(x) = \lambda P_{0,\iota} e^{-(\mu-\lambda)x}, x > 0.$$

Due to the assumed Poisson arrivals, the actual-wait G/M/1 results agree with the forms of P_0 and $f(x), x > 0$ for the virtual wait, and number in system as they must, in $M_\lambda/M_\mu/1$, a necessary condition (see (2.14) in Chapter 2), where the Poisson arrival stream implies:

$$P_{0,\iota} = P_0 = 1 - \frac{\lambda}{\mu},\ f_\iota(x) = f(x), x > 0,$$

$$P_{n,\iota} = P_n = \left(\frac{\lambda}{\mu}\right)^n P_0,$$

in agreement with PASTA Section (2.1.2).

5.1.14 Sojourn of Extended Age Process Above or Below a Level in a Standard G/M/1 Queue

Focus on the extended-age process "$\{Vt)\}_{t\geq 0}$" . We determine the expected sojourn above x ($E[a_x]$), and on the expected sojourn below x "$E[b_x]$" at the state-space level x (see Figure 5.2 in Section 5.1.1).

5.1.14.1 Inter-upcrossing Time "u_x" of State-space Level x

We first define "u_x" to be the "inter-upcrossing time" of level x (i.e., time between two successive upcrossings of level x). Consider two cases separately: Case1: level $x \geq 0$; and Case 2: level $x < 0$. Note that the SP motion above $x = 0$ differs from the SP motion below level $x = 0$ (see Figure 5.2).

5.1.14.2 Case 1: Level $x \geq 0$

For $x \geq 0$, upcrossings of x are regenerative points, since both the service times and the remaining service times after upcrossing x are $\stackrel{d}{=}$ Exp_μ, due to the memoryless property of the service time. By the elementary renewal theorem in Section 1.8.3 in Chapter 1 and the basic level crossing theorem,

$$E[u_x] = \frac{1}{\frac{\lim_{t\to\infty} \mathcal{U}_t(x)}{t}} = \frac{1}{f(x)} = \frac{1}{Ke^{-\gamma x}} = \frac{e^{\gamma x}}{K}, x \geq 0, \tag{5.52}$$

where γ and K are given in formulas (5.16) and (5.26) respectively. (To compute K, we may use C_γ given in formula (5.29).)

For $x \geq 0$, $a_x \stackrel{d}{=}$ busy period $\mathcal{B}$ "independent of x", due to exponential service times and the memoryless property of $\text{Exp}_{(.)}$. By formula (5.12)

$$E[a_x] = E[\mathcal{B}] = \frac{1}{\gamma}, x \geq 0. \tag{5.53}$$

Case 2: Level $x < 0$

If $x < 0$ then $-x$ is the "time until the next customer arrives". A new customer arrives when a sample path of extended age $\{V(t)\}_{t\geq 0}$ hits level 0 from below with slope +1. Upcrossings of x are regenerative points, since the remaining time $-x$ is followed immediately by a service time starting at level $0 \stackrel{d}{=} \text{Exp}_\mu$. The elementary renewal theorem and the basic LC theorem for upcrossings and formula (5.18) together, give E [inter-upcrossing time] denoted by

$$E[u_x] = \frac{1}{\lim_{t\to\infty} \frac{\mathcal{U}_t(x)}{t}} = \frac{1}{h(x)} = \frac{1}{\mu K \int_{y=0}^{\infty} \overline{A}(y-x)e^{-\gamma y}dy}, x < 0. \quad (5.54)$$

Let $a_x :=$ sojourn of $\{V(t)\}_{t\geq 0}$ above level $x \geq 0$. Since $u_x = a_x + b_x$,

$$\begin{aligned} E[a_x] &= E[u_x] - E[b_x] = \frac{1}{h(x)} - E[b_x] \\ &= \frac{1}{\mu K \int_{y=0}^{\infty} \overline{A}(y-x)e^{-\gamma y}dy} \\ &- \int_{z=0}^{\infty} \left[\frac{\int_{y=0}^{\infty} \overline{A}(y-x+z)Ke^{-\gamma y}dy}{\int_{y=0}^{\infty} \overline{A}(y-x)Ke^{-\gamma y}dy} \right] dz, x < 0, \end{aligned} \quad (5.55)$$

where $b_x :=$ sojourn below level x, and $E[b_x]$, $x < 0$, is derived in Case 2–Level $x < 0$. in Proposition 5.9 immediately below.

5.1.14.3 Sojourn b_x below Level x in Standard G/M/1 Queue

We now derive the expected SP sojourn below level x.

5.1.14.4 Case 1–Level $x \geq 0$

Since $u_x = a_x + b_x$, applying formula (5.52) for $E[u_x]$ and (5.53) for $E[a_x]$, gives

$$E[b_x] = E[u_x] - E[a_x] = \frac{e^{\gamma x}}{K} - \frac{1}{\gamma}, x \geq 0. \quad (5.56)$$

5.1.14.5 Case 2–Level $x < 0$

For $x < 0$, $E[b_x]$ is given by the following proposition.

Proposition 5.9 *The expected sojourn time of* $\{V(t)\}_{t\geq 0}$ *below level* x *is*

$$E[b_x] = \int_{z=0}^{\infty} \left[\frac{\int_{y=0}^{\infty} \overline{A}(y-x+z)e^{-\gamma y}dy}{\int_{y=0}^{\infty} \overline{A}(y-x)e^{-\gamma y}dy} \right] dz, x < 0. \quad (5.57)$$

Proof *We present two proofs, which complement each other.*

(1). Consider an SP vertical downward jump that ends below level $x < 0$ *(all vertical downward jumps begin above level* 0*). Let* $r_x :=$ *excess of a jump ending below level* x*. Since a sample path of* $\{V(t)\}_{t\geq 0}$ *increases steadily at rate* +1 *when* $V(t) < 0$, $b_x \stackrel{d}{=} r_x$, *and* $E[b_x] = E[r_x]$*. Using Bayes' rule (see*

Section 1.5.2 in Chapter 1) condition on the event "a jump downcrosses level x, yielding:

$$\begin{aligned}
P(r_x > z) &= P(a\ jump\ downcrosses\ level\ x - z | the\ jump\ downcrosses\ level\ x) \\
&= \frac{P(a\ jump\ downcrosses\ level\ x - z\ and\ downcrosses\ level\ x)}{P(the\ jump\ downcrosses\ level\ x)} \\
&= \frac{P(a\ jump\ downcrosses\ level\ x - z)}{P(the\ jump\ downcrosses\ level\ x)} \\
&= \frac{\int_{y=0}^{\infty} \overline{A}(y - x + z) K e^{-\gamma y} dy}{\int_{y=0}^{\infty} \overline{A}(y - x) K e^{-\gamma y} dy}, x < 0, z > 0.
\end{aligned}$$

Thus

$$\begin{aligned}
E[b_x] = E[r_x] &= \int_{z=0}^{\infty} P(r_x > z) dz \\
&= \int_{z=0}^{\infty} \left[\frac{\int_{y=0}^{\infty} \overline{A}(y - x + z) e^{-\gamma y} dy}{\int_{y=0}^{\infty} \overline{A}(y - x) e^{-\gamma y} dy} \right] dz, x < 0.
\end{aligned}$$

(2). An alternative approach is to denote the cdf of the excess r_x by $F_{r_x}(z)$, $z > 0$. Using rate balance gives

$$(1 - F_{r_x}(z)) \cdot h(x) = \mu \int_{y=0}^{\infty} \overline{A}(y - x + z) K e^{-\gamma y} dy,$$

where the left and right sides are two different expressions for the SP down-crossing rate of level $x - z$. (A similar argument is used in the derivation of $E[a_x]$ in Variant 3 of the M/G/1 queue with bounded virtual wait, given in formula (2.82) in Chapter 3.) Since

$$h(x) = \mu \int_{y=0}^{\infty} \overline{A}(y - x) K e^{-\gamma y} dy,$$

and $b_x = r_x$,

$$1 - F_{r_x}(z) = \frac{\mu \int_{y=0}^{\infty} \overline{A}(y - x + z) K e^{-\gamma y} dy}{h(x)} = \frac{\mu \int_{y=0}^{\infty} \overline{A}(y - x + z)\ \not{K} e^{-\gamma y} dy}{\mu \int_{y=0}^{\infty} \overline{A}(y - x)\ \not{K} e^{-\gamma y} dy},$$

$$E[b_x] = \int_{z=0}^{\infty} (1 - F_{r_x}(z))\, dz$$

$$= \int_{z=0}^{\infty} \left[\frac{\int_{y=0}^{\infty} \overline{A}(y - x + z) e^{-\gamma y} dy}{\int_{y=0}^{\infty} \overline{A}(y - x) e^{-\gamma y} dy} \right] dz,\ x < 0, \tag{5.58}$$

which is in agreement with Proof (1) above.

Example 5.1.1 *Assume G/M/1 is an $M_\lambda/M_\mu/1$ queue, so that the arrival stream is Poisson. From (5.57)*

$$E[b_x] = E[r_x] = \int_{z=0}^{\infty} \left[\frac{\int_{y=0}^{\infty} e^{-\lambda(y-x+z)} e^{-\gamma y} dy}{\int_{y=0}^{\infty} e^{-\lambda(y-x)} e^{-\gamma y} dy} \right] dz$$

$$= \int_{z=0}^{\infty} e^{-\lambda z} dz = \frac{1}{\lambda},\ x < 0,$$

by factoring out $e^{\lambda z}$ in the numerator of the integrand of $\int_{z=0}^{\infty} \cdots dz$.

5.1.15 Events During a Sojourn a_x above a Level in G/M/1 Queue

5.1.15.1 Number of Completed System Times $N^\sigma_{a_x}$ in a_x in G/M/1 Queue

See the discussion of system time σ in the extended age process $\{V(t)\}_{t\geq 0}$ in Section 5.1.10. Consequently, σ = wait before service + service time = $W_{q,\iota} + S$. For each customer, σ completes at the end-of-service instant (at departure instant from the system, i.e., at the corresponding "sample-path peak" of $\{V(t)\}_{t\geq 0}$). The $W_{q,\iota}$s complete "just before" the start of service instants, i.e., at sample-path troughs. (See Figures 5.2 and 5.3.)

Level $x \geq 0$

Let $N^\sigma_{a_x}$:= number of service completions (departures) during sojourn a_x, $x \geq 0$, including the departure that that ends a_x (i.e., the length of a "run of system times $> x$", or number of sample-path peaks $> x$ in a_x). Let S_i and T_i, $i = 1, 2,...$, denote the service times and immediately following inter-arrival times, respectively, counting from the instant a sample path of $\{V(t)\}_{t\geq 0}$ upcrosses level x (starts a_x). If $x = 0$, then S_1 is a full service time. If $x > 0$, then S_1 is the remaining service time measured from level x, and $S_1 \stackrel{d}{=} \text{Exp}_\mu$, due to the memoryless property. Then (see Figure 5.2)

$$N^\sigma_{a_x} = \min\left\{ n | \sum_{i=1}^{n} (S_i - T_i) \leq 0 \right\}, x \geq 0. \tag{5.59}$$

$N^\sigma_{a_x}$ is a stopping time for the sequence $\{S_i - T_i\}_{1=1,2,\ldots}$ (e.g., pp. 678–679 in [52]) for both sequences $\{S_i - T_i\}_{i=1,2,\ldots}$ and $\{S_i\}_{i=1,2,\ldots}$. Since $a_x = \sum_{i=1}^{N^\sigma_{a_x}} S_i$ and $a_x \stackrel{d}{=} \mathcal{B}$ for all $x \geq 0$, p. 679 in [52])

$$E[a_x] = E[N^\sigma_{a_x}] \times E[S_i],$$

$$E[N^\sigma_{a_x}] = \frac{E[a_x]}{E[S_i]} = \frac{E[\mathcal{B}]}{E[S]}, \tag{5.60}$$

independent of x. Substituting from (5.53) into (5.60), gives

$$E\left[N^{\sigma}_{a_x}\right] = \frac{\frac{1}{\gamma}}{\frac{1}{\mu}} = \frac{\mu}{\gamma} = \frac{1}{P_{0,\iota}}, x \geq 0, \tag{5.61}$$

where $P_{0,\iota}$ is given in formula (5.44). Formula (5.61) is interesting in that it is similar to formula (2.39) in Chapter 2. A similar formula holds for the standard M/G/1 queue (see formula (3.88) on p. 89 in Chapter 3 in [22] (2017)).

From (5.61) $E\left[N^{\sigma}_{a_x}\right] > 1$ since $\mu > \gamma$. This agrees with intuition, because there must be at least one departure (sample-path peak) counted in a_x (e.g., the "last departure in a_x").

5.1.15.2 Number Served in a Busy Period $\mathcal{B}$ in G/M/1 Queue

Let $N^{\sigma}_{\mathcal{B}}$:= number of system-time completions in $\mathcal{B}$ (:= number of "sample-path peaks", i.e., number of "customers served", in $\mathcal{B}$). Since $\mathcal{B} = a_0$, $N^{\sigma}_{\mathcal{B}} = N^{\sigma}_{a_0}$, and from (5.61), $E\left[N^{\sigma}_{a_x}\right] = \frac{\mu}{\gamma}$, independent of x. From (5.37), we obtain a notable formula for G/M/1,

$$E\left[N^{\sigma}_{\mathcal{B}}\right] = \frac{\mu}{\gamma} = \frac{1}{P_{0,\iota}}. \tag{5.62}$$

5.1.16 Number of Completed Waiting Times (Service Starts) $N^{w}_{a_x}$ in a Sojourn a_x in G/M/1 Queue

Consider the extended age process $\{V(t)\}_{t\geq 0}$. We assume $x \geq 0$, since service starts can occur only when $V(t) \geq 0$. Let $N^{w}_{a_x}$ denote the number of customers that start service during a_x, $x \geq 0$. Then $N^{w}_{a_x}$ is the number of customers that wait $> x$ time units (strictly greater than x) before starting service during a_x (excluding the upcrossing that initiates a_x). Alternatively viewed, $N^{w}_{a_x}$:= number of sample-path "troughs" during a_x. Figure 5.2 shows that $N^{w}_{a_x} = N^{\sigma}_{a_x} - 1$, $x \geq 0$, since the count of service starts with waits $> x$ during a_x is one less than the total count of service completions in a_x (which includes the completion which ends a_x). Hence

$$E\left[N^{w}_{a_x}\right] = E\left[N^{\sigma}_{a_x}\right] - 1 = \frac{\mu}{\gamma} - 1 > 0, x \geq 0. \tag{5.63}$$

5.1.17 Events Occurring above Level x During a Busy Period $\mathcal{B}$ in G/M/1 Queue

A busy period $\mathcal{B}$ of the extended age $\{V(t)\}_{t\geq 0}$ starts with an SP upcrossing of level 0. It may contain several sojourns above level $x > 0$. For example, in Figure 5.2 in Section 5.1.1, the first three busy periods contain 2, 2, and 1 sojourn(s) above level x, respectively.

5.1.18 Expected Number of SP Sojourns above Level $x > 0$ During Busy Period $\mathcal{B}$ of Extended Age Process $\{V(t)\}_{t\geq 0}$ in G/M/1 Queue

Let $N_{a_x}^{soj}(\mathcal{B}) :=$ number of sojourns above level $x \geq 0$ during $\mathcal{B}$. Then $N_{a_x}^{soj}(\mathcal{B}) \in \{0, 1, 2, ...\}$. We first obtain $E\left[N_{a_x}^{\text{soj}}(\mathcal{B})\right]$.

Let $\mathcal{C}$ denote a busy cycle. Let $\mathcal{U}_{\mathcal{C}}(x)$ denote the number of SP upcrossings of level x during $\mathcal{C}$, all of which occur during the corresponding contained busy period $\mathcal{B}$ (not in idle period $\mathcal{I}$). There is a one-to-one correspondence between the starts of sojourns a_x and SP upcrossings of x, implying that $N_{a_x}^{soj}(\mathcal{B}) = \mathcal{U}_{\mathcal{C}}(x)$ (see Figure 5.2). Each busy cycle is a regenerative cycle of a regenerative process By the renewal reward theorem,

$$\frac{E\left[N_{a_x}^{soj}(\mathcal{B})\right]}{E[\mathcal{C}]} = \frac{E\left[\mathcal{U}_{\mathcal{C}}(x\right]}{E[\mathcal{C}]} = \lim_{t\to\infty} \frac{\mathcal{U}_t(x)}{t} = f(x) = Ke^{-\gamma x}, \ x \geq 0. \tag{5.64}$$

In formula (5.64), $\mathcal{U}_t(x)$ is the number of upcrossings in the time interval $(0, t)$ and the third equality is an expression of the basic LC theorem for SP upcrossings of level x. By the elementary renewal theorem

$$E\left[\mathcal{C}\right]) = \frac{1}{f(0)} = \frac{1}{K}.$$

From the first and last terms in (5.64)

$$E\left[N_{a_x}^{soj}(\mathcal{B})\right] = \frac{1}{K} \times Ke^{-\gamma x} = e^{-\gamma x}, x \geq 0. \tag{5.65}$$

A necessary condition when $x = 0$ in (5.65) is met, i.e., $E\left[N_{a_0}^{soj}(\mathcal{B})\right] = 1$, which means that $\mathcal{B}$ consists of exactly one sojourn a_0 above level 0.

5.1.19 Further Comparison of the G/M$_\mu$/1 Queue and the M/M$_\mu$/1 Queue

Consider the standard M/M/1 queue of Chapter 2 in the light of the G/M/1 results in Sections 5.1.15–5.1.17. Let "G" in G/M/1 := Exp$_\lambda$. From equation (5.62) in G/M/1,

$$\begin{aligned} E\left[N_{a_x}^{w}(\mathcal{B})\right] &= \left(\frac{\mu}{\gamma} - 1\right) e^{-\gamma x} = \left(\frac{\mu}{\mu - \lambda} - 1\right) e^{-\gamma x} \\ &= \left(\frac{1}{1 - \frac{\lambda}{\mu}} - 1\right) e^{-\gamma x} = \left(\frac{1}{P_0} - 1\right) e^{-\gamma x}, x \geq 0, \end{aligned} \tag{5.66}$$

where $\gamma = \mu - \lambda$, and $P_0 = 1 - \frac{\lambda}{\mu}$ in M/M/1. When $x = 0$ formula (5.66) reduces to

$$E\left[N_{a_0}^{w}(\mathcal{B})\right] = \frac{1}{P_0} - 1.$$

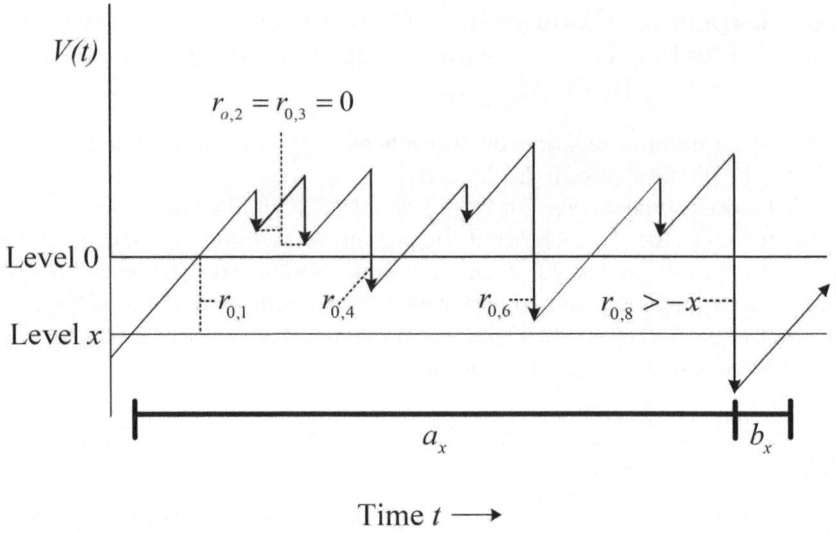

FIGURE 5.4
a_x and b_x in G/M/1 queue for $x < 0$. Illustrates vertical downeard jump "overshoots": $r_{0,1} := -x$; $r_{0,i} = 0$ if jump ends above 0; $0 < r_{0,i} < -x$ iff a_x continues; if $r_{0,i} > -x$ then a_x ends and b_x begins.

In M/M/1 (also in standard M/G/1),

$$E\,[\text{number served in a "busy period"}] := E\,[N_{\mathcal{B}}] = E\,[N_{\mathcal{B}}^{\sigma}]$$

$$= E[\text{number of customers served in a "busy cycle"}] = \frac{1}{P_0},$$

since zero cutomers are served in an idle period $\mathcal{I}$ (see formula (2.39) in Chapter 2). The customer initiating busy period $\mathcal{B}$ waits 0; all others served in $\mathcal{B}$ wait > 0. This explains why $E\left[N_{a_0}^{w}(\mathcal{B})\right] = E\,[N_{\mathcal{B}}^{\sigma}] - 1$.

In M/M/1 (and in M/G/1) the proportion of customers that wait > 0 in a busy period is

$$\frac{E\left[N_{a_0}^{w}(\mathcal{B})\right]}{E\,[N_{\mathcal{B}}^{\sigma}]} = \frac{\frac{1}{P_0} - 1}{\frac{1}{P_0}} = 1 - P_0 = \frac{\lambda}{\mu} = \rho,$$

where "ρ" is the traffic intensity. This agrees with the result for G/M/1 given in formula (5.32).

5.1.19.1 Related Results for $M_\lambda/M_\mu/1$

Similarly as in the analyses above for G/M/1, we obtain the following results for $M_\lambda/M_\mu/1$ (see Figure 5.4). The expected number of system times completed in a sojourn a_x is

$$E\left[N^{\sigma}_{a_x}\right] = E\left[N^{\sigma}_{\mathcal{B}}\right] = \frac{1}{P_0} = \frac{\mu}{\mu - \lambda}, x \geq 0, \tag{5.67}$$

since $N^{\sigma}_{a_x} \overset{d}{=} N_{\mathcal{B}}$ in both M/M/1 and G/M/1, for all $x \geq 0$. Also, $E\left[N^{\sigma}_{a_x}\right] > 1$ since $\mu > \mu - \lambda$ (i.e., for stability $0 < \lambda < \mu$).

E [number that wait $> x$ during a_x] is

$$E\left[N^{w}_{a_x}\right] = \frac{\mu}{\mu - \lambda} - 1 = \frac{1}{P_0} - 1 = \frac{\lambda}{\mu - \lambda} > 0. \tag{5.68}$$

E [number of sojourns above level x in $\mathcal{B}$] is, using (5.65),

$$E\left[N^{soj}_{a_x}(\mathcal{B})\right] = e^{-\gamma x} = e^{-(\mu - \lambda)x}, x \geq 0; \tag{5.69}$$

if $x = 0$ then

$$E\left[N^{soj}_{a_x}(\mathcal{B})\right] = e^0 = E\left[N^{soj}_{a_0}(\mathcal{B})\right] = 1.$$

The busy perod $\mathcal{B}$ consists of exactly one sojourn above level 0. $\mathcal{B}$ may contain a random number of sojourns above an arbitrary fixed level $x > 0$.

The number of system times (service completions, i.e., sample-path peaks) above level x in $\mathcal{B}$ is

$$N^{\sigma}_{a_x}(\mathcal{B}) = N^{\sigma}_{a_x}(\mathcal{C}) = \sum_{i=1}^{N^{soj}_{a_x}(\mathcal{C})} N^{\sigma}_{a_x i},$$

where $N^{\sigma}_{a_x i} = N^{\sigma}_{a_x}$ independent of x. By (5.67) and (5.69),

$$\begin{aligned} E\left[N^{\sigma}_{a_x}(\mathcal{B})\right] &= E\left[N^{\sigma}_{a_x}\right] = E\left[N^{\sigma}_{\mathcal{B}}\right] \times E\left[N^{soj}_{a_x}(\mathcal{C})\right] \\ &= \frac{\mu}{\mu - \lambda} \times e^{-(\mu - \lambda)x}, \geq 0. \end{aligned} \tag{5.70}$$

E[number of waiting times $> x$ in $\mathcal{B}$] is

$$\begin{aligned} E\left[N^{w}_{a_x}(\mathcal{B})\right] &= \left(\frac{\mu}{\mu - \lambda} - 1\right) \times e^{-(\mu - \lambda)x} \\ &= \frac{\lambda}{\mu - \lambda} \times e^{-(\mu - \lambda)x}, x \geq 0; \end{aligned} \tag{5.71}$$

if $x = 0$, then $E\left[N^{w}_{a_x}(\mathcal{B})\right] = \frac{\lambda}{\mu - \lambda}$ = expected number that wait > 0 in $\mathcal{B}$.

The "proportion' of customers that wait > 0 in $\mathcal{B}$ is

$$\frac{E\left[N^{w}_{a_0}(\mathcal{B})\right]}{E\left[N^{\sigma}_{\mathcal{B}}\right]} = \frac{\frac{\lambda}{\mu-\lambda}}{\frac{\mu}{\mu-\lambda}} = \frac{\lambda}{\mu} = 1 - P_0,$$

where $N_{\mathcal{B}}^{\sigma}$:= number served in $\mathcal{B}$. The explanation of the last formula is that the long-run proportion of customers that wait > 0 is $1 - P_0$ (the regeerative cycle C is a probabilistic replica of the entire time line $\{t|t > 0\}$. All arrivals take place during the $\mathcal{B}$ embedded in C).

Next, let $N_{b_x}^{\sigma}(B)$:= number of σs $\leq x$ in $\mathcal{B}$ (i.e., occur in a sojourn b_x below level x, where $x > 0$).

Proposition 5.10 *In M/M/1 E[number of system times $\leq x$ in a busy period $\mathcal{B}$] is*

$$E\left[N_{b_x}^{\sigma}(\mathcal{B})\right] = \frac{\mu}{\mu-\lambda} - \frac{\mu}{\mu-\lambda}e^{-(\mu-\lambda)x}$$
$$= \frac{\mu}{\mu-\lambda}(1 - e^{-(\mu-\lambda)x}), x \geq 0. \tag{5.72}$$

Proof *The total number of σs (sample-path peaks) corresponding to service completions in a busy period is, computed from (5.67),*

$$E\left[N_{b_x}^{\sigma}(\mathcal{B})\right] + E\left[N_{a_x}^{\sigma}(\mathcal{B})\right] = E\left[N_{\mathcal{B}}^{\sigma}\right] = \frac{\mu}{\mu-\lambda}.$$

Then (5.72) follows from (5.67) and (5.70).

Proposition 5.11 *For M/M/1 E[number of waiting times $\leq x$ in a busy period $\mathcal{B}$] is*

$$E[N_{b_x}^{w}(\mathcal{B})] = \frac{\mu}{\mu-\lambda} - \frac{\lambda}{\mu-\lambda}e^{-(\mu-\lambda)x}, x \geq 0. \tag{5.73}$$

Proof *In $\mathcal{B}$, the number of customers with waiting times $\leq x$ plus the number with waiting times $> x$, is equal to the number served in $\mathcal{B}$, namely $N_{\mathcal{B}}^{\sigma}$. By (5.67),*

$$E[N_{b_x}^{w}(\mathcal{B})] + E[N_{a_x}^{w}(\mathcal{B})] = E[N_{\mathcal{B}}^{\sigma}] = \frac{\mu}{\mu-\lambda}.$$

Thus, (5.73) follows from (5.68) and (5.71).

Remark 5.12 *In $M_\lambda/M_\mu/1$:*

If $x = 0$ then $E\left[N_{b_x}^{\sigma}(\mathcal{B})\right] = 0$.

If $x \to \infty$ then $E\left[N_{b_x}^{\sigma}(\mathcal{B})\right] \to \mu/(\mu-\lambda) = 1/P_0$.

If $x = 0$ then $E\left[N_{b_x}^{w}(\mathcal{B})\right] = 1$ (initiator of $\mathcal{B}$ waits 0).

If $x \to \infty$ then $E\left[N_{b_x}^{w}(\mathcal{B})\right] \to \frac{\mu}{\mu-\lambda} = 1/P_0$.

5.1.20 Boundedness of PDF of Wait $f(x)$, $x > 0$ in G/M/1 Queue

In G/M/1 let $S = \text{Exp}_\mu$; let the inter-arrival time have cdf $A(y)$, $y > 0$, with $A(0^+) = 0$. Assume the stationary pdf of wait $f(x)$, $x > 0$, exists.

The stationary pdf of the virtual wait, $f(x) = Ke^{-\gamma x}$, $x > 0$, is decreasing and convex on $(0, \infty)$. Thus

$$f(x) < K < \gamma < \mu, \ x > 0,$$

by (5.26), and by (5.17), in Section 5.1.5.

The pdf of the actual wait, $f_\iota(x) = K_\iota e^{-\gamma x}$, $x > 0$, has similar properties. Thus

$$f_\iota(x) < K_\iota = \gamma\left(1 - \frac{\gamma}{\mu}\right) < \gamma < \mu, \ x > 0,$$

where K_ι is given in formula (5.35) in Section 5.1.8.

Proposition 5.13 derives boundedness of $f(x)$ from "first principles" without drawing on the result $f(x) = Ke^{-\gamma x}$, $x > 0$. We include it for ideas that may be useful to obtain bounds on the pdf of wait in variants of G/M/1 (or random variables in other models), from basic LC considerations.

Proposition 5.13 *In G/M/1 with the foregoing assumptions*

$$f(x) < \mu\,(1 - F(x)) < \mu, x > 0. \tag{5.74}$$

Proof *An alternative form of the LC integral equation for G/M/1 (formula (5.8)) is*

$$f(x) = \mu(1 - F(x)) - \mu \int_{y=x}^{\infty} A(y - x) f(y) dy, x > 0. \tag{5.75}$$

The subtracted term in (5.75) is such that

$$\mu \int_{y=x}^{\infty} A(y - x) f(y) dy < \mu \int_{y=x}^{\infty} 1 \times f(y) dy = \mu\,(1 - F(x))\,, x > 0,$$

so that (5.74) follows.

Example 5.1.2 *The $M_\lambda/M_\mu/1$ is a special case of G/M//1 in which $\rho = \lambda/\mu < 1$ for stability, with $F(x) = 1 - \rho e^{-(\mu-\lambda)x}$, $f(x) = \lambda P_0 e^{-(\mu-\lambda)x}$, and $E[\mathcal{B}] = \frac{1}{\mu-\lambda}$. In the $M_\lambda/M_\mu/1$ queue, a little algebra shows that*

$$f(x) = (1 - F(x)) \times (\mu - \lambda) = \mu\left(1 - \frac{\lambda}{\mu}\right)(1 - F(x))$$

$$= \mu P_0\,(1 - F(x)) < \mu\,(1 - F(x))\,,$$

which satisfies (5.74), since $P_0 < 1$.

5.2 Zero-Waits Receive Special Service in G/M/1 Queue

Queues where "zero-wait customers receive special service" (aka "...exceptional service") have many real-world applications. This motivates deriving the stationary pdf $\{P_0, f(x)\}_{x>0}$ of wait in this G/M/1 queue using the "method of Lines and Pages" (for the method see Section 4.1.1 and Figure 4.2 in Section 4.2). Formula (2.61) in Section 2.5 in Chapter 2, derives $\{P_0, f(x)\}_{x>0}$ in the M/M/1 queue where zero-waits receive special service. We present an alternative derivation here, and will check that the two derivations give the same result, below.

We utilize the duality between the virtual wait and the extended age (see Section 5.1.2). Let $\{V(t), M(t)\}_{t\geq 0}$ denote the extended age process, where $V(t)$:= time in system of the customer who is in service at time t. In the "discrete auxiliary process", $M(t) = 0$ if a zero-wait customer (the initiator of a new busy period) is in service, or the system is empty, at time t; and $M(t)$ $= 1$ if a positive-wait customer is in service at time t (see Figure 5.5). Let S_0 := service time of zero-waits and S_1 := service time of positive-waits, where $S_i \stackrel{d}{=} \text{Exp}_{\mu_i}$, $i = 0, 1$. Assume the arrival stream is a general "renewal process" with inter-arrivals having cdf $A(\cdot)$, ccdf $\overline{A}(\cdot) = 1 - A(\cdot)$, and mean $1/a$. (In Section 5.2.2 we will assume the arrival process is Poisson with rate λ, and get

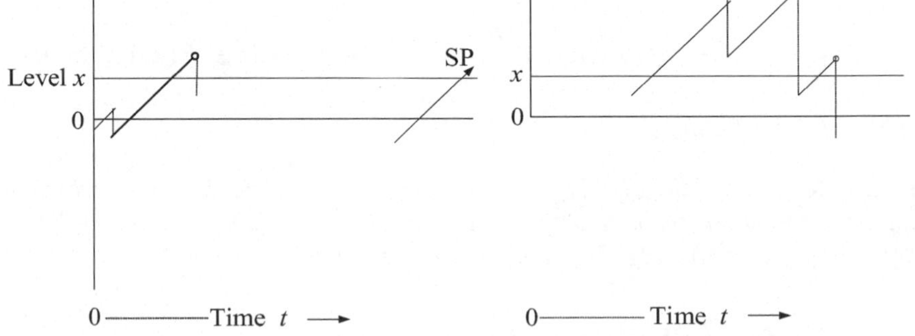

FIGURE 5.5
Sample path of $\{V(t), M(t)\}_{t\geq 0}$ in G/M/1 where zero-waits get special service. Pages have same time axis. Page 0 overlays page 1. The section of sample path on Page 1 fills in the gap in the segments of the sample path on Page 0.

the same stataionary pdf of wait as for M/M/1 in Section 2.5.3—a necessary condition for the correctness of the G/M/1 derivation using the method of Lines and Pages given here.

Let $f_0(x)$, $x > 0$, denote the stationary pdf of the extended age $V(t)$ as $t \to \infty$ when a "zero-wait customer ($i = 0$) is in service". Let $f_1(x)$, $x > 0$, denote the stationary pdf of the extended age $V(t)$ as $t \to \infty$ when a positive-wait customer ($i = 1$) is in service. Let $h(x)$, $x < 0$, denote the pdf of "remaining time until the next arrival occurs" (i.e., $-x$), for a zero-wait customer. The domains (and ranges) of $f_0(x)$, $h(x)$ and $f_1(x)$ are respectively: $\{(0, \infty), 0\}$, $\{(-\infty, 0), 0\}$, and $\{(0, \infty), 1\}$, in which the right-most numbers after the ";"s, are $M = 0$ and $M = 1$, denoting the Pages where they are in effect. Let $P_0 := P$(an arrival waits 0).

5.2.1 Integral Equations for $f_0(x)$, $f_1(x)$ and $h(x)$ in G/M/1 Queue

Figure 5.5 illustrates a sample path of $\{V(t), M(t)\}_{t\geq 0}$, which aids in deriving the integral equations.

Equation for $f_0(x), x > 0$

Consider state set $\{(x, \infty), 0\}_{x\geq 0}$. The SP rate into is $\lim_{t\to\infty} \mathcal{U}_t(x)/t = f_0(x)$; the SP rate out is $\lim_{t\to\infty} \mathcal{D}_t(x)/t = \mu_0 \int_x^\infty f_0(y)dy$ because the exponential rate μ_0 at which the service ends is independent of how long the customer has been in service (memoryless property of Exp_{μ_0}). The ends of service times at sample-path peaks of zero-waits, initiate downward jumps (size $\stackrel{d}{=} \text{Exp}_\lambda$) ending in $\{(-\infty, 0), 0\}$ or in $\{(0, \infty), 1\}$. Rate balance across level x on Page 0 gives

$$f_0(x) = \mu_0 \int_x^\infty f_0(y)d\dot{y}, x > 0. \tag{5.76}$$

Converting (5.76) to a differential equation

$$f_0'(x) + \mu_0 f_0(x) = 0,$$

and solving, gives the solution

$$f_0(x) = \alpha e^{-\mu_0 x}, x > 0, \tag{5.77}$$

where α is a constant to be determined.

Equation for $f_1(x), x > 0$

Consider state set $\{(x, \infty), 1\}_{x\geq 0}$. The rate-balance equation across level $x > 0$ on Page 1 is:

$$\text{entrance rate into } \{(x, \infty), 1\}_{x\geq 0} = \text{exit rate from } \{(x, \infty), 1\}_{x\geq 0}$$

$$f_1(x) + \mu_0 \int_{y=x}^\infty A(y - x) f_0(y)dy = \mu_1 \int_{y=x}^\infty \overline{A}(y - x) f_1(y)dy, x > 0.$$

Transposing terms and substituting from (5.77) yields

$$f_1(x) - \mu_1 \int_x^{\infty} \overline{A}(y-x) f_1(y) dy = -\mu_0 \boldsymbol{\alpha} \int_x^{\infty} A(y-x) e^{-\mu_0 y} dy, x > 0. \quad (5.78)$$

Equation for $h(x), x < 0$

Consider state-space set $\{(-\infty, x), 0\}_{-\infty < x \le 0}$. The rate-balance equation is

$$h(x) = \mu_0 \int_0^{\infty} \overline{A}(y-x) \boldsymbol{\alpha} e^{-\mu_0 y} dy + \mu_1 \int_0^{\infty} \overline{A}(y-x) f_1(y) dy, x < 0, \quad (5.79)$$

where the left side is the exit rate ($\lim_{t\to\infty} \mathcal{U}_t(x)/t$) and the right side is the entrance rate (the first term is "SP from" $\{(0,\infty),0\}$, the second term is "SP isfrom" $\{(0,\infty),1\}$).

Note that

$$P_0 = \int_{-\infty}^{0} h(x) dx. \quad (5.80)$$

The normalizing condition is

$$P_0 + \int_{x=0}^{\infty} f_0(x) dx + \int_{x=0}^{\infty} f_1(x) dx = 1. \quad (5.81)$$

We shall not solve equations (5.77)–(5.81) completely here "for a general $A(\cdot)$". Section 5.2.2 gives a full solution when the arrival rate is Poisson at rate λ, which serves as an example of the method of Lines and Pages for single-server queues, and a check on a necessary condition for the model in this Section.

5.2.2 M/M/1 as Special Case of G/M/1

We now relax the "G" in G/M/1 to be Exp_λ, to determine if we get the corresponding necessary M/M/1 result in formula (2.61) in Section 2.5 in Chapter 2 which is well-known and tested. This procedure confirms a necessary condition that the G/M/1 solution derived here is valid. (This is a necessary, but not a sufficient condition for its correctness. Fulfillment of the necessary condition increases "belief" in its correctness.)

Assume cdf $A(y) = 1 - e^{-\lambda y}$, $y > 0$, ccdf $\overline{A}(y) = e^{-\lambda y}$, $y \ge 0$. Observe that the SP upcrossing rate of level 0 is equal to $h(0) = f_0(0) =$ "arrival rate of busy-period initiators. These SP upcrossings on Page 0 are for all zero-wait customers only. There are no SP upcrossings of level 0 on Page 1. All troughs in sojourns on Page 1 are for starts of service of positive-wait customers. Therefore, the SP rate of the Idle period completions is $f_0(0) = \lambda P_0$. This implies that $\boldsymbol{\alpha} = \lambda P_0$ and

$$f_0(x) = \lambda P_0 e^{-\mu_0 x}.$$

Substituting in equation (5.78), yields

$$f_1(x) - \mu_1 \int_x^{\infty} e^{-\lambda(y-x)} f_1(y)dy = -\mu_0 \lambda P_0 \int_x^{\infty} \left[1 - e^{-\lambda(y-x)}\right] e^{-\mu_0 y} dy,\ x > 0 \tag{5.82}$$

or

$$f_1(x) - \mu_1 \int_x^{\infty} e^{-\lambda(y-x)} f_1(y)dy = -\frac{\lambda^2 P_0}{\lambda + \mu_0} e^{-\mu_0 x}, x > 0,$$

Applying differential operator $\langle D - \lambda \rangle$ to both sides results in the differential equation:

$$f_1'(x) + (\mu_1 - \lambda) f_1(x) = \lambda^2 P_0 e^{-\mu_0 x}.$$

Using the condition that $f_1(0) = 0$ (because the upcrossing rate of level 0 on Page 1, is 0), the solution of the differential equation is

$$f_1(x) = \frac{\lambda^2 P_0}{\mu_1 - \lambda - \mu_0} e^{-\mu_0 x} - \frac{\lambda^2 P_0}{\mu_1 - \lambda - \mu_0} e^{-(\mu_1 - \lambda)x},$$

assuming that $\mu_0 \neq \mu_1 - \lambda$.

The total pdf of wait is the sum

$$f(x) = f_0(x) + f_1(x)$$

$$= \frac{(\mu_1 - \mu_0)\lambda P_0}{\mu_1 - \lambda - \mu_0} e^{-\mu_0 x} - \frac{\lambda^2 P_0}{\mu_1 - \lambda - \mu_0} e^{-(\mu_1 - \lambda)x}.$$

The normalizing equation (5.81) gives

$$P_0 \left[1 + \frac{\lambda}{\mu_1 - \lambda - \mu_0} \left(\frac{\mu_1 - \mu_0}{\mu_0} - \frac{\lambda}{\mu_1 - \lambda}\right)\right] = 1,$$

which leads to

$$P_0 = \frac{1 - \rho_1}{1 - \rho_1 + \rho_0}, \tag{5.83}$$

where $\rho_i = \lambda/\mu_i, i = 0, 1-$ the same as for M/M/1 in Section 2.5.3 and elsewhere in the open literature. Formula (5.83) is a widley known formula in Queueing theory.

Exercises

Exercise 5.2.1 *Check the calculations between formula (5.82) and formula (5.83) inclusive, in Section 5.2.21.*

Exercise 5.2.2 *Study the G/M/1 queue with "tnterarrival time depending on waiting time" in the publication [15](1988).*

Exercise 5.2.3 *For the $D/M/1$ Queue, where $D > 0$ is the constant interarrival time, derive the equations for obtaining the stationary pdf of the waiting time before service. (Use "duality between the virtual wait and the extended age" in Section 5.1.2.)*

6

Inventories

6.1 $\langle s, S\rangle$ Inventory with No Product Decay

Consider an $\langle s, S\rangle$ inventory, having "inter-demand" times $\overset{d}{=} \text{Exp}_\lambda$; "demand sizes" $\overset{d}{=} \text{Exp}_\mu$; no product decay; and no lead time. The SP of the corresponding sample path of net inventory describes its motion in the state space $(-\infty, S]$ over time. Let $\{I(t)\}_{t\geq 0}$ denote the net inventory at time $t \geq 0$. Note that $\{I(t)\}_{t\geq 0}$ is a "regenerative process", as we next show. Once the SP "enters a fixed level" in $(s, S]$, the SP remains at that level for a time $\overset{d}{=} \text{Exp}_\lambda$, until the next "demand" instant (see Figure 6.1). The SP returns to level S at ends of a sequence of "ordering cycles". These cycles are iid r.v.s due to the identical probabilistic structure of each cycle. Therefore, the stationary mixed pdf $\{\Pi_S, f(x)\}_{s<x<S}$ exists (see, e.g., [55] (1993) or [57] (2003)). The term "Π_S" is an "atom" at level S (positive probability). The term $f(x), 0 < x < S$, is the continuous part of the pdf. Every level $x \in (s, S)$ is a continuous point (not an atom) of $f(x)$, because the probability of entering any $x \in (s, S)$ is zero due to continuous demand sizes $\overset{d}{=} \text{Exp}_{\mu}.$. Equating the SP vertical down- and upcrossing rates of level $x \in (s, S)$ yields an integral equation for $f(x)$:

$$\lambda \Pi_S e^{-\mu(S-x)} + \lambda \int_{y=x}^{S} e^{-\mu(y-x)} f(y)dy$$
$$= \lambda \Pi_S e^{-\mu(S-s)} + \lambda \int_{y=s}^{S} e^{-\mu(y-s)} f(y)dy = \lambda \Pi_S, \tag{6.1}$$

or more simply

$$\lambda \Pi_S e^{-\mu(S-x)} + \lambda \int_{y=x}^{S} e^{-\mu(y-x)} f(y)dy = \lambda \Pi_S, s < x < S. \tag{6.2}$$

Equation (6.2) employs the second equality in (6.1), which expresses rate balance across levl x, $0 < x < S$

SP downcrossing rate of level x = SP upcrossing rate of level x.

The normalizing condition is

$$\Pi_S + \int_{x=s}^{S} f(x)dx = 1. \tag{6.3}$$

DOI: 10.1201/9780429297601-6

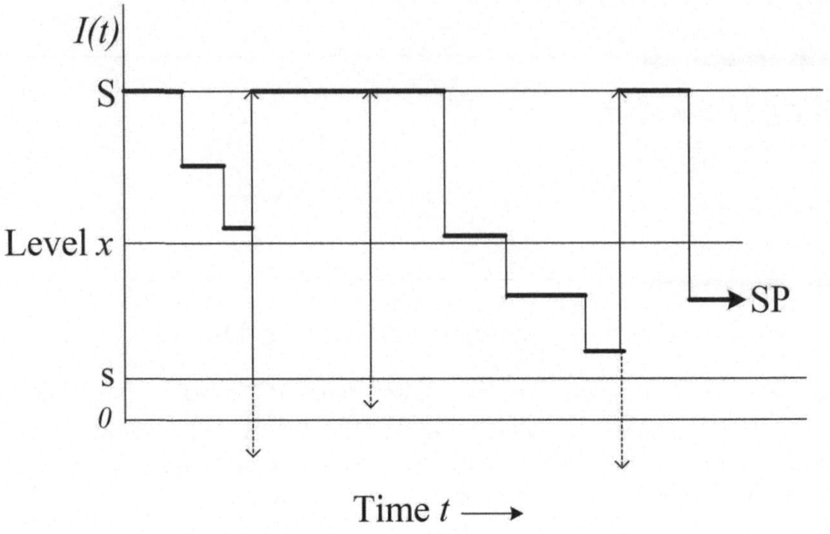

FIGURE 6.1
Sample path of $\{I(t)\}_{t\geq 0}$ in $\langle s, S\rangle$ inventory with no product decay. The SP stays at a fixed level for a time $\stackrel{d}{=}$ Exp$_\lambda$, before making a vertical downward jump with size $\stackrel{d}{=}$ Exp$_\mu$. When a vertical jump downcrosses the reorder point level s, a new replenishment order is initiated which generates an SP vertical upward jump into level $\dot{S}$.

6.1.1 Solution for Stationary Pdf of Net Inventory

Some algebra, and using (6.3), gives $\{\Pi_S, f(x)\}_{x\in(s,S)}$ in terms of the input parameters. Taking $\frac{d}{dx}$ on both side of (6.2) and using differentiation under the integral sign, gives

$$\mu \times \lambda \times \Pi_S - \lambda f(x) = 0, s < x < S,$$

$$f(x) = \mu \Pi_S, s < x < S.$$

and

$$\begin{aligned} \Pi_S &= \tfrac{1}{1+\mu(S-s)}, \\ f(x) &= \tfrac{\mu}{1+\mu(S-s)}, s < x < S. \end{aligned} \tag{6.4}$$

That is, $f(x)$ is partially uniformly distributed on state-space interval (s, S), and there is an atom at level S. In formula (6.4): if $\mu < 1$ then $\Pi_S > f(x)$, $x \in (s, \dot{S})$; if $\mu > 1$ then $\Pi_S < f(x)$, $x \in (s, \dot{S})$; and if $\mu = 1$ then $\Pi_S = f(x)$, $x \in (s, \dot{S})$.

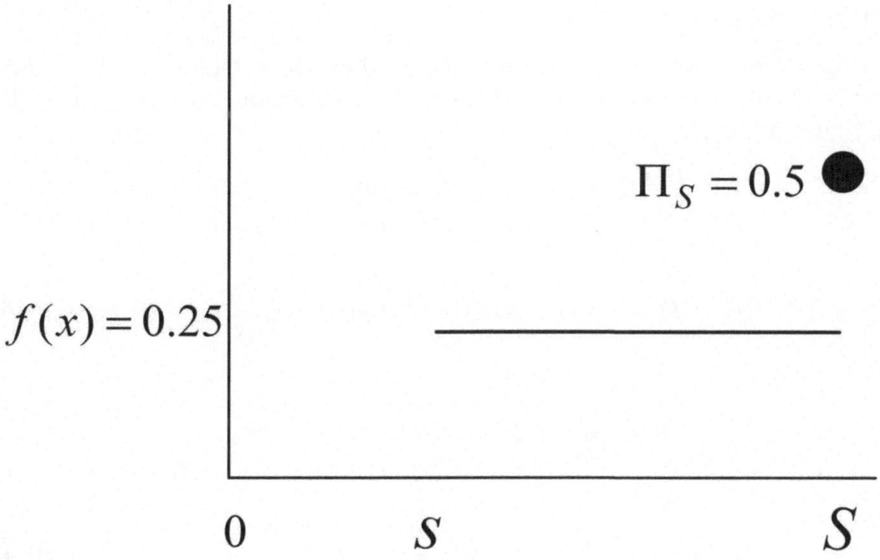

FIGURE 6.2
PDF $\{\Pi_S, f(x)\}_{s<x<S}$ where $s = 1$, $S = 3$, $S - s = 2$, $\mu = 0.5$; $\Pi_S = 0.5$, $f(x) = 0.25$, $s < x < S$.

Example 6.1.1 *Suppose* $s = 1$, $S = 3$, $\mu = 0.5$. *Then* $\Pi_S > f(x)$, $x \in (s, S)$. $S - s = 2$, $\Pi_S = 0.5$, $f(x) = 0.25$, $s < x < S$. *See Figure 6.2.*

6.1.2 Sojourn above and Sojourn below a Level x in $\langle s,S\rangle$ Inventory with No Decay

Referring to Figure 6.1 and Equation (6.2), the rate of replenishments up to level S is the total SP downcrossing rate of the reorder point at level s, namely

$$\lambda\Pi_S e^{-\mu(S-s)} + \lambda\int_{y=s}^{S} e^{-\mu(y-s)} f(y)dy = \lambda\Pi_S.$$

The size of all inventory replenishments include (1) the excess below s of "the final demand in the ordering cycle" plus "the amount $S - s$". Let $d_S :=$ time "between any two successive replenishments" up to level S ($d_S :=$ duration of ordering cycle). By the elementary renewal theorem (see Section 1.8.3 in Chapter 1).

$$E\left[d_S\right] = \frac{1}{\lambda\Pi_S e^{-\mu(S-s)} + \lambda\int_{y=s}^{S} e^{-\mu(y-s)} f(y)dy} = \frac{1}{\lambda\Pi_S}. \tag{6.5}$$

6.1.2.1 Sojourns a_x and b_x

Let a_x := "sojourn above" level $x \in (s, S)$. Let b_x := "sojourn below" level $x \in (s, S)$. By the renewal reward theorem (see Section 1.8.4 in Chapter 1), and formula (6.5),

$$\frac{E\left[a_x\right]}{E\left[d_S\right]} = 1 - F(x), x \in [s, S],$$

$$E\left[a_x\right] = (1 - F(x)) \times E\left[d_S\right] = \frac{1 - F(x)}{\lambda \Pi_S}. \tag{6.6}$$

Similarly,

$$\frac{E\left[b_x\right]}{E\left[d_S\right]} = F(x), x \in [s, S],$$

$$E\left[b_x\right] = F(x) \times E\left(d_S\right) = \frac{F(x)}{\lambda \Pi_S}, \tag{6.7}$$

where

$$F(x) = \int_{y=s}^{x} f(y)dy = \frac{\mu(x-s)}{1+\mu(S-s)}, s < x \leq S.$$

The normalizing condition (law of total probability) gives the necessary condition,

$$F(S) = \frac{\mu(S-s)}{1+\mu(S-s)} + \Pi_S = \frac{\mu(S-s)}{1+\mu(S-s)} + \frac{1}{1+\mu(S-s)} = 1.$$

6.1.3 Some Ordering Characteristics in $\langle s, S \rangle$ Inventory with No Decay

This section focuses on the "cost rate" of the $\langle s, S \rangle$ inventory with no product decay, when there is "no lead time" and "no backlogging".

6.1.3.1 Ordering Rate in $\langle s, S \rangle$ Inventory with No Decay

The total ordering rate is the right hand side of (6.2), i.e., $\lambda \Pi_S = \frac{\lambda}{1+\mu(S-s)}$, i.e., the rate at which the SP exits level S (same as rate at which the SP vertically downcroses level s).

6.1.3.2 Expected Order Size in $\langle s, S \rangle$ Inventory with No Decay

Denote the order size by $ORD_{\mathrm{Si}\,ze}$. All orders are signaled by demands ending at or below the reorder point at level s. By the memoryless property of the demand sizes which downcross level s, the expected order size is

$$E\left[ORD_{\mathrm{Si}\,ze}\right] = S - s + \frac{1}{\mu}. \tag{6.8}$$

6.1.3.3 Expected Number of Demands in an Ordering Cycle in $\langle s,S\rangle$ Inventory with No Decay

Let N_{d_S} := number of demands in an ordering cycle of duration d_S. By the renewal reward theorem,

$$\frac{E\left[N_{d_S}\right]}{E\left[d_S\right]}=\lim_{t\to\infty}\frac{E[\text{number of demands in }(0,t)]}{t}=\frac{\lambda t}{t}=\lambda,$$

$$E\left[N_{d_S}\right]=\lambda E\left[d_S\right]=\frac{\lambda}{\lambda\Pi_S}=\frac{1}{\Pi_S}=1+\mu(S-s). \tag{6.9}$$

An alternative derivation of (6.9) using stopping times is of interest. Let $\{D_i\}_{i=1,2,...}$ denotes the sequence of demand sizes starting when the net inventory is at level S. Then

$$N_{d_S}=\min\left\{n|\sum_{i=1}^{n}D_i>S-s\right\}. \tag{6.10}$$

Thus N_{d_S} is a "stopping time" for the sequence $\{D_i\}_{i=1,2,...}$. Also

$$d_S=\sum_{i=1}^{N_{d_S}}T_i,$$

where $T_i\stackrel{d}{\equiv}T, i=1,2,...,$ and $T\stackrel{d}{=}\text{Exp}_\lambda$. N_{d_S} is also a stopping time for the sequence $\{T_i\}_{i=1.2....}$. By Wald's equation,

$$E\left[d_S\right]=E\left[N_{d_S}\right]\times E\left[T_i\right]=E\left[N_{d_S}\right]\times\frac{1}{\lambda}.$$

and it follows that

$$E\left[N_{d_S}\right]=\frac{E\left[d_S\right]}{E\left[T\right]}=\frac{\frac{1}{\lambda\Pi_S}}{\frac{1}{\lambda}}=\frac{1}{\Pi_S}=1+\mu(S-s), \tag{6.11}$$

corroborating (6.9). (For the concepts "stopping timé" and "Wald's equation", see Problem 13, p. 486, in [52] (2010).)

6.1.4 Cost Rate in $\langle s,S\rangle$ Inventory with No Product Decay

The cost rate has three components: (1) Setup cost per order denoted by C_{setup}; (2) Holding cost in inventory per order per unit time denoted by C_{hold}. The total average cost rate is denoted by $C_{\cos t}$ where

$$C_{\cos t}=C_{setup}\times(\text{ordering rate})+C_{hold}\times\int_{x=s}^{S^+}xf(x)dx.$$

The ordering rate is $\lambda\Pi_S = \lambda/\left(1+\mu(S-s)\right)$. The average stock on hand is

$$\int_{x=s}^{S^+} xf(x)dx = S\Pi_S + \int_{x=s}^{S} \frac{\mu x}{1+\mu(S-s)}dx$$

$$= \frac{S}{1+\mu(S-s)} + \frac{\mu(S^2-s^2)}{2(1+\mu(S-s))}$$

$$= \frac{2S+\mu(S^2-s^2)}{2(1+\mu(S-s))}.$$

Thus

$$C_{\cos t} = \frac{\lambda}{1+\mu(S-s)} \times C_{setup} + \frac{2S+\mu(S^2-s^2)}{2(1+\mu(S-s))} \times C_{hold}. \tag{6.12}$$

Remark 6.1 *In the standard $\langle s, S\rangle$ inventory with no product decay, suppose the inter-demand times form a "renewal process" (not necessarily a Poisson process). Then the results will be the same as in this section, but only for the "arrival-point mixed pdf " denoted by $\{\Pi_{S,\iota}, f_\iota(x)\}_{x\in(s,S)}$, because PASTA would not apply. Then, the integral equation for the "continuous part of the arrival-point pdf $f_\iota(x)$" would be the same as (6.2), where λ represents the "renewal rate" of the demand process. The arrival rate λ cancels out of the equation. Thus the formulas for $\Pi_{S,\iota}$ and $f_\iota(x)$, $x \in (s,S)$ in (6.4) are independent of λ. The underlying reason for this property is that all demands are made at the ends (and starts) of inter-demand times. When the SP jumps up to level S, the time until the next demand is a full inter-demand time. Hence, $\{d_{S,i}\}$ is a renewal process, where $d_{S,i} \stackrel{d}{=} d_S$.*

Remark 6.2 *For exposition, we apply LC here to three basic $\langle s, S\rangle$ inventory systems only. We emphasize that LC equally applies to a large number of other inventory models as well, e.g., $\langle r, nQ\rangle$ models; variations of EOQ models; models with lead time; models with backlogging; production-inventory models of various complexity; models with a variety of state-dependent control policies (e.g.,[4] (1986), [5]) (1992)).*

6.2 The $\langle s, S\rangle$ Inventory with Linear Product Decay

Denote the stationary pdf of net inventory by $f(x)$, $s < x < S$. Note that there are "no atoms in the stationary pdf". The slope of product decay does not permit any particular state-space level to have a positive probability. In a sample path of $\{I(t)\}_{t\geq 0}$, the SP moves constantly downward with slope $-k$;

or the SP makes a vertical downward jump due to a new demand; or the SP makes a vertical upward jump at end of each ordering cycle. The SP is never at a fixed level for a finite duration. The rate of demands is assumed to be Poisson with rate λ. The demand sizes are assumed to be $\stackrel{d}{=} \text{Exp}_\mu$. The order cycles (aka replenishment cycles) can end in two ways: (1) SP downcrossing of the reorder point s "with slope $-k$", and (2) SP downcrossing of the reorder point s by a "vertical downcrossing jump" due to a demand. At the end of each order cycle, the SP jumps vertically upward to level S and starts immediately to move downward with slope $-k$. That is, the sample path is tangent to level S from below, the SP does not enter level S (see Section 1.4.5 in Chapter 1). Note that $\{I(t)\}_{t\geq 0}$ is a "regenerative process"; the order cycles are regenerative cycles (see Section 1.8.2 in Chapter 1).

6.2.1 Integral Equation for Stationary Pdf $f(x)$ in $\langle s, S\rangle$ Inventory with Linear Product Decay

Consider the sample path of $\{I(t)\}_{t\geq 0}$ in Figure 6.3. Fix an arbitrary level $x \in (s, S]$. The SP downcrossing rate of level x consists of two types: (1) those

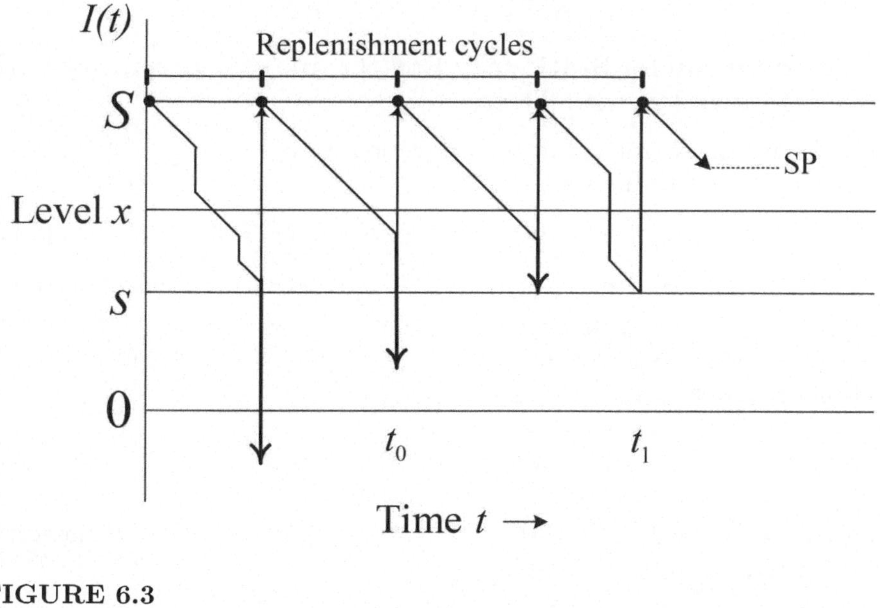

FIGURE 6.3
Sample path of $\langle s, S\rangle$ inventory with linear product decay with slope $k = -1$. Demands occur in a Poisson process with rate λ. Demand sizes are $\stackrel{d}{\equiv} \text{Exp}_\mu$, shown by vertical downward jumps. Shows ordering (replenishment) cycles, i.e., times between SP vertical upward jumps into the order up-to level S (same as times between SP vertical downcrossings of the reorder point s).

with slope "$-k$" due to product decay and (2) vertical downward jumps due to a new demand which downcross level s. By the basic LC crossing theorem the total SP downcrossing rate of level x is

$$kf(x) + \lambda \int_{y=x}^{S} e^{-(y-x)} f(y)dy. \tag{6.13}$$

The total SP upcrossing rate of level x is equal to the total downcrossing rate of level s

$$kf(s) + \lambda \int_{s}^{S} e^{-\mu(y-s)} f(y)dy = kf(S^-). \tag{6.14}$$

By rate balance across level x (6.13) and formula (6.14), we obtain the integral equation

$$k\times f(x)+\lambda \int_{y=x}^{S} e^{-\mu(y-x)} f(y)dy = k\times f(s)+\lambda \int_{s}^{S} e^{-\mu(y-s)} f(y)dy = k\times f(S^-). \tag{6.15}$$

The associated normalizing condition is

$$\int_{s}^{S} f(y)dy = 1. \tag{6.16}$$

6.2.2 Solution for Stationary Pdf $f(x)$ in $\langle s, S\rangle$ Inventory with Linear Product Decay

We now give the solution of integral Equation (6.15).

Taking $\frac{d}{dx}$ in (6.15) gives

$$kf'(x) + \mu\left[kf(S) - kf(x)\right] - \lambda f(x) = 0. \tag{6.17}$$

In (6.17), using the second equality in Equation (6.35) we replace the term $\mu\lambda \int_{y=x}^{S} e^{-\mu(y-x)} g(y)dy$ by $\mu\left[kg(S) - kf(x)\right]$. The term $-\lambda f(x)$ results from taking the derivative under the integral sign. Simplifying (6.17) gives the differential equation

$$f'(x) - \left(\mu + \frac{\lambda}{k}\right) f(x) = -\mu f(S). \tag{6.18}$$

Multiplying both sides of (6.18) by the integrating factor $e^{-\left(\mu+\frac{\lambda}{k}\right)x}$, integrating with respect to x, and simplifying gives

$$f(x) = \frac{\mu f(S)}{\mu + \frac{\lambda}{k}} + Ce^{-\left(\mu+\frac{\lambda}{k}\right)x}, \tag{6.19}$$

where C is a constant. Setting $x = S$ in (6.19) yields

$$f(S) = \frac{\mu + \frac{\lambda}{k}}{\frac{\lambda}{k}} \cdot Ce^{-\left(\mu+\frac{\lambda}{k}\right)S};$$

which substituted back into (6.19) gives

$$f(x) = \frac{\mu k}{\lambda} \times Ce^{-(\mu+\frac{\lambda}{k})S} + Ce^{-(\mu+\frac{\lambda}{k})x}, s < x \leq S. \tag{6.20}$$

Applying the normalizing condition (6.36) leads to

$$\frac{\mu k}{\lambda} \times Ce^{-(\mu+\frac{\lambda}{k})S}(S-s) + C\frac{e^{-(\mu+\frac{\lambda}{k})S} - e^{-(\mu+\frac{\lambda}{k})s}}{\mu+\frac{\lambda}{k}} = 1,$$

$$\text{or} \quad \frac{1}{C\frac{\mu k}{\lambda}e^{-(\mu+\frac{\lambda}{k})S}} = (S-s) + \frac{\lambda}{\mu k}\frac{1-e^{-(\mu+\frac{\lambda}{k})(S-s)}}{\mu+\frac{\lambda}{k}}.$$

Factoring (6.20) results in

$$f(x) = C\frac{\mu k}{\lambda}e^{-(\mu+\frac{\lambda}{k})S}\left(1 + \frac{\lambda}{\mu k}e^{-(\mu+\frac{\lambda}{k})(S-x)}\right), s < x \leq S.$$

Letting

$$A := C\frac{\mu k}{\lambda}e^{-(\mu+\frac{\lambda}{k})S}, \tag{6.21}$$

gives

$$f(x) = A\left(1 + \frac{\lambda}{\mu k}e^{-(\mu+\frac{\lambda}{k})(S-x)}\right), s < x \leq S, \tag{6.22}$$

where (applying (6.36)), results in

$$A = \left[(S-s) + \frac{\lambda}{\mu k\left(\mu+\frac{\lambda}{k}\right)}\left(1 - e^{-(\mu+\frac{\lambda}{k})(S-s)}\right)\right]^{-1}. \tag{6.23}$$

In (6.22) $f(x)$ is convex and increasing on (s, S) (i.e., $f'(x) > 0$, $f''(x) > 0$). This property agrees with intuition which suggests that the stock resides a large proportion of time at high levels closer to S and a smaller proportion of time near s, for every $k > 0$. This accumulation of inventory near S is a consequence of the re-order policy, which instantaneously replenishes the stock up to level S at each replenishment instant, since there is no lead time. (See the numerical example in Section 6.2.7 and Figures 6.4 and 6.5.)

6.2.3 Sojourns above and below Level x

Let a_x:= sojourn time above x, and b_x := sojourn time at or below x, $x \in (s, S]$, in $\{I(t)\}_{t\geq 0}$. Every instant $t \geq 0$ such that $I(t) = S$ is a regenerative point of $\{I(t)\}_{t\geq 0}$. The regenerative property holds whether replenishments up to S are due to SP smooth decays into level s from above, or due to SP jumps that end at or below level s as a result of a demand. For example, consider Figure 6.6. At time points like t_1 the SP makes a left-continuous hit of level s from above, and jumps upward to level S. The Poisson arrival process for demands ensures that the excess arrival time until the next demand is $\stackrel{d}{=}$ Exp_λ (memoryless property).

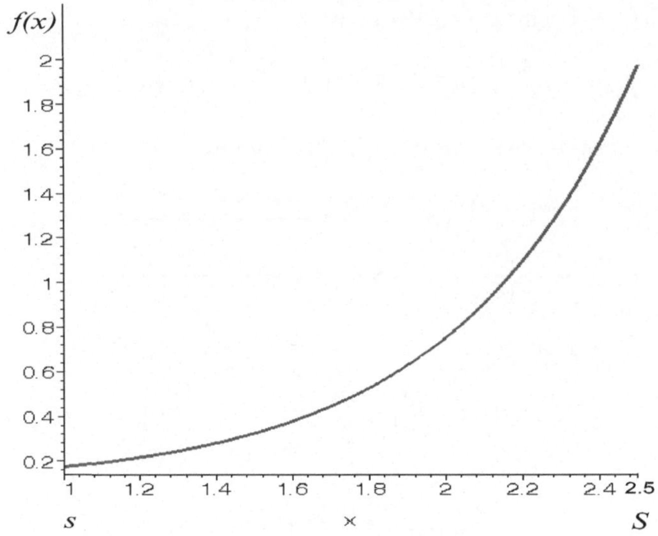

FIGURE 6.4
PDF $f(x)$ in $\langle s, S \rangle$ inventory with linear decay rate k. $\lambda = 2, \mu = 0.10$, $k = 1$, $S = 2.5$, and $s = 1$. Note that $f(x) = 0 \begin{cases} x < 1, \\ x > 2.5. \end{cases}$

6.2.3.1 Expected Time between Successive SP Hits of Level S from below in $\langle s, S \rangle$ Inventory with Linear Product Decay

The times between successive instants when $I(t) = S$, form a renewal process. Let $d_S :=$ inter-level-S time. From (6.35), the total SP downcrossing rate of level s is identical to the

$$\text{replenishment rate } = kf(s) + \lambda \int_{y=s}^{S} e^{-\mu(y-s)} f(y)dy = kf(S^-).$$

The value of $f(S)$ is obtained from (6.22). Thus, with A as in (6.22),

$$E\left[d_S\right] = \frac{1}{kf(S)} = \frac{\mu}{A \times (k\mu + \lambda)}. \tag{6.24}$$

6.2.3.2 Expected Value of a_x in $\langle s, S \rangle$ Inventory with Linear Product Decay

The proportion of time the SP spends above level x is $1 - F(x)$, which is equal to $\frac{E[a_x]}{E[d_S]} = E\left[a_x\right] kf(S)$ (by the renewal reward theorem $E\left[d_S\right] = \frac{1}{kf(S)}$). Thus

$$E\left[a_x\right] = \frac{1 - F(x)}{kf(S)} = \frac{\mu(1 - F(x))}{A(k\mu + \lambda)}, \tag{6.25}$$

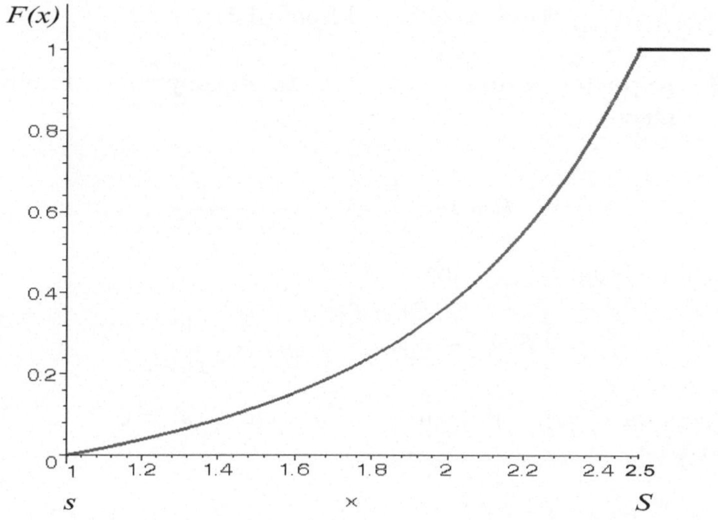

FIGURE 6.5
CDF $F(x)$ in $\langle s, S\rangle$ inventory with linear decay rate k. $\lambda = 2, \mu = 0.10$, $k = 1$, $S = 2.5$, and $s = 1$.

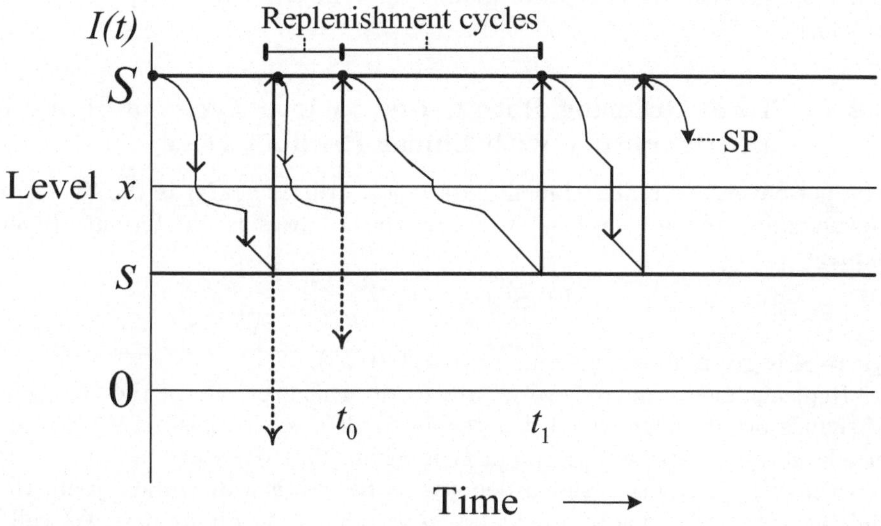

FIGURE 6.6
Sample path $\{I(t)\}_{t\geq 0}$ in $\langle s, S\rangle$ inventory with general decay. The SP motion when the net inventory has a general decay pattern is similar to the SP motion when the net inventory has a "linear decay" pattern.

where $F(x) = \int_{y=s}^{x} f(y)dy$ is obtained from (6.22).

6.2.3.3 Expected Value of b_x in $\langle s, S\rangle$ Inventory with Linear Product Decay

Similarly,

$$E[b_x] = \frac{F(x)}{kf(S)} = \frac{\mu F(x)}{A(k\mu+\lambda)}. \tag{6.26}$$

We can also obtain $E[b_x]$ using (6.25) and

$$E[a_x] + E[b_x] = E[d_x] = \frac{1}{kf(S)}.$$

A check on $E[a_x]$, $E[b_x]$ when $x = s$, using $F(s) = 0$, $1 - F(s) = 1$, and (6.24), is

$$E[a_s] = \frac{1-F(s)}{kf(S)} = \frac{1}{kf(S)} = \frac{\mu}{A(k\mu+\lambda)} = E[d_S],$$

the expected replenishment cycle, as intuitively expected. Also $E[b_s] = F(s)/(kf(S)) = 0$, which agrees with the SP spending zero time below level s. (State-space jumps, including jumps that end below s, occur not in Time because they are perpendicular to the time axis. (See Chapter 2 in [22] (2017).)

6.2.4 Total Ordering Rate Is due to Two Types of Signal in $\langle s, S\rangle$ Inventory with Linear Product Decay

The net-onventory replenishment rate (total ordering rate) is the SP "total downcrossing rate of level s", which is the "right side" in formula (6.35), namely

$$kf(S^-) = A \times \frac{k\mu+\lambda}{\mu},$$

where A is given above in formulas (6.21)–(6.23).

Replenishments up to level S are made whenever one of the two types of signals occurs. One signal denotes an SP left continuous decay at rate k into level s from above (time point t_1 in Figure 6.3). We call it a "type-c" for "continuous". The other signal denotes an SP vertical downward jump that ends at or below s, due to a demand (time point t_0 in Figure 6.3). We call it a "type-j" for "jump". An order cycle (same as replenishment cycle) is the time between two successive instants when an order is received, namely, "d_S". The net inventory $\{\mathcal{I}(t)\}_{t\geq 0}$ is a regenerative process with regenerative cycle $\stackrel{d}{=} d_S$. Due to Poisson arrivals of demands, the sequence $\{d_{S,i}\}_{i=1,2,...}$ with $d_{S,i} \stackrel{d}{=} d_S$, is a renewal process. An order initiating d_S is due either to a type-c

or type-j SP downcrossing of level s. Let $P_{\boldsymbol{c}} := P(\text{an order is type-}c)$, $P_{\boldsymbol{j}} := P(\text{an order is type-}j)$. Then $P_c + P_j(s) = 1$.

We now determine $P_{\boldsymbol{c}}$ and $P_{\boldsymbol{j}}$. The counting process $\{D_t^c(s)\}_{t\geq 0}$ is a renewal process due to Poisson arrivals of demand. Since there is exactly 1 type-c or exactly 1 type-j signal for an order in an ordering cycle,

$$E[\text{number of type-}\boldsymbol{c}\text{ orders in } d_S] = 1\cdot P_{\boldsymbol{c}} + 0\cdot P_{\boldsymbol{j}} = P_{\boldsymbol{c}},$$

$$E[\text{number of type-}\boldsymbol{j}\text{ orders in } d_S] = 0\cdot P_{\boldsymbol{c}} + 1\cdot P_{\boldsymbol{j}} = P_{\boldsymbol{j}}.$$

By the renewal reward theorem,

$$\frac{E[\text{number of type-}\boldsymbol{c}\text{ orders in } d_S]}{E\left[d_S\right]} = \lim_{t\to\infty}\frac{E[\mathcal{D}_t^c(s)]}{t} = kf(s),$$

$$\frac{P_{\boldsymbol{c}}}{E\left[d_S\right]} = kf(s).$$

Since $E\left[d_S\right] = \frac{1}{kf(S^-)}$,

$$P_{\boldsymbol{c}} = kf(s)\times E\left[d_S\right] = \frac{kf(s)}{kf(S^-)} = \frac{f(s)}{f(S^-)}. \tag{6.27}$$

In the second right most term of (6.27) the numerator is the rate of type-c generated orders; the denominator is the total rate of orders.

Also, $\{\mathcal{D}_t^j(s)\}_{j\geq 0}$ is a renewal process (due to Poisson arrivals of demands). Therefore,

$$\frac{E\left[\text{number of type-}\boldsymbol{j}\text{ orders in } d_S\right]}{E\left[d_S\right]} = \frac{1\cdot P_{\boldsymbol{j}} + 0\cdot P_{\boldsymbol{c}}}{E\left[d_S\right]}$$

$$= \frac{P_{\boldsymbol{j}}}{E\left[d_S\right]} = \lim_{t\to\infty}\frac{E\left[\mathcal{D}_t^j(s)\right]}{t},$$

and (see Section 1.8.4 in Chapter 1),

$$P_{\boldsymbol{j}} = \lim_{t\to\infty}\frac{E\left[\mathcal{D}_t^j(s)\right]}{t}E\left[d_S\right] = \frac{\lambda\int_{y=s}^{S} e^{-\mu(y-s)}f(y)dy}{kf(S^-)}. \tag{6.28}$$

In (6.28) the numerator is the rate of type-j orders; the denominator is the total rate of orders.

Formulas (6.27) and (6.28), with A as in (6.22), imply that

$$P_{\boldsymbol{c}} = \frac{k\mu + \lambda e^{-(\frac{k}{\lambda}+\mu)(S-s)}}{k\mu+\lambda}, \tag{6.29}$$

$$P_{\boldsymbol{j}} = \frac{\mu\lambda A\int_{y=s}^{S} e^{-\mu(y-s)}\left(1+\frac{\lambda e^{-\left(\frac{\lambda}{k}+\mu\right)(S-y)}}{k\mu}\right)dy}{k\mu+\lambda}. \tag{6.30}$$

6.2.5 Expected Order Size in $\langle s, S\rangle$ Inventory with Linear Product Decay

Denote the replenishment order by R. If an order is generated by a type-c signal then $R = S - s$. If an order is generated by a type-j signal then $R = S - s + r_s$ where $r_s :=$ excess demand below s. Since the demand sizes are $\stackrel{d}{=}$ Exp_μ then $r_s \stackrel{d}{=} \text{Exp}_\mu$ by memoryless property of Exp_μ. Since $P_{\boldsymbol{c}} + P_{\boldsymbol{j}} = 1$, the expected order size is

$$E\,[R] = (S - s)P_{\boldsymbol{c}} + \left(S - s + \frac{1}{\mu}\right) P_{\boldsymbol{j}} = S - s + P_{\boldsymbol{j}}\frac{1}{\mu}, \tag{6.31}$$

where $P_{\boldsymbol{j}}$ is given in (6.30).

6.2.6 Cost Rate in $\langle s, S\rangle$ Inventory with Linear Product Decay

Since there is no backlogging or lead-time costs in the $\langle s, S\rangle$ model considered here, the cost function accounts only for costs of "setup" of "placing orders", and for "holding inventory". Let $C_{total} :=$ total average cost rate, $C_{\boldsymbol{s}} :=$ setup cost per type-c generated order, $C_{\boldsymbol{j}}, :=$ setup cost per type-j generated order. Let C_{Hold} be the holding cost per unit per unit time. Then

$$C_{total} = \mathcal{C}_{\boldsymbol{s}} \times (\text{type-}\boldsymbol{c}\text{ ordering rate})$$

$$+\mathcal{C}_{\boldsymbol{j}} \times (\text{type-}\boldsymbol{j}\text{ ordering rate}) + C_{Hold} \times \int_{x=s}^{S} x f(x)dx$$

$$= \mathcal{C}_{\boldsymbol{s}} k f(s) + \mathcal{C}_{\boldsymbol{j}}\lambda \int_{x=s}^{S} e^{-(x-s)\mu} f(x)dx + C_{Hold} \int_{x=s}^{S} x f(x)dx, \tag{6.32}$$

where $f(x)$ is given by (6.22).

6.2.7 Numerical Example in $\langle s, S\rangle$ Inventory with Linear Product Decay

In $\langle s, S\rangle$ with linear decay rate k and all demand sizes $\stackrel{d}{=}$ Exp_μ, assume $\lambda = 2$, $\mu = 0.10$, $k = 1$, $S = 2.5$, and $s = 1$. Calculations give $A = 0.094200$ in formula (6.22). The stationary pdf of inventory is

$$f(x) = 0.094200 + 1.88400e^{(-5.250+2.10x)}, 1 < x \leq 2.5,$$

$$f(1) = 0.1749, \qquad f(2.5) = 1.9782.$$

The cdf of inventory is

$$F(x) = -0.132645 + 0.094200x + 0.897144e^{(-5.25+2.1x)}, 1 < x \leq 2.5,$$

$$F(1) = 0, \qquad F(2.5) = 1.0.$$

Functions $f(x)$ and $F(x)$ are plotted in Figures 6.4 and 6.5, which demonstrate "convexity" and the probability "massed towards level S".

6.3 The $\langle s,S\rangle$ Inventory with More General Product Decay

Consider an $\langle s,S\rangle$ inventory system with reorder point $s \geq 0$, and order-up-to the level $S > s$. Assume that demands for stock (net inventory) occur at a Poisson rate λ. The demand amounts are denoted by $D_i, i = 1, 2, ...$ These amounts are iid random variables with common cdf $B(x)$, and ccdf $\overline{fB}(x) = 1 - B(x)$, $x \geq 0$. Denote the "net inventory" process by $\{I(t)\}_{t\geq 0}$. Assume the net inventory decays at rate $dI(t)/dt = -r(I(t))$, where $r(I(t))$ is a non-negative continuous function of $I(t) \in (s,S]$. The ordering policy is similar to that in Section 6.2. That is, if the net inventory either: (1) decays continuously to level s from above, or (2) jumps vertically downward to level s or below level s due to a demand of stock then an order is placed and received immediately. This order replenishes the net inventory up to level S^-. In a sample path of $\{I(t)\}_{t\geq 0}$, the SP makes a vertical upward jump, which is "tangent to S" from below (Section 1.4.5 in Chapter 1). The "regular" state space is (s,S) since all probability is concentrated on (s,S). The "wide-sense" state space is $(-\infty,S)$, which accounts for all SP downward jumps ending below s, and immediately jumping up to be tangent at S ("double jumps") (see Figure 6.6).

Let $f(x)$, $s < x < S$, and $F(x)$, $s<x < S$, denote, respectively, the stationary pdf and cdf of $I(t)$ as $t \to \infty$. Assume each order size $\stackrel{d}{=} \text{Exp}_\mu$.

6.3.1 PDF of Net Inventory with More General Product Decay Rate $r(x) = kx$

We illustrate the derivation of the pdf of net inventory using a decay rate "$r(x) = kx$" (generalized from "$r(x) \equiv k$" in Section 6.2) (see [19] (2009)). We will derive an integral equation for the stationary pdf of net inventory $f(x), x \in (s,S)$ in Section 6.3.2.

In the present $\langle s,S\rangle$ model, the time between successive demands is $\stackrel{d}{\equiv} \text{Exp}_\lambda$. Thus, $\lambda :=$ rate of SP vertical downward jumps (see Figure 6.7). The demand sizes are $\stackrel{d}{\equiv} \text{Exp}_\mu$. Thus, the cdf of all the demand sizes is $B(x) = 1 - e^{-\mu x}$, $x > 0$. The SP upward jumps are due to replenishments (prescribed

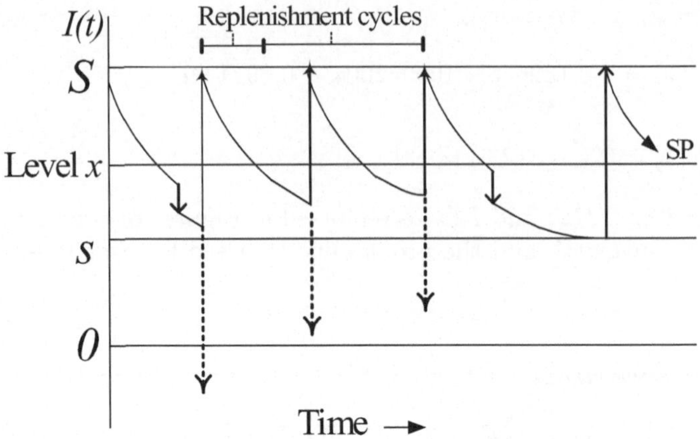

FIGURE 6.7
Sample path $\{I(t)\}_{t\geq 0}$ in $\langle s, S\rangle$ inventory with the specific decay pattern $r(x) = kx$. Here $\{I(t)\}_{t\geq 0}$ resembles the case where the net inventory has "linear decay" in Figure 6.3. There is a distinctly convex curvature between vertical SP jumps in the present figure.

jumps, i.e., policy jumps, up to level S). The upward jumps occur whenever a demand causes the SP to vertically downcross the reorder point s, or to glide smoothly into level s. (The study in [19] (2009) discusses the model when "$r(x) = kx$".)

6.3.2 Integral Equation for Stationary Pdf of Net Inventory When $r(x) = kx$

We use LC to develop an integral equation for $f(x)$, $x \in (s, S)$ based on simple LC limit theorems (Section 1.4 in Chapter 1). Consider a sample path of $\{I(t)\}_{t\geq 0}$ (Figures 6.6 and 6.7). Fix level $x \in (s, S)$. The rate at which the SP "decays into" level x from above (due to left-continuous downcrossings of x) is

$$\lim_{t\to\infty} \frac{E\left[\mathcal{D}_t^c(x)\right]}{t} = r(x)f(x) = kxf(x).$$

(We use the terms "rate" for $\lim_{t\to\infty} \frac{\mathcal{D}_t^c(x)}{t}(a.s.)$ and "expected rate" for $\lim_{t\to\infty} \frac{E[\mathcal{D}_t^c(x)]}{t}$ synonymously.) The SP smooth decay rate into the reorder point level s from above is

$$\lim_{t\to\infty} \frac{E\left[\mathcal{D}_t^c(s)\right]}{t} = r(s^+)f(s^+) \equiv r(s)f(s) = kxf(s).$$

The rate at which the SP vertically jump downcrosses level $x \in (s, S)$ due to a demand is

$$\lim_{t\to\infty} \frac{E\left[\mathcal{D}_t^j(x)\right]}{t} = \lambda \int_{y=x}^{S} \overline{B}(y-x)f(y)dy = \lambda \int_{y=x}^{S} e^{-\mu(y-x)}f(y)dy$$

(downcrossing jumps start at level $y \in (x, S)$ and are greater than $y - x$). The total SP downcrossing rate of level $x \in (s, S)$ is the sum

$$\lim_{t\to\infty} \frac{E\left[D_t^c(x)\right]}{t} + \lim_{t\to\infty} \frac{E\left[\mathcal{D}_t^j(x)\right]}{t} = r(x)f(x) + \lambda \int_{y=x}^{S} \overline{B}(y - xf(y)dy$$

$$= kxf(x) + \lambda \int_{y=x}^{S} e^{-\mu(y-x)}f(y)dy, x \in (s, S].$$

The "total SP downcrossing" rate of the reorder point s is

$$\lim_{t\to\infty} \frac{E\left[\mathcal{D}_t(s)\right]}{t} = r(s)f(s) + \lambda \int_{y=s}^{S} \overline{B}(y-s)f(y)dy$$

$$= kxf(s) + \lambda \int_{y=s}^{S} e^{-\mu(y-s)}f(y)dy,$$

where we have considered a left-continuous hit of the reorder point level s from above to be an "SP downcrossing of s". (See generalized basic LC theorem in Section 1.4.3 in Chapter 1.)

The SP total downcrossing rate of level s is equal to the SP downward rate from the "order-up-to level S". This equality is due to the ordering policy, which replenishes net inventory up to level S with: (1) each left continuous hit of level s from above, or (2) each vertical jump downcrossing of level s. (There is a one-to-one correspondence between SP egresses from S below, and total downcrossings of level s.) SP "rate balance into and out of" state $\{S^-\}$ results in the equality:

$$r(s)f(s) + \lambda \int_{y=s}^{S} \overline{B}(y-s)f(y)dy = r(S^-)f(S^-),$$

$$ksf(s) + \lambda \int_{y=s}^{S} e^{-\mu(y-s)}f(y)dy = kSf(S^-), \tag{6.33}$$

where $r(S) = r(S^-)$ and $f(S) := f(S^-)$ without any change in probabilities.

A simplifying feature of this $\langle s, S\rangle$ model is the total SP upcrossing rate of every level $x \in (s, S)$ is equal to the total downcrossing rate of level s

(replenishment rate), because net inventory declines at rate $r(S)$ just after it is replenished to level S until it eventually downcrosses level s again by a continuous motion or by a vertical downward jump. Applying Equation (6.33) and rate balance across level x, yields an integral equation for $f(x)$:

$$\text{downcrossing rate of } x = \text{upcrossing rate of } x$$

$$r(s)f(x) + \lambda \int_{y=x}^{S} \overline{B}(y-x)f(y)dy = r(s)f(s) + \lambda \int_{y=s}^{S} \overline{B}(y-s)f(y)dy$$

$$= r(S)f(S), x \in (s, S]. \tag{6.34}$$

If $r(x) \equiv kx$, then

$$kxf(x) + \lambda \int_{y=x}^{S} e^{-\mu(y-x)}f(y)dy = kxf(s) + \lambda \int_{y=s}^{S} e^{-\mu(y-s)}f(y)dy.$$

$$kxf(x) + \lambda \int_{y=x}^{S} e^{-\mu(y-x)}f(y)dy = kxf(S^{-}), x \in (s, S]. \tag{6.35}$$

The state space has no atoms, i.e., there is no state where the SP spends a positive sojourn time. The probability distribution of net inventory is concentrated on (s, S). The normalizing condition is

$$\int_{x=s}^{S} f(x)dx = 1. \tag{6.36}$$

Exercises

Exercise 6.3.1 *Determine the physical "dimensions" of A, C, and k in formula (6.21).*

Exercise 6.3.2 *Check that all three terms in equation (6.35) have the "same dimension".*

Exercise 6.3.3 *Give an expression for E[number of demands in an ordering cycle] in the $\langle s, S\rangle$ inventory with no product decay (use the renewal reward theorem in Section 1.8.4 in Chapter 1 for this and next two exercises).*

Remark 6.3 *For the following four Exercises, consider the calculations in Sections 6.1.3–6.1.4.*

Exercise 6.3.4 *Give an expression for E[number of demands in an ordering cycle] in the $\langle s, S\rangle$ inventory with linear product decay $r(x) = k$.*

Exercise 6.3.5 *Give an expression for E[number of demands in an ordering cycle] in the $\langle s, S\rangle$ inventory with more general product decay $r(x) = kx$, $x \in (s, S)$.*

Exercise 6.3.6 *Give an expression for the cost rate in $\langle s, S\rangle$ inventory with linear decay $r(x) = k$.*

Exercise 6.3.7 *Give an expression for the cost rate in $\langle s, S\rangle$ inventory with non-linear decay $r(x) = kx$.*

7

Dams

This chapter introduces and analyzes several types of stationary (aka steady-state) "dams" where the rate of inputs is a Poisson process and input sizes are $\stackrel{d}{=} \text{Exp}_\mu$. More general dams where the rate of inputs is a non-Markovian process (i.e., not a Poisson process) are given in Chapters 6 and 8 in [22], "Time-dependent" (aka transient) analyses of dams using LC are analyzed in Chapter 6 in [22] (2017), in [12] (1988), and in [13] (1988).

7.1 M/M/r(·) Dam

A dam may be perceived as having a "liquid" content, e.g., volume measured in cubic meters of water, or mass measured in kilograms, etc. A typical "sample path" is given in Figure 7.1. We assume that new liquid is added to the content by finite amounts $\stackrel{d}{\equiv} L$, at discrete time points. Denote the "content" of the dam at time t by $\{C(t)\}_{t\geq 0}$. Let C_{cap}:= capacity level of the dam. Two time points of interest are when the content in the dam "overshoots" level C_{cap} due to an input; or decreases to level 0 initiating an "empty" period. Assume that all inputs L are $\stackrel{d}{=} \text{Exp}_\mu$, which occur in a Poisson process with rate λ. Thus $E[L] = \frac{1}{\mu}$. Assume $C(t)$ decreases continuously at rate $r(x)$ when $0 < x < C_{cap}$. When the content hits level 0 it remains there for a finite time until the next input increases the content to a positive amount.

7.2 M/M/r(·) Dam When $r(x) = kx + m$

Let "$r(x) = kx + m$", where k and m are positive constants. Thus, $r(x) > 0$, $0 < x < C_{cap}$. The dam is considered to be stable if the content returns to level 0 (empty state $\{0\}$), with probability 1. In this model of a dam, we assume that there are no vertical downward jumps of content.

A typical sample path of the M/M/r(·) Dam is given in Figure 7.1.

DOI: 10.1201/9780429297601-7

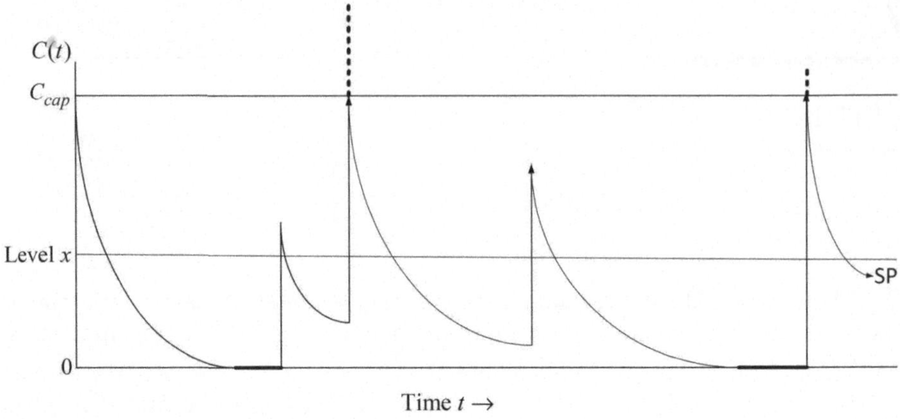

FIGURE 7.1
Sample path of M/M/$r(x)$ dam where $r(x) = kx + m$, and $k > 0$, $m > 0$. Shows 2 SP "empty" periods at level 0; 5 "sojourns" a_x (above x) and 4 full "sojourns" b_x (below x); 3 "non-empty" periods and 2 "empty" periods, alternating; and 2 "excess" amounts of liquid over level C_{cap}, which flow away into an independent reservoir.

7.3 Equation for Stationary PDF of Content in M/M/r(·) Dam

Denote the stationary pdf of the content of the dam by $\{P_0, f(x)\}_{0<x<C_{cap}}$, where $P_0 = P(\text{content} = 0)$ and $f(x)$ is the continuous part of the pdf. We derive the balance equation for the pdf of the dam using LC, like Equation (6.34) for the more general $\langle s, S\rangle$ inventory in Section 6.3.2 of Chapter 6. The integral equation for the present M/M/r(·) dam is

$$r(x)f(x) = \lambda P_0 e^{-\mu x} + \lambda \int_{y=0}^{x} e^{-\mu(x-y)} f(y)dy, 0 < x < C_{cap}, \tag{7.1}$$

$$f(x) = \frac{\lambda}{r(x)} \left(P_0 e^{-\mu x} + \int_{y=0}^{x} e^{-\mu(x-y)} f(y)dy \right), 0^+ < x < C_{cap}. \tag{7.2}$$

A brief explanation of Equation (7.1) follows. The left side is the SP "smooth downcrossing rate" of level x, $0 < x < C_{cap}$. The right side is the upcrossing rate of level x by "vertical upward" jumps of content inputs. SP "rate balance" across level x gives equation (7.1).

7.4 Solution of Equation (7.1) for PDF of Content in M/M/r(·) Dam

Assume $P_0 > 0$. (That is, state $\{0\}$ is positive recurrent; the content returns in a finite time to level 0, with probability 1.) Applying differential operator $\langle D + \mu \rangle$ to both sides of (7.1), leads to the differential equation for $f(x)$,

$$\frac{f'(x)}{f(x)} = -\frac{r'(x) + \mu r(x) - \lambda}{r(x)}, 0 < x < C_{cap},$$

$$\frac{f'(x)}{f(x)} + \frac{r'(x)}{r(x)} = -\frac{\mu r(x) - \lambda}{r(x)}, 0 < x < C_{cap},$$

$$\frac{d}{dx} \ln\left(r(x)f(x)\right) = -\mu + \frac{\lambda}{r(x)}, 0 < x < C_{cap}, \tag{7.3}$$

by using the properties: (1) "derivative of a product of two functions" and (2) "derivative of the logarithm of a function" (both standard in calculus texts). Integating both sides of (7.3) with respect to x gives

$$\ln\left(r(x)f(x)\right) = -\mu x + \lambda \int_{x=0}^{C_{cap}} \frac{1}{r(x)} dx + c,$$

where $c :=$ constant.

$$f(x) = \frac{\lambda P_0}{r(x)} e^{-\mu x + \lambda \int_{y=0}^{x} \frac{1}{r(y)} dy}, 0 < x < C_{cap}, \tag{7.4}$$

upon applying the "initial condition" $r(0^+)f(0) = \lambda P_0$ (i.e., "SP rate into level 0 by a smooth content decline" = "SP rate out of level 0 by a vertical jump"). (Recall that $r(0) = m > 0$.)

Substituting $f(x)$ from (7.4) into the normalizing condition

$$P_0 + \int_{x=0}^{C_{cap}} f(x) dx = 1, \tag{7.5}$$

gives the solution for P_0:

$$P_0 = \frac{1}{1 + \lambda \int_{x=0^+}^{C_{cap}} \frac{1}{r(x)} e^{-\mu x + \lambda \int_{y=0}^{x} \frac{1}{r(y)} dy} dx}. \tag{7.6}$$

Figure 7.2 plots the cdf $F(x)$, pdf $f(x)$ and ccdf $\overline{F}(x)$ for an arbitrary set of the model's parameters. To check a "necessary condition" for formula

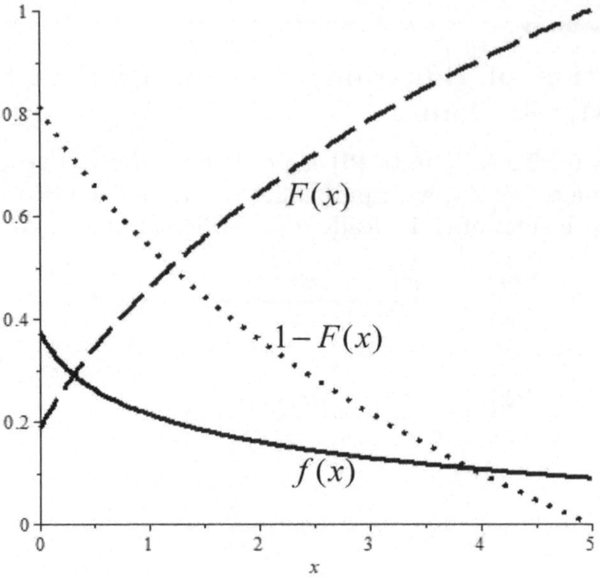

FIGURE 7.2
Figure shows cdf $F(x)$, pdf $f(x)$, ccdf $\overline{F}(x) \equiv 1 - F(x)$, $0 < x < C_{cap}$; $r(x) = kx + m$, $k = 3$, $m = 1$, $\lambda = 2$, $\mu = 0.1$, $Ccap = 5$.

(7.6) to be correct, set $r(x) \equiv k > 0$, $m = 0$, and set $C_{cap} = \infty$. Then, from Equation (7.6) some algebra and integration gives:

$$P_0 = \frac{1}{1 + \frac{\lambda}{k\left(\mu - \frac{\lambda}{k}\right)}} = \frac{k\mu - \lambda}{k\mu} = 1 - \frac{\lambda}{k\mu}. \tag{7.7}$$

If $k = 1$ in formula (7.7) then P_0 for the foregoing "relaxed" M/M/r(·) dam, has the same value of the stationary P_0 of the virtual wait $\{W(t)\}_{t \geq 0}$ in the standard M/M/1 queue (Chapter 2). This is a "necessary" condition for formula (7.6) to hold when $r(x) \equiv 1$ and $C_{cap} = \infty$. (This is a "mild check" because it is not a "necessary and sufficient" condition. The mild check is useful because it increases "confidence" that the computations leading to formula (7.6) in the present section are "correct".)

Example 7.4.1 *Assume the model parameters are* $r(x) = kx + m$, $k = 3$, $m = 1$, $\lambda = 2$, $\mu = 0.1$, $Ccap = 5$. *Computation shows that*

$$\begin{aligned}
f(x) &= \frac{\lambda P_0 e^{-\mu x + ((1/3)\lambda ln(3x+1))}}{kx+m}, 0 < x < 5, \\
\frac{f(x)}{P_0} &= \frac{2e^{-0.1x}}{(3x+1)^{1/3}}, 0 < x < 5, \\
P_0 &= \frac{1}{1 + \int_{x=0}^{5} \frac{2e^{-0.1x}}{(3x+1)^{1/3}} dx}, \\
P_0 &= 0.186908,
\end{aligned} \tag{7.8}$$

where "ln(y)" denotes the natural logarithm if y. (Note that "$e^{ln(y)} = y$", $y > 0$.)

7.5 Sojourn Times above and below a State-space Level in the M/M/r(·) Dam with $r(x) = kx + m$

Let a_x:= SP sojourn "above" level x, $0 \le x < C_{cap}$. Let b_x := SP sojourn "below" level x, $0 \le x < C_{cap}$. Let d_x:= time between any two "successive" SP downcrossings of level x. Here, a "downcrossing" of level 0 means "left limit" (aka "left continuous") entrance into level 0 at "rate" m, since state $\{0\}$ is the lower boundary of the state space of $\{C(t)\}_{t\ge 0}$. (See "left limit", "entrance from above" in Figure 1.2 in Section 1.4.5 in Chapter 1.)

Since $P_0 > 0$ in (7.6), when $x = 0$, E["total cycle"] $= E\left[d_0\right]$, where

$$d_0 = \text{"empty period"} + \text{"non-empty period"}$$

$$= \text{"empty period"} + a_0,$$

where a_0 = "non-empty period". The sequence of d_xs forms a "regenerative process" in Time, for each x, $0 \le x < C_{cap}$. Thus, $E\left[a_0\right] := E$["non-empty" period]. By the "elementary renewal theorem" (Section 1.8.3 in Chapter 1),

$$E\left[d_x\right] = \frac{1}{r(x)f(x)}, \; 0 \le x < C_{cap}. \tag{7.9}$$

In particular

$$E\left[d_0\right] = \frac{1}{mf(0)} = \frac{1}{\lambda P_0}.$$

When the dam is stable, the content returns in a finite time to state $\{0\}$ with probability 1. In sample paths of $\{C(t)\}_{t\ge 0}$:

$$\text{"SP rate into level 0"} = \text{"SP rate out of level 0"}.$$

By the "renewal reward" theorem (Section 1.8.4 in Chapter 1),

$$\frac{E\left[a_0\right]}{E\left[d_0\right]} = 1 - P_0,$$

$$E\left[a_0\right] = E\left[d_0\right] \times \left(1 - P_0\right),$$

$$E\left[a_0\right] = \frac{1 - P_0}{mf(0)} = \frac{1 - P_0}{\lambda P_0}, \tag{7.10}$$

where P_0 is given in formula (7.6).

In the M/M/r(·) dam, all upward jumps are $\stackrel{d}{=} \text{Exp}_\mu$. By the memoryless property, the "excess" of a vertical jump over any level x is also $\stackrel{d}{=} \text{Exp}_\mu$. Generally the sojourn a_x depends on x (i.e., generally, $r(x) \neq$ "constant"). The dependence of a_x on x when $r(x) \equiv kx+m$ usually differs from a_x for other models considered in this book. For example, the virtual wait $\{W(t)\}_{t\geq 0}$ in the standard M/M/1 queue has slope -1; the content $\{C(t)\}_{t\geq 0}$ in the M/M/r(·) dam with $r(x) \equiv m$ (see the next section); the net inventory $\{I(t)\}_{t\geq 0}$ in the $\langle s.S\rangle$ inventory with linear constant product decay $r(x) \equiv k > 0$, $x > 0$. Thus, generally in the M/M/r(·) dam, a_x depends on $f(y)$ and $r(y)$, $0 \leq x < y < C_{cap}$. We can determine $E[d_x]$, $E[a_x]$, and $E[b_x]$ provided we can solve for $\{P_0, f((x)\}_{0>x<C_{cap}}$ as in formulas (7.3)–(7.6).

7.5.1 M/M/r(·) Dam with $r(x) \equiv m$ (Constant)

Assume that the "efflux" rate (:= rate of liquid flow "out of" the dam) is $r(x) \equiv m$, $m > 0$, for all x, $0 < x < C_{cap}$. The sample-path structure is similar to the sample path of the virtual wait in the M/M/1 queue (see Section 2.1.4 in Chapter 2). Thus, from (7.7),

$$P_0 = 1 - \frac{\lambda}{m\mu},$$

$$E[d_x] = \frac{m\mu}{m\mu - \lambda}, 0 \leq x < C_{cap},$$

$$E[a_x] = \frac{\frac{1}{\mu}}{m\left(1 - \frac{\lambda}{m\mu}\right)} = \frac{1}{m\mu - \lambda}, x \geq 0. \tag{7.11}$$

Exercises

Exercise 7.5.1 *Compare the "similarities" and "differences" between the* $\langle s, S\rangle$ *inventory with product decay* $r(x) = kx$ *and the M/M/r(·) dam where the efflux rate is* $r(x) = kx + m$. *(Compare the "sample path" in Figure 6.7 in Section 6.3.1 in Chapter 6 with the "sample path" in Figure 7.1 in the present chapter).*

Exercise 7.5.2 *Check the computations of the steps in formula (7.8).*

Exercise 7.5.3 *In the M/M/r(·) dam with* $r(x) = kx + m$, *give the "dimensions" of* $r(x)$, k, *and* m.

Exercise 7.5.4 *As a mild check on the correctness of formula (7.11), compute the dimensions of the associated quantities:* P_0, $E[d_x]$ *and* $E[a_x]$.

Exercise 7.5.5 *Explain why the sequence of* $d_x s$ *form a regenerative process for each* x*, in formulas (7.9) and (7.10).*

Exercise 7.5.6 *Prove the validity of formula (7.6) by deriving it from formula (7.5).*

Exercise 7.5.7 *Prove the validity of formula (7.7) by deriving it from formula (7.6).*

Exercise 7.5.8 *In the* $M/M/r(\cdot)$ *dam in Section 7.2 the content may be* 0 *(empty) for a positive time, i.e.,* $P_0 > 0$*. To ensure that there is always some liquid present (i.e., always non-empty), it is possible to model the dam so that* $P_0 = 0$*. One way to accomplish this is to decide on a content level, say* $\omega > 0$*, as follows. Whenever the content decreases to level* ϖ*, a special input to the dam occurs, from an independent reservoir. The size of the special input may be* $\stackrel{d}{=}$ *all other inputs. The special input is triggered in the same fashion as an order in an* $\langle s, S\rangle$ *inventory when the net inventory hits the reorder point* s *due to product decay. The objective is to give the analysis of the* $M/M/r(.)$ *where* $\omega < C(t) < C_{cap}$*. (See Section 6.3.2 in Chapter 6.)*

8

Actuarial Risk Model

This chapter discusses the classical "standard" Cramér-Lundberg ("C-L") risk model, which tracks the insurer's risk reserve fund (aka "surplus") up to the time of ruin, i.e., when the risk reserve drops to level zero or less. The chapter also studies a variant of it. The variant introduces a "dividend barrier" and extends the active time period beyond the first ruin time, by restarting the process from the same initial risk reserve amount, repeatedly.

8.1 Standard Cramér-Lundberg Actuarial Risk Model

The aim of the standard Cramér-Lundberg Actuarial risk model is to determine the expected time until the "surplus" of an insurance company decreases to ≤ 0 (aka "time until ruin"), and the probability that ruin occurs in a finite time.

Let $\{X(t)\}_{0\leq t\leq \tau_{rn}}$ denote the "surplus" in a standard "C-L" risk model with initial reserve $x_0 > 0$ (see, e.g., pp. 22–28 in [32] (2003); or Chapter 1 in [7] (1999)). Assume that "insurance claims" occur in a Poisson process at rate λ, with claim "amounts" which are iid r.v.s denoted by $\{Z_i\}_{i=1,2,\ldots}$ with common cdf $B(x)$, $x \geq \dot{0}$. Then, the surplus at time t is

$$X(t) = x_0 + ct - \sum_{i=1}^{\mathbf{N}(t)} Z_i, 0 \leq t \leq \tau_{rn}, \tag{8.1}$$

where $c > 0$ is the "premium" rate at which funds are paid by the insurees into the Insurance company's "surplus"; and $\{\mathbf{N}(t)\}_{t\geq 0}$ is a Poisson process having rate λ. (The premim rate "c" is assumed to be constant.) Denote the time until "ruin" by τ_{rn} (subscript "rn" stands for "$ruin$"). In the dtandard C-L model, τ_{rn} is the first time after time $t = 0$ when $X(t)$ downcrosses level 0. In mathematical notation,

$$\tau_{rn} = \inf\{t > 0 \mid X(t) \ \leq 0\}, \tag{8.2}$$

where "inf" (abbreviation of "infimum") means "greatest lower bound" of t when the condition "$X(t) \leq 0$" occurs. (We may also the terminology "$\min\{t > 0 \mid X(t) \leq 0\}$".)

A sample path of the "standard" C-L model is given in Figure 8.1.

DOI: 10.1201/9780429297601-8

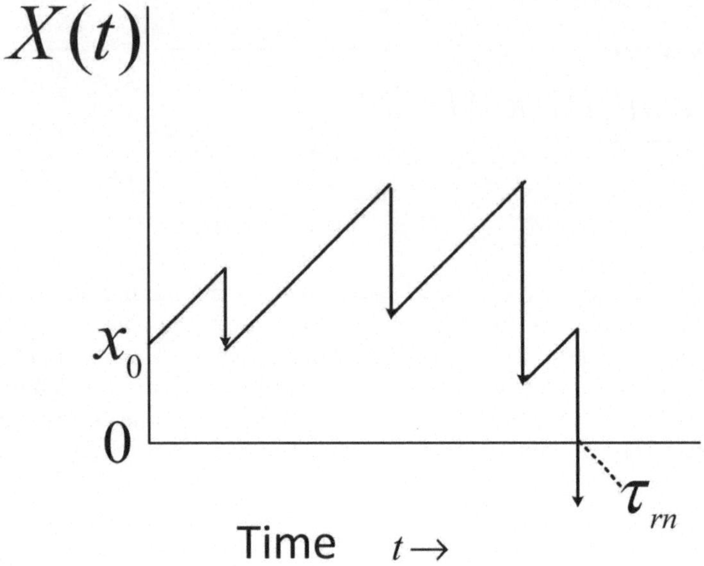

FIGURE 8.1
Sample path of $\{X(t)\}_{0\leq t\leq \tau_{rn}}$ in standard C-R risk model. Shows "surplus" $X(0) = x_0$; $X(t)$, $t \in (0, \tau_{rn})$; and $X(\tau_{rn}) \leq 0$.

8.2 A Variant of the Standard Cramér-Lundberg Model

In this chapter, Section 8.3 analyzes a "variant" of the standard "Cramér–Lundberg" ("C-L") risk process $\{X(t)\}_{t\geq 0}$, based on Model 1, pp. 289–302 in [27] (2011). In the variant, the time axis extends indefinitely beyond the first "ruin" time. The process $\{X(t)\}_{t\geq 0}$ restarts at level x_0 immediately after the first ruin time (Figure 8.1); x_0 is the same level as at time 0, i.e., $X(0) = x_0$. The durations between any two "successive" ruin times form the "regenerative cycles" of a "regenerative process".

8.3 C-L Risk Model with Dividend Barrier and Reinvestment Strategy

Here, we consider the variant of the standard C-L model (characterized by formula (8.3), with the following properties. There is a horizontal "dividend" barrier at level $M > x_0$, and a special "reinvestment strategy". Whenever $\{X(t)\}_{t\geq 0}$ reaches the barrier M due to the paid premiums, a portion of

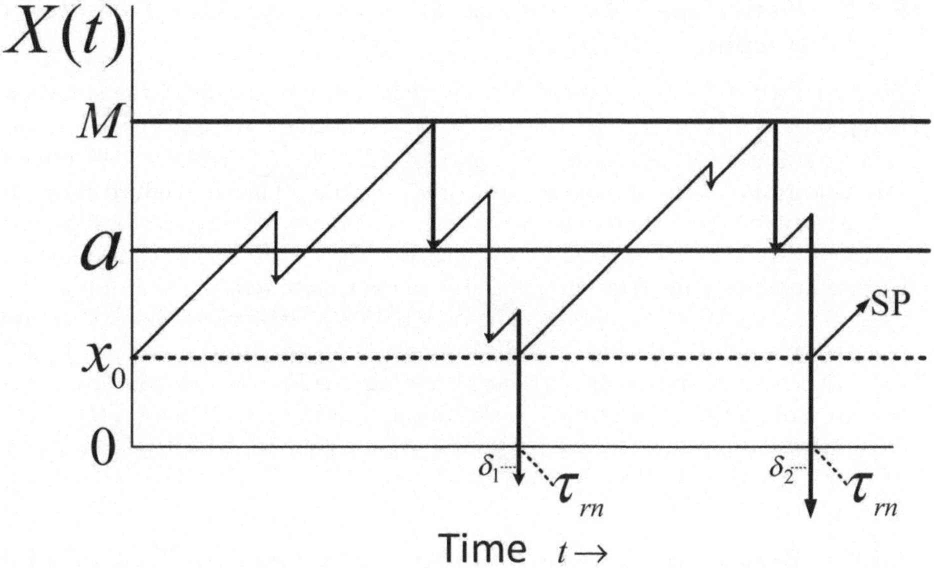

FIGURE 8.2
Sample path of surplus $\{X(t)\}_{t\geq 0}$ in the variant of the C-R model with dividend barrier M and reinvestment strategy. Figure shows 2 "ruin cycles" starting at level x_0 and ending at a level ≤ 0. Shows M; reinvestment level a; 2 reinvestment amounts $:= M - a$; restart initial amount x_0; 2 "deficits" δ_1, δ_2 at ruin times. Figure shows 2 "ruin" cycles; intervals $(0, 1^{\text{st}}\ \tau_{rn})$ and $\left(1^{\text{st}}\ \tau_{rn}, 2^{\text{nd}}\ \tau_{rn}\right)$.

net gain in surplus, $M - a$, is transferred out immediately and moved into alternative investments (e.g., a conservative investment portfolio). Specifically, if $X(t^-) = M$ then immediately $X(t) = a$, where $x_0 < a < M$. Every instant (say t_M) when the surplus hits level M from below, i.e., $t_M \in \{t\ X(t^-) = M,\ t \geq 0\}$ is a regenerative point of $\{X(t)\}_{t\geq 0}$. (There can be several different regenerative processes associated with a single sample path.) On reaching each t_M the surplus jumps vertically downward by the "reinvestment amount $M - a$" to level a and starts increasing at the average premium rate c (see Figure 8.2). Let k be the number of instants t_M which occur in $(0, t)$. The following formulas characterize the surplus $X(t)$ at time t for this variant:

$$\begin{aligned} &X(t) = x_0 + ct - \textstyle\sum_{i=1}^{\mathbf{N}(t)} Z_i - k(M-a),\ 0 < X(t) < M, t \in (0, \tau) \\ &X(t) = a \text{ if } X(t^-) = M, t \in (0, \tau), \end{aligned} \tag{8.3}$$

where "τ" is the first ruin time of $\{X(t)\}_{t\geq 0}$. The characteristics in formula (8.3) are repeated in each interval $(\tau_n, \tau_{n+1})\,, n = 1, 2, ...$

8.3.1 Extending $\{X(t)\}$ from $[0,\tau)$ to $[0,\infty)$ and Deficits of Surplus

We create a regenerative process whose cycles are iid replicas of the standard R-L process $\{X(t)\}_{t\in[0,\tau)}$, as follows. At the ruin time τ, we restore the surplus up to the initial value x_0, and the process $\{X(t)\}_{t\in[0,\tau)}$ restarts. We repeat this restoration at each successive "ruin" instant. This procedure forms a regenerative process, denoted by $\{X(t)\}_{t\geq 0}$. Figure 8.2 shows the first two cycles of $\{X(t)\}_{t\geq 0}$. The regenerative process has the advantage containing an infinite number of iid "ruin cycles", and enables us to analyze the ruin model using its regenerative properties. The time points τ_1, $\tau_1+\tau_2$, $\tau_1+\tau_2+\tau_3$, ..., are regenerative points at which the SP makes a "double jump" – one downward to, or below level 0 causing a "deficit" ending a ruin cycle, and immediately one upward to the initial level x_0, starting the next cycle. Since $\{X(t)\}_{t\geq 0}$ is a regenerative process, its limiting pdf exists as $t\to\infty$, which we denote by $f(x)$, $x\in(0,M)$ (see Section 1.8.2 in Chapter 1).

8.3.2 Expected Time of Ruin in Variant of C-L Risk Model

Consider the regenerative process $\{X(t)\}_{t\geq 0}$ in the inervals $\{\tau_n-\tau_{n-1}\}_{n=1,2,\ldots}$, where each $\tau_n-\tau_{n-1}\overset{d}{=}\tau_{rn}$ (:= time of ruin starting from level x_0). Let $N_\tau(t)$:= number of renewals (i.e., number of "ruins") in $(0,t)$ where t is anarbitrary time point. Let $\mathcal{D}_t(x)$:= number of SP downcrossings of level $x\geq 0$ during $(0,t)$ (see Figure 8.2). From the "elementary renewal theorem" and the simple basic LC theorems in Section 1.4, Chapter 1, we obtain:

$$\lim_{t\to\infty}\frac{E\left[\mathcal{D}_t(x)\right]}{t}=\lambda\int_{y=x}^{M}\overline{B}(y-x)f(y)dy, 0\leq x\leq M,$$

$$\lim_{t\to\infty}\frac{E\left[\mathcal{D}_t(0)\right]}{t}=\lim_{t\to\infty}\frac{N_\tau(t)}{t}=\lim_{t\to\infty}\frac{E\left[N_\tau(t)\right]}{t}=\frac{1}{E\left[\tau\right]},\tag{8.4}$$

where $\overline{B}(\cdot)=1-B(\cdot)$. In particular the expected time of ruin starting from level x_0 is given by formula (8.5):

$$\frac{1}{E\left[\tau\right]}=\lambda\int_{y=0}^{M}\overline{B}(y)f(y)dy,$$

$$E\left[\tau\right]=\frac{1}{\lambda\int_{y=0}^{M}\overline{B}(y)f(y)dy}.\tag{8.5}$$

Let $\mathcal{U}_t(x)$:= number of SP "upcrossings" of level x in $(0,t)$. All upcrossings of any level x, $0<x<M$, occur when the sample path is continuous and rising at the average premium rate c, except at the "ruin" instants, when

upcrossings of level x are jumps over level x, $0 < x < x_0$, ending at level x_0(the "restart" amount). From LC theory and formula (8.5),

$$\lim_{t\to\infty} \frac{E\left[\mathcal{U}_t(x)\right]}{t} = cf(x), x_0 \leq x < M;$$

$$\lim_{t\to\infty} \frac{E\left[\mathcal{U}_t(x)\right]}{t} = cf(x) + \lambda \int_0^M \overline{B}(y)dF(y), 0 < x < x_0;$$

$$= cf(x) + \frac{1}{E\left[\tau\right]}, 0 < x < x_0.$$

8.3.3 Cdf and Pdf of the Deficit at Ruin in Variant C-L Risk Model

A random variable of interest in risk theory is the "deficit" at ruin, denoted by $\boldsymbol{\delta}$, with cdf denoted by $F_{\boldsymbol{\delta}}(y)$, and pdf denoted by $f_{\boldsymbol{\delta}}(y), y > 0$ (Figure 8.2). The random variable $\boldsymbol{\delta}$ is equal to the "excess" of an SP downward jump ending below level 0 due to a claim.

Theorem 8.1 *The cdf and pdf of the deficit at ruin $\boldsymbol{\delta}$ are given by*

$$F_{\boldsymbol{\delta}}(x) = 1 - \lambda E\left[\tau\right] \int_{y=0}^{M} \overline{B}(y+x)f(y)dy, x > 0, \tag{8.6}$$

$$f_{\boldsymbol{\delta}}(x) = \lambda E\left[\tau\right] \int_{y=0}^{M} b(y+x)dF(y), x > 0, \tag{8.7}$$

where $E\left[\tau\right]$ is given by (8.5), and where $b(y) = \frac{dB(y)}{dy}$, $y > 0$ is the pdf of the claim size.

Proof *Fix level $x < 0$. By formula (8.5), $1/E\left[\tau\right]$ is the downcrossing rate of level 0. The term $1 - F_{\boldsymbol{\delta}}(|x|)$ is the probability that a downward jump of level 0, also downcrosses level x, where $|x| :=$ absolute value of x. Thus*

$$\frac{1}{E\left[\tau\right]}(1 - F_{\boldsymbol{\delta}}(|x|))$$

is the SP downcrossing rate of level x. (See Bayes' rule in Section 1.5.2 in Chapter 1.) Since downcrossings of level 0 are caused by insurees' claims only, we have the following equation for $F_{\boldsymbol{\delta}}(|x|)$

$$\frac{1}{E\left[\tau\right]}(1 - F_{\boldsymbol{\delta}}(|x|)) = \lambda \int_0^M \overline{B}(y-x)f(y)(y), x < 0, \tag{8.8}$$

which is equivalent to (8.6) (since $x < 0$). Taking d/dx on both sides of (8.6) gives (8.7).

8.3.4 Further Analysis of the Variant of the Standard Cramér-Lundberg Model with Dividend Barrier and Reinvestment Strategy

Let us define expressions the pdf $f(x)$ as the sum of three distinct functions $f_0(x)$, $f_1(x)$, and $f_2(x)$, as follows:

$$f(x) := f_0(x)\boldsymbol{I}_{[0,x_0)}(x) + f_1(x)\boldsymbol{I}_{[x_0,a)}(x) + f_2(x)\boldsymbol{I}_{[a,M)}(x),$$

where, for a given set A, $\boldsymbol{I_A}(x) = 1$ if $x \in \boldsymbol{A}$ and $\boldsymbol{I_A}(x) = 0$ if $x \notin \boldsymbol{A}$.

Define $f_2(M) := f_2(M^-)$ for convenience; this does not affect any probability measures. By using LC and rate balance across level x (see explanations immediately after formula (8.13)), we obtain equations (8.9)–(8.11). The left side of each equal sign is the SP upcrossing rate of level x; the right side is the SP downcrossing rate of level x.

$$cf_2(x) = cf_2(M^-) + \lambda \int_x^M \overline{B}(y-x) f_2(y)\, dy, a \le x < M, \tag{8.9}$$

$$cf_1(x) = \lambda \int_x^a \overline{B}(y-x) f_1(y)\, dy + \lambda \int_a^M \overline{B}(y-x) f_2(y)\, dy, x_0 \le x < a, \tag{8.10}$$

$$cf_0(x) + \frac{1}{E[\tau]} = \lambda \int_x^{x_0} \overline{B}(y-x) f_0(y)\, dy + \lambda \int_{x_0}^a \overline{B}(y-x) f_1(y)\, dy$$

$$+\lambda \int_a^M \overline{B}(y-x) f_2(y)\, dy, 0 < x < x_0, \tag{8.11}$$

where

$$\frac{1}{E[\tau]} = \lambda \int_0^{x_0} \overline{B}(y) f_0(y)\, dy + \lambda \int_{x_0}^a \overline{B}(y) f_1(y)\, dy + \lambda \int_a^M \overline{B}(y) f_1(y)\, dy. \tag{8.12}$$

The normalizing condition is

$$\int_0^{x_0} f_0(y)\, dy + \int_{x_0}^a f_1(y)\, dy + \int_a^M f_2(y)\, dy = 1. \tag{8.13}$$

8.3.4.1 Further Explanation of (8.9)–(8.11).

In (8.9), the left side is the upcrossing rate of level x, $a < x < M$. On the right side, the first term is the downcrossing rate of level x, due to hits of level M from below causing immediate jumps downward to level a. The second term is the downcrossing rate of level x due to claims occurring when the surplus is in the interval (x, M). Equation (8.10) is explained similarly.

In equation (8.11), the left side is the upcrossing rate of level x, $0 < x < x_0$. The first term is the upcrossing rate at continuous points of the sample path. The second term is the total downcrossing rate of level 0, by (8.12), resulting in immediate upcrossings of level x (a "double jump"). The right side of (8.11) is the downcrossing rate of x caused by claims when the SP is above x.

Rate balance across level x yields equations (8.9), (8.10), and (8.11).

Remark 8.2 *Consider the hit and egress rates of level x_0. We obtain the following rate balance equation*

$$c\, f_0(x_0^-) + \frac{1}{E\,[\tau]} = c\, f_1(x_0). \tag{8.14}$$

In (8.14), the left side is the SP total hit rate of x_0 from below; the right side is the egress rate out of x_0 to above (Figure 8.2). The term $cf_0(x_0^-)$ is the hit rate of level x_0 from below at continuous sample-path points. Since downward jumps ending below level 0 cause ruin and immediate SP jumps up to level x_0, $1/E\,[\tau]$ is also the hit rate of level x_0 from below. Finally, on the right side, $cf_1(x_0)$ is the egress rate out of level x_0 above.

From (8.14), $f(x)$ has a jump discontinuity at x_0, given by

$$f_1(x_0) - f_0(x_0^-) = \frac{1}{cE\,[\tau]}. \tag{8.15}$$

Similar reasoning for level a gives

$$f_2(a) - f_1(a^-) = f_2(M) \tag{8.16}$$

(The derivations of (8.15) and (8.16) are examples of how LC can lead to analytical properties of the pdf in an intuitive manner based on simple LC theorems. These formulas also serve as a check on integral equations (8.9)–(8.11).)

8.3.5 Solution of Variant when Claim Amounts are Exponentially Distributed

We now solve equations (8.9)–(8.11) and (8.13) for $f_0(x), f_1(x)$, and $f_2(x)$ on their respective state-space intervals, when the claim sizes are $\stackrel{d}{=} \mathrm{Exp}_\mu$, to illustrate the solution technique. Thus, $e^{-\mu(\cdot)}$ is substituted for $\overline{B}(\cdot)$ in (8.9)–(8.11). There are two cases: (1) $c \neq \frac{\lambda}{\mu}$, (2) $c = \frac{\lambda}{\mu}$.

8.3.5.1 General Case $c \neq \frac{\lambda}{\mu}$

Taking d/dx on both sides of (8.9)–(8.11) and solving the resulting differential equations for $f_2(\cdot)$, $f_1(\cdot)$, and $f_0(\cdot)$ yields

$$\begin{cases} f_0(x) = A'_{10} + B'_{10} e^{(\mu - \frac{\lambda}{c})x}, 0 \le x < x_0, \\ \\ f_1(x) = A'_{11} e^{(\mu - \frac{\lambda}{c})x}, x_0 \le x < a, \\ \\ f_2(x) = f_2(a) \frac{\mu - \left(\frac{\lambda}{c}\right) e^{-(\mu - \frac{\lambda}{c})(M-x)}}{\mu - \left(\frac{\lambda}{c}\right) e^{-(\mu - \frac{\lambda}{c})(M-a)}}, a \le x < M, \end{cases} \tag{8.17}$$

where A'_{10}, B'_{10}, A'_{11} are constants.

Substituting $f_2(\cdot)$, $f_1(\cdot)$, and $f_0(\cdot)$ into equations (8.9)–(8.11) and (8.13) yields

$$\begin{cases} A'_{10} = \left(C'_{11}\right)^{-1} \lambda \mu e^{x_0 \mu} \left(e^{\frac{\lambda}{c} M + a\mu} - e^{\frac{\lambda}{c} a + M\mu}\right), \\ \\ B'_{10} = -A'_{10}, \\ \\ A'_{11} = \left(C'_{11}\right)^{-1} \left(\frac{\lambda}{c}\right) \left(e^{\frac{\lambda}{c} M + a\mu} - e^{\frac{\lambda}{c} a + M\mu}\right) \left(\lambda e^{\frac{\lambda}{c} x_0} - c\mu e^{x_0 \mu}\right) \\ \\ f_2(a) = \left(C'_{11}\right)^{-1} c e^{a(\mu - \frac{\lambda}{c})} \left(\frac{\lambda}{c} e^{\frac{\lambda}{c} x_0} - \mu e^{x_0 \mu}\right) \left(\frac{\lambda}{c} e^{\frac{\lambda}{c} M + a\mu} - \mu e^{\frac{\lambda}{c} a + M\mu}\right), \end{cases} \tag{8.18}$$

where

$$C'_{11} = (a - M)\, c\mu e^{(M+a)\mu} \left(\frac{\lambda}{c} e^{\frac{\lambda}{c} x_0} - \mu e^{x_0 \mu}\right)$$

$$+ \lambda(1 + x_0 \mu) e^{x_0 \mu} \left(e^{\frac{\lambda}{c} M + a\mu} - e^{\frac{\lambda}{c} a + M\mu}\right).$$

The expected time until ruin is, from (8.12),

$$E[\tau] = C'_{11} \lambda^{-1} (\lambda - c\mu)^{-1} e^{-x_0 \mu} \left(e^{\frac{\lambda}{c} M + a\mu} - e^{\frac{\lambda}{c} a + M\mu}\right)^{-1}. \tag{8.19}$$

8.3.5.2 Expected Number of Claims before Ruin

Let N_{Clms} be the number of claims before ruin. From $E[\tau]$ given in (8.19), we obtain

$$E[N_{Clms}] = \lambda E[\tau] = C'_{11} (\lambda - c\mu)^{-1} e^{-x_0 \mu} \left(e^{\frac{\lambda}{c} M + a\mu} - e^{\frac{\lambda}{c} a + M\mu}\right)^{-1}. \tag{8.20}$$

8.3.5.3 Special Case $c = \frac{\lambda}{\mu}$

By a similar analysis, we can show that when $c = \lambda/\mu$,

$$\begin{cases} f_0(x) = B''_{10}x, 0 \le x < x_0, \\ \\ f_1(x) = A''_{11}, x_0 \le x < a, \\ \\ f_2(x) = A''_{12} + B''_{12}x, a \le x < M, \end{cases} \tag{8.21}$$

where

$$B''_{10} = 2\mu^2 \left(C''_{11}\right)^{-1}, \qquad A''_{11} = 2\mu \left(1 + x_0\mu\right) C''^{-1}_{11},$$

$$B''_{12} = \left(\frac{1}{a - M}\right) A''_{11}, \qquad A''_{12} = -\left(M + \frac{1}{\mu}\right) B''_{12}$$

and $C''_{11} = 2 - x_0^2\mu^2 + \mu\left(a + M\right)\left(1 + x_0\mu\right)$.

The expected time until ruin is

$$E\left[\tau\right] = \left(\frac{1}{2\lambda}\right) C''_{11}. \tag{8.22}$$

8.3.5.4 Expected Capital Transferred out of Surplus to Investments before Ruin

Let N_M be the number of hits of level M and let CAP be the capital transferred out, before ruin. We give a method to derive $E\left[CAP\right]$, since each illustrates useful intuitive notions.

$E\left[CAP\right]$

Using the renewal reward theorem, we get

$$\frac{E\left[N_M\right]}{E\left[\tau\right]} = \lim_{t \to \infty} \frac{E\left[\mathcal{U}_t(M)\right]}{t} = cf_2(M)$$

$$\Longrightarrow E\left[N_M\right] = cE\left[\tau\right] f_2(M). \tag{8.23}$$

When $c \neq \lambda/\mu$, substituting $f_2(M)$ and $E\left[\tau\right]$ from (8.17) and (8.19) into (8.23) yields

$$E\left[CAP\right] = (M - a)\, E\left[N_M\right] = (M - a)\left(\frac{c}{\lambda}\right)\left(\frac{\left(\frac{\lambda}{c}\right) e^{\left(\frac{\lambda}{c} - \mu\right)x_0} - \mu}{e^{\left(\frac{\lambda}{c} - \mu\right)M} - e^{\left(\frac{\lambda}{c} - \mu\right)a}}\right).$$

When $c = \lambda/\mu$, $E\left[\tau\right]$ is given by (8.22), and $f_2(M)$ is obtained from (8.21). Then it can be shown that

$$E[CAP] = x_0 + \frac{1}{\mu}.$$

Note that when $c = \lambda/\mu$, the expected capital transferred out before ruin into alternative investments is independent of the barrier M.

For additional analytical details, numerical examples, and applications of LC to other risk models, see [27] (2011).

Exercises

Exercise 8.3.1 *Give the dimensions of all terms in equations (8.9)–(8.12). Is the necessary condition satisfied about the correctness of the equations?*

Exercise 8.3.2 *Explain the validity of equation (8.1).*

Exercise 8.3.3 *Give the dimension of each term in equation (8.1).*

Exercise 8.3.4 *Explain the validity of equation (8.3).*

Exercise 8.3.5 *Give the dimension of each term in equation (8.3).*

Exercise 8.3.6 *Explain the validity of equation (8.8).*

Exercise 8.3.7 *Give the dimension of each term in equation (8.8).*

Exercise 8.3.8 *In Figure 8.2 above, the intervals between successive ruin instants are cycles of a regenerative process. Find and explain any other regenerative precesses in Figure 8.2.*

9

Observation of a Piecewise Cyclic Trajectory at a Random Time Point

A "piecewise cyclic trajectory" is a "single sample path" of a regenerative process, with regenerative cycles. We assume that the cycles may derive from probabilistic models or from deterministic models. Many stochastic models have sample paths with this property. Examples are: the virtual wait in M/M/1 queues; the "content" in an M/M/r(·) dam; the "surplus" in a C-L Insurance risk model; the "net inventory" in an $\langle s, S\rangle$ inventory model; the serum "concentration of a drug" in a pharmacokinetic model with multiple dosing; the position of a point moving in simple harmonic motion ("SHM"), with examples: the position of the weight at the end of a swinging pendulum about a center point, as in a grandfather clock; the position of a weight at the bottom end of a vertical spring, which is fixed in position at the top end–like a bungee jump). Note: these positions are not the same as time-dependent (transient) results.

9.1 Observation of the Virtual Wait in an M/M/1 Queue at a Random Time Point

Consider an $M_\lambda/M_\mu/1$ with virtual-wait process $\{W(t)\}_{t\geq 0}$ (see Sections 2.1.4 and 2.1.5 in Chapter 2).

Formulas (2.14) and (2.15) in Section 2.1.10 in Chapter 2 give the stationary pdf of $\{W(t)\}_{t\geq 0}$ as $t \to \infty$, denoted by $\{P_0, f(x)\}\ x > 0$, repeated here for easy access:

$$\begin{aligned} f(x) &= \lambda P_0 e^{-(\mu-\lambda)x}, x > 0, \\ P_0 &= 1 - \tfrac{\lambda}{\mu} \end{aligned} \tag{9.1}$$

Suppose that an "observer" arrives at an $M_\lambda/M_\mu/1$ queue (e.g., makes a phone call to a medical office; or enters a store where there is only one server; i.e., "joining a waiting line" where there is a single server.) The observer questions how long a time she/he must wait before starting service. To answer this question, we will use formula (9.1).

Let us assume further that the observer has some data which suggests that the values $\lambda = 2$, $\mu = 3$, are reasonable in this situation. Then formula (9.1)

DOI: 10.1201/9780429297601-9

becomes, since $\mu - \lambda = 1$,

$$f(x) = 2P_0 e^{-x}, x > 0, \qquad P_0 = 1 - \tfrac{2}{3} = \tfrac{1}{3}. \tag{9.2}$$

Let us explore the "median" waiting time, denoted by "x_{mdn}". Then x_{mdn} is the value such that the "probability of waiting $< x_{mdn}$" is $= \frac{1}{2} =$ "probability of waiting. $> x_{mdn}$". To derive x_{mdn} we must solve the equation (9.3):

$$\int_{x_{mdn}}^{\infty} f(y)dy = \frac{1}{2}$$

$$\int_{x_{mdn}}^{\infty} \frac{2}{3}e^{-y}dy = \frac{1}{2} \tag{9.3}$$

The solution of (9.3) for x_{mdn} is obtained using computational software (e.g., [46] (2017)), which gives

$$x_{mdn} = .287682 \text{ time units.}$$

This analysis provides information to the observer. She/he has probability $\frac{1}{2}$ of waiting less than x_{mdn} time units.

(Note: This analysis requires that $P_0 < \frac{1}{2}$, i.e., $1 - \frac{\lambda}{\mu} < \frac{1}{2} \iff \frac{\lambda}{\mu} > \frac{1}{2}$. The above analysis used $\frac{\lambda}{\mu} = \frac{2}{3}$. The reason for assuming $P_0 < \frac{1}{2}$ is: if $P_0 > \frac{1}{2}$ then the median $x_{mdn} = 0$, and the probability of waiting zero time would be $> \frac{1}{2}$.)

A related analysis can answer questions like: what is the probability of waiting $< .100000$ time units when $1 - P_0 = \frac{\lambda}{\mu} > \frac{1}{2}$? (see Exercise 9.6.3).

9.2 Observation of a Dam at a Random Time Point

Consider the M/M/r($\cdot$) dam where $r(x) = kx + m$, $0 < x < C_{cap}$, k and m are constants > 0, and a sample path in Figure 7.1 (see Section 7.2 in Chapter 7).

Assume, as in Example 7.4.1 that the model parameters are $r(x) = kx+m$, $k = 3$, $m = 1$, $\lambda = 2$, $\mu = 0.1$, $Ccap = 5$. For convenience we repeat the formula for the mixed statuionary pdf of the content $\{P_0, f(x)\}$ of the dam given in formula (7.8) in Chapter 7, namely,

$$f(x) = \frac{\lambda P_0 e^{-\mu x} \times (3x+1)^{(1/3)\lambda}}{kx+m}, 0 < x < 5,$$

$$\frac{f(x)}{P_0} = \frac{2e^{-0.1x} \times (3x+1)^{2/3}}{(3x+1)^{3/3}}, 0 < x < 5,$$

$$\frac{f(x)}{P_0} = 2e^{-0.1x}(3x+1)^{-\frac{1}{3}}, 0 < x < 5, \tag{9.4}$$

$$P_0 = \frac{1}{1+\int_{x=0}^{5} 2e^{-0.1x}(3x+1)^{-\frac{1}{3}}dx},$$

$$P_0 = 0.186908.$$

Thus,

$$f(x) = \frac{1}{3}(0.461560 \times 2) \times e^{-.1x} \times (3x+1),$$

(see Figure 9.1)

Our immediate goal is to determine the median of the content of the dam.denote the meian by x_{mdn}. Then x_{mdn} is the solution of the equation

$$\int_{x_{mdn}}^{5} f(y)dy = \frac{1}{2}$$

$$\int_{y=x_{mdn}}^{5} 0.373816 \times e^{-0.1y(3y+1)^{1/3}} dy = \frac{1}{2} \tag{9.5}$$

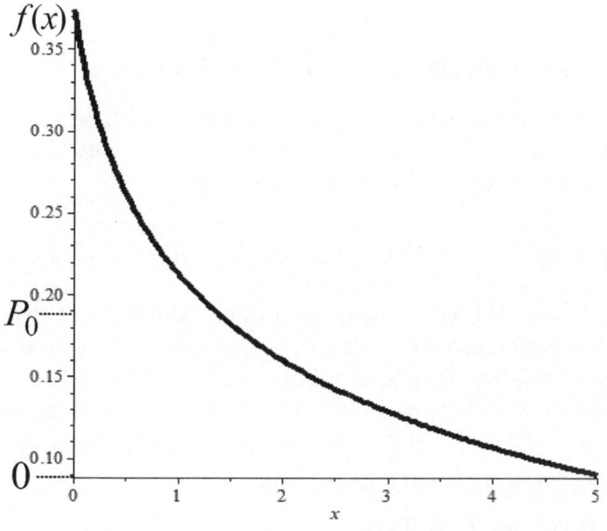

FIGURE 9.1
Plot of mixed pdf of content in dam, $\{P_0, f(x)\}_{0<x<5.0}$, when $r(x) = kx + m$, $k = 3$, $m = 1$, $\lambda = 2$, $\mu = 0.1$, $Ccap = 5.0$, where $P_0 :=$ probability that content is empty.

The solution for the median content in formula (9.5) is: "x_{mdn}= 1.202803" liquid units. An arbitrary observer at a random time point would see a content greater than 1.202803 units with probability 1/2. The observer would find the dam empty with probabobility P_0 (= 0.186908). When x_{mdn} = 1.202803, computation shows that $\int_{y=xmdn}^{5.0} f(y)dy = 0.500000$.

9.3 Pharmacokinetic Model with Multiple Poisson Dosing and Exponential Dose Amounts

This Section outlines an LC approach to analyzing "multiple dosing" in pharmacokinetics. It deals with a simplified one-compartment model. We assume bolus dosing, i.e., a full dose of a drug is absorbed into the blood stream immediately at each dosing instant. Also, inter-dose times are $\stackrel{d}{=} \text{Exp}_\lambda$. Thus doses occur in a Poisson process at rate λ. This assumption may be valid outside of a closely controlled environment like a clinic. In an uncontrolled environment, many patients may take certain medications over time in a Poisson process (see [23] (1980)).

We assume first-order kinetics. That is, the "concentration" of the drug in the blood stream decays at a rate which is proportional to the concentration. This is equivalent to the "sample path" of the concentration over time having a negative exponential shape between doses (similar to Figure 9.2).

9.3.1 Pharmacokinetic Model Description

Let $\{W(t)\}_{t\geq 0}$, denote the drug concentration in the blood at time t. Denote the dosing times by $\{\tau_n\}$, $\tau_n < \tau_{n+1}$, $n = 0, 1, 2, ...$. Let the decay rate of concentration between doses due to drug elimination be

$$\frac{dW(t)}{dt} = -kW(t), \tau_n \leq t < \tau_{n+1}, n = 0, 1, 2, ..., \tag{9.6}$$

with constant $k > 0$. We shall use a simplified "measure of concentration" , called the "mass concentration", i.e., "$\frac{\text{mass of solute}}{\text{volume of solution}}$", where each instilled drug dose is the "solute" (mass in milligrams) and the blood volume is the solution (volume in litters or milliliters). The dimension of the concentration $W(t)$ is equal to $\left[\frac{Mass}{Volume}\right] = \left[\frac{M}{L^3}\right]$, where M := Mass and L := Length. The dimension of the decay rate (due to treatment use and elimination) is $\left[\frac{dW(t)}{dt}\right]$ $= \left[\frac{M}{L^3}\right] \times \left[T^{-1}\right]$, where T := Time.

Let $\{P_0, f(x)\}_{x>0}$, denote the stationary pdf of mass concentration (aka concentration). We assume that $P_0 = 0$ due to the negative exponential shape of the sample path (decay over time) between doses (see Figure 9.2, and Sections 6.2.4, 6.2.5, and 6.4 in Chapter 6 in [22] (2017)). In theory, the concen-

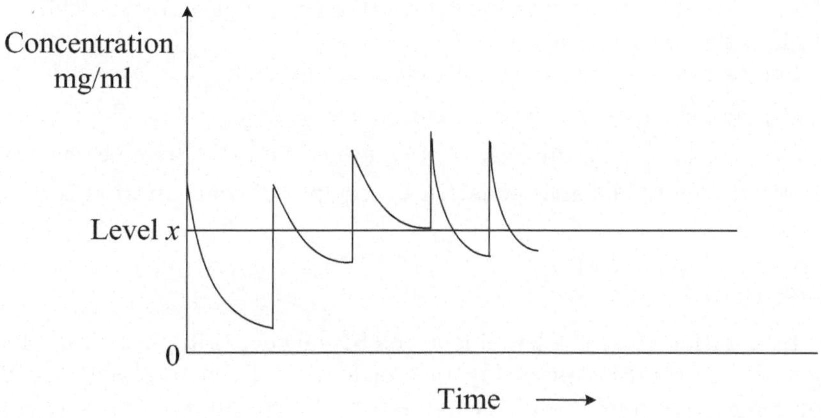

FIGURE 9.2
Sample path of drug concentration $\{W(t)\}_{t\geq 0}$ in one-compartment model with bolus multiple dosing and first-order kinetics.

tration of the drug never vanishes. In practice, the concentration eventually "tends to 0", or is "negligible". (We are not discussing the treatment effects of multiple dosing. We are discussing only the concentration dynamics of the drug concentration.)

9.3.2 Dose Amounts Exponentially Distributed in Pharmacokinetic Model with Poisson Multiple Dosing

We first analyze a model which assumes a Poisson process for instillation of doses, and "exponentially distributed dose sizes". Some small samples of multiple dosing have shown that the hypothesis of Poisson arrivals cannot be rejected statistically in an uncontrolled environment; see [23] (1980). Dose amounts are assumed to be $= \text{Exp}_\mu$. This assumption may be valid if the amount of each dose absorbed is affected randomly by the dosing environment (e.g., acidity, presence of enzymes, interaction with other medications, etc.). Another instance could occur when eye drops are instilled by a patient, say "approximately" every six hours. Often, the sizes of the individual drops may vary considerably, due to using a hand-squeezed medicine dropper. The location on the cornea of the instillation may vary from dose to dose, possibly affecting absorption. This could create "random increases in concentration" with the successive doses during a dosing regimen. Similar remarks apply to fast-acting sprays, such as nitrolingual pump sprays (which open blood vessels for more oxygen delivery to cells), or to nasal sprays. (For certain medications it may be feasible to study the effect of using "random dose sizes" as an exponential random variable "inherently in a prescription", in order to test whether it will decrease the long-run variance of concentration while staying within a therapeutic range.)

Let us assume the bolus dose amounts in milligrams are such that the corresponding increase in concentration are $\stackrel{d}{\equiv} \text{Exp}_\mu$, where the dimension of μ is $\left[\frac{1}{\mu}\right] = \left[\frac{M}{L^3}\right]$, i.e., dimension of concentration. We assume that $P_0 = 0$. Therefore, the LC balance equation for the pdf of concentration is

$$kxf(x) = \lambda \int_{y=0}^{x} e^{-\mu(x-y)} f(y)dy, x > 0. \tag{9.7}$$

In equation (9.7) the left side is the SP downcrossing rate of level x. The right side is the SP upcrossing rate of level x. Each total side has dimension $\left[\frac{1}{Time}\right]$. To understand this, note that dimension $[kxf(x)] = \left[\frac{1}{Time}\right]$ (see Theorem 1.6 in Section 1.4.3 in Chapter 1).

The constant k has dimension $[k] = \left[\frac{1}{Time}\right]$ because:

$$\begin{aligned} [xf(x)] = [x]\,[f(x)] = \tfrac{[x]}{[x]} = [1] \quad &\text{(dimensionless integer)} \\ \text{and } [k] \times [xf(x)] = \left[\tfrac{1}{Time}\right] \quad &\text{(same as } [\lambda]\,). \end{aligned} \tag{9.8}$$

That is, $[k] \times [\text{constant } 1] = \left[\frac{1}{Time}\right]$, implying that $[k] = \left[\frac{1}{Time}\right]$.

9.3.2.1 Brief Note on the Dimension of "$\frac{1}{\mu}$" in Pharmacokinetic Model with Dose Amounts Exponentially Distributed

"Prior to instillation", a dose has mass (measured in mg.s), i.e., having dimension $[M]$. "Immediately after" installation the dose is dissolved in the solvent (blood) and it immediately causes an increases in the concentration of the drug, since the volume of the solvent remains approximately unchanged. Our analysis focuses on the dynamics of the "concentration" over time. The posterior (just after instillation) effect of the dose increases the concentration. Therefore it is more appropriate to assign

$$\begin{aligned} \text{dimension } \left[\tfrac{1}{\mu}\right] &= \left[ML^{-3}\right] = [\text{concentration}], \\ \text{dimension } [\mu] &= \left[\tfrac{L^3}{M}\right] = \tfrac{1}{[\text{concentration}]}. \end{aligned} \tag{9.9}$$

9.3.3 Solving for Stationary Pdf of Concentration of Blood Serum with Poisson Multiple Instillations and Posterior Dose Amounts Exponentially Distributed

To solve equation (9.7) for $f(x), x > 0$, we first take $\frac{d}{dx}$ of both sides. This yields

$$kxf'(x) + kf(x) = -\mu kxf(x) + \lambda f(x),$$

$$kxf'(x) = (-k - \mu kx + \lambda)\, f(x),$$

$$\frac{f'(x)}{f(x)} = \frac{-k - \mu kx + \lambda}{kx},$$

$$\frac{d \ln f(x)}{dx} = -\frac{1}{x} - \mu + \frac{\lambda}{kx},$$

$$\ln f(x) = -\ln x - \mu x + \frac{\lambda}{k} \ln x + \text{constant}$$

$$f(x) = Ae^{-\ln x - \mu x + \frac{\lambda}{k}\ln x}, x > 0,$$

$$f(x) = Ae^{-\mu x} x^{\frac{\lambda}{k}-1}, x > 0,$$

where A is a constant. Using the normalizing condition to compute A,

$$A = \frac{1}{\int_{x=0}^{\infty} e^{-\mu x} x^{\frac{\lambda}{k-1}}\, dx};$$

therefore

$$f(x) = \frac{e^{-\mu x} x^{\frac{\lambda}{k}-1}}{\int_{x=0}^{\infty} e^{-\mu x} x^{\frac{\lambda}{k}-1} dx}, x > 0. \tag{9.10}$$

Multiplying numerator and denominator by $\mu^{\frac{\lambda}{k}}$ gives

$$f(x) = \frac{e^{-\mu x} (\mu x)^{\frac{\lambda}{k}-1} \mu}{\int_{x=0}^{\infty} e^{-\mu x} (\mu x)^{\frac{\lambda}{k}-1} \mu dx}$$

$$f(x) = \frac{e^{-\mu x} (\mu x)^{\frac{\lambda}{k}-1} \mu}{\int_{x=0}^{\infty} e^{-\mu x} (\mu x)^{\frac{\lambda}{k}-1} d\mu x}, x > 0. \tag{9.11}$$

Formula (9.11) can be expressed in terms of the "Gamma function". In the denominator as follows, make the transformation: "$t = \mu x$". If $x = 0$ then $t = 0$; if $x = \infty$ then $t = \infty$. Thus

$$f(x) = \frac{e^{-\mu x} (\mu x)^{\frac{\lambda}{k}-1} \mu}{\int_{t=0}^{\infty} e^{-t} (t)^{\frac{\lambda}{k}-1} dt}$$

$$f(x) = \frac{e^{-\mu x} (\mu x)^{\frac{\lambda}{k}-1} \mu}{\Gamma\left(\frac{\lambda}{k}\right)}, x > 0. \tag{9.12}$$

where the "Gamma function" is defined by

$$\Gamma(y) \equiv \int_{t=0}^{\infty} e^{-t} t^{y-1} dt, y > 0, \tag{9.13}$$

is the well-known. (For the definition of $\Gamma(y)$, and for useful properties of $\Gamma(y)$ see Problem 25, p. 327 in [54] (1967); or other Calculus books; or [1] (1964).)

The dimension of $\Gamma(y)\Gamma(y)$ is [1], i.e., a real number. The left side of Equation (9.12) has dimension $\left[\frac{L^3}{M}\right] = \frac{1}{[\text{concentration}]}$, and the right side has the same dimension (see formula (9.9). Let "W" denote the "stationary" concentration of $W(t)$ as $t \to \infty$, having the pdf in formula (9.12). The first and second moments of W are

$$E[W] = \frac{\lambda}{k\mu}, \quad E\left[W^2\right] = \frac{\lambda}{k\mu^2}\left(\frac{\lambda}{k}+1\right). \tag{9.14}$$

The variance of W is

$$Var(X) = E\left[W^2\right] - (E[W])^2 = \frac{\lambda}{k\mu^2}, \tag{9.15}$$

(see p. 58 in [37] (1995) for statistical definitions and properties; or the "Brief review of statistics" in Chapter 1 in [53] (2013)).

We can find the probability that the concentration W is between two threshold limits, say $\alpha < \beta$, using the formula

$$P(\alpha < W < \beta) = \int_{x=\alpha}^{\beta} \frac{1}{\Gamma\left(\frac{\lambda}{k}\right)} \mu(\mu x)^{\left(\frac{\lambda}{k}-1\right)} e^{-\mu x} dx. \tag{9.16}$$

The information in formula (9.16) can be useful when multiple dosing continues for a long time, e.g., when a person is taking the blood thinner comedian (warfarin). If the concentration is $<\alpha$ the blood becomes too thick and the comedian dose is not effective for the intended treatment. If the concentration is $>\beta$ the blood becomes too thin for the intended treatment and undesirable bleeding may occur. The interval (α, β) is called here, "therapeutic range".

The type of analysis outlined briefly here can be extended to various pharmacokinetic models of varying complexity.

9.3.4 Observation of the Drug Concentration of the Blood Serum in a Pharmacokinetic Model at a Random Time Point

Let x_{mdn} denote the median of the pdf $f(x), x > 0$, in formula (9.12). Then, x_{mdn} is the solution of the equation,

$$\int_{y=x_{mdn}}^{\infty} \frac{1}{\Gamma\left(\frac{\lambda}{k}\right)} (\mu y)^{\left(\frac{\lambda}{k}-1\right)} e^{-\mu y} \mu dy = \frac{1}{2}. \tag{9.17}$$

For a numerical example, let $\lambda = 3$, $k = 2$, $\mu = 1$. Then $\Gamma\left(\frac{\lambda}{k}\right) = \frac{1}{2}\sqrt{\frac{\lambda}{k}} =$ 0.88622692, by computation. The solution of Equation (9.17) using mathematical software is (to six decimals) is given by

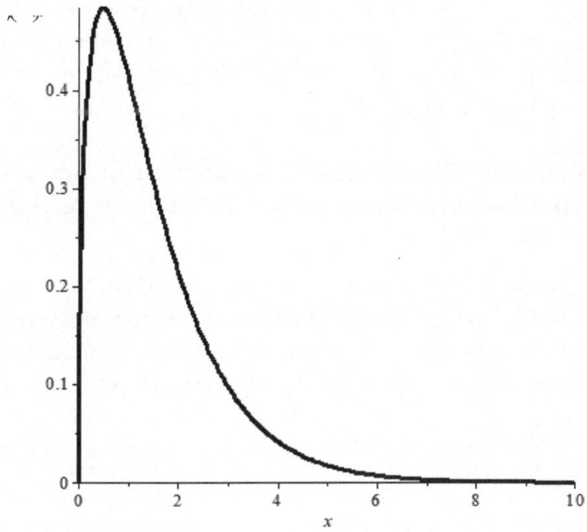

FIGURE 9.3
Stationary pdf of the concentration (mg/L), in multiple dosing pharmacokinetic model. Therapeutic range is $(2.0, 5.0)$ and its probability is the area between $x = 2.0$ and $x = 5.0$, under the curve in the figure.

$$x_{mdn} = 1.182986 \text{ units}.$$

This means: when a blood sample is taken at a random time point there is a probability of 1/2 that the observed concentration will be greater than 1.182986 units, and probability 1/2 that the concentration will be less than 1.182986 units.

Similar computations can determine the probability that the multiple dosing regimen is in the therapeutic range (α, β) when observed at a random time. In this "theoretical" example, let $\alpha = 2.0$ mg/L, and $\beta = 5.0$ mg/L. The probability that the multiple dosing regimen will be therapeutic is

$$\int_{x=2}^{5} \frac{1}{\Gamma\left(\frac{\lambda}{k}\right)} (\mu x)^{\left(\frac{\lambda}{k}-1\right)} e^{-\mu x} \mu dx = 0.242898.$$

Figure 9.3 shows the stationary pdf when $\lambda = 3$, $k = 2$, $\mu = 1$. The therspeutic range is $2.0 < x < 5.0$.

9.3.5 Pharmacokinetic Model with Dose Amounts Deterministic

Next, suppose the "prior dose“ amounts (before instillation) are all the same size, 1.e., the same number of units in mg.s with dimension $[M]$ (M $\equiv$ mass).

Also, assume that the "posterior" dose amounts after intillation are all the same amounts in "mgs per liter of blood serum", say D, with dimension $\left[\frac{M}{L^3}\right]$ ($L. \equiv$ lenth). This model is equivalent to an M/D/$r(\cdot)$ dam (see, e.g., p. 504ff, Section 11.6.3 in Chapter 11 in [22] (2017)).

9.3.5.1 Equation for Stationary Cdf and Pdf of Serum Concentration in Pharmacokinetic Model with Dose Amounts Deterministic

Consider a sample path of $\{W(t)\}_{t\geq 0}$ (similar to Figure 9.2 with all upward jump amounts due to prior doses which cause equal posterior doses to be $D > 0$). That is, all posterior doses increase the concentration in the blood serum by the same amount immediately (or very soon) after the instillation.

In the sample path, fix level $x > 0$. The SP downcrossing rate of level $x > 0$ is $\lim_{t\to\infty} \frac{\mathcal{D}_t(x)}{t}$ ="$kxf(x)$". The SP upcrossing rate of level x is equal to $\lim_{t\to\infty} \frac{\mathcal{U}_t(x)}{t}$ ="$\lambda F(x) - \lambda F(x-D)$" where $F(x) = \int_{y=0}^{x} f(y)dy$ (see Section 3.4.3 in Chapter 3). The principle of "rate balance across level x" gives the "analytic equation" for the pdf $f(x)$ and cdf $F(x)$ of W, namely

$$kxf(x) = \lambda F(x) - \lambda F(x - D), x > 0. \tag{9.18}$$

Due to all upward jumps being of the same size D, we derive $f(x)$ and $F(x)$ on "successive non-overlapping intervals of length D" as follows. Denote the pdf and cdf of W by $f_0(x)$ and $F_0(x)$ if $x \in (0, D]$, and by $f_n(x)$ and $F_n(x)$ if $nD < x < (n+1)D$, $n = 1, 2,...$

Note that equation (9.18) is a "dfifferential equation" for $F(x)$ since $f(x) = \frac{dF(x)}{dx} \equiv F'(x)$, $x > 0$. Note that $x - D < 0$ and $F(x - D) \equiv 0$ for all $x \in (0, D)$. Thus,

$$kxf_0(x) = \lambda F_0(x), 0 < x < D,$$

implying

$$\frac{F_0'(x)}{F_0(x)} \equiv \frac{d}{dx} \ln F_0(x) = \frac{\lambda}{kx}, 0 < x < D,$$

with solution given by

$$\frac{d \ln F_0(x)}{dx} = \frac{\lambda}{kx},$$

$$\ln F_0(x) = \frac{\lambda}{k} \ln(x) + \text{constant},$$

$$\ln F_0(x) = \ln(x^{\frac{\lambda}{k}}) + \text{constant}$$

$$F_0(x) = e^{\ln(x^{\frac{\lambda}{k}}) + \text{constant}}$$

or

$$F_0(x) = Ax^{\frac{\lambda}{k}}, 0 < x < D, \tag{9.19}$$

where $e^{\ln(x^{\frac{\lambda}{k}})} = x^{\frac{\lambda}{k}}$, and $A = e^{\text{constant}}$ is a constant to be determined.

For $x \in [D, 2D)$, substituting from formula (9.19) into Equation (9.18), gives

$$kxf_1(x) = \lambda F_1(x) - \lambda F_0(x - D), D \leq x < 2D,$$

$$f_1(x) = \frac{\lambda}{kx} F_1(x) - \frac{\lambda}{kx} A (x - D)^{\frac{\lambda}{\kappa}}, D \leq x < 2D,$$

$$F_1'(x) - \frac{\lambda}{kx} F_1(x) = -\frac{\lambda}{kx} A (x - D)^{\frac{\lambda}{\kappa}}, D \leq x < 2D. \tag{9.20}$$

To solve the differential Equation (9.20) multiply both sides by the "integrating factor" $e^{-\int_D^x \frac{\lambda}{ku} du}$ and apply the continuity of the CDF at level D, i.e.,

$$F_1(D^+) = F(D_0^-),$$

to obtain

$$F_1'(x) e^{-\int_D^x \frac{\lambda}{ku} du} - \frac{\lambda}{kx} F_1(x) e^{-\int_D^x \frac{\lambda}{ku} du} = -\frac{\lambda}{kx} A (x - D)^{\frac{\lambda}{\kappa}} e^{-\int_D^x \frac{\lambda}{ku} du},$$

$$\frac{d}{dx} F_1(x) e^{-\int_D^x \frac{\lambda}{ku} du} = -\frac{\lambda}{kx} A (x - D)^{\frac{\lambda}{\kappa}} e^{-\int_D^x \frac{\lambda}{ku} du},$$

$$F_1(x) = -e^{+\frac{\lambda}{k} \ln(x/D)} \int_D^x \frac{\lambda}{ky} A (y - D)^{\frac{\lambda}{\kappa}} e^{-\frac{\lambda}{k} \ln(y/D)} dy + C_1 e^{+\frac{\lambda}{k} \ln(x/D)},$$

where C_1 is a constant to be determined. Setting $x = D$ in the last equality immediately above, gives $F_1(D) = C_1 = AD^{\frac{\lambda}{k}} = F_0(D^-)$, from (9.19), since $\ln(1) = 0$. This leads to

$$F_1(x) = Ax^{\frac{\lambda}{k}} \left[-\frac{\lambda D}{k} \int_D^x \left(\frac{y}{D} - 1 \right)^{\frac{\lambda}{\kappa}} \left(\frac{y}{D} \right)^{-(\frac{\lambda}{k}+1)} dy + 1 \right], x \in [D, 2D), \tag{9.21}$$

where A is defined just afterformula (9.19). Setting $x = D$ in (9.21), checks with $F_1(D^+) = AD^{\frac{\lambda}{k}} = F_0(D^-)$.

9.3.5.2 Mathematical Induction for Cdf in Pharmacokinetic Model with Dose Amounts Deterministic

The solution for $F_n(x)$, $x \in [nD, (n + 1)D)$, $n = 2, 3, \ldots$, can be obtained by a similar "recursive procedure" using "mathematical induction". We now give the induction step. Suppose we know $F_n(x)$, $x \in [nD, (n+1)\,D)$. Then, for $x \in [(n+1)\,D, (n+2)\,D)$, abbreviating the integrating factor to $\mathbf{\Phi}(x) := e^{-\frac{\lambda}{k}\int_{(n+1)D}^{x}\frac{1}{u}du}$, we get:

$$F'_{n+1}(x)\mathbf{\Phi}(x) - \frac{\lambda}{kx}F_{n+1}(x)\mathbf{\Phi}(x) = -\frac{\lambda}{kx}F_n(x-D)\mathbf{\Phi}(x),$$

$$\frac{d}{dx}F_{n+1}(x)\mathbf{\Phi}(x) = -\frac{\lambda}{kx}F_n(x-D)\mathbf{\Phi}(x),$$

$$F_{n+1}(x)\left(\frac{x}{(n+1)\,D}\right)^{-\frac{\lambda}{k}} = -\frac{\lambda}{k}\int_{(n+1)D}^{x}\frac{1}{y}F_n(y-D)\left(\frac{y}{(n+1)\,D}\right)^{-\frac{\lambda}{k}} + C_{n+1},$$

$$F_{n+1}(x) = -\frac{\lambda}{k}\left(\frac{x}{(n+1)\,D}\right)^{+\frac{\lambda}{k}}\int_{(n+1)D}^{x}\frac{1}{y}\left(\frac{y}{(n+1)\,D}\right)^{-\frac{\lambda}{k}} F_n(y-D)dy + C_{n+1}\left(\frac{x}{(n+1)\,D}\right)^{+\frac{\lambda}{k}}.$$

Letting $x = (n+1)\,D$ gives $C_{n+1} = F_{n+1}((n+1)\,D) = F_n((n+1)\,D^-)$ which is known on the assumption that we know $F_n(x)$, $x \in [nD, (n+1)\,D)$. Moreover, C_{n+1} will be in terms of the factor A. In principle, we can derive $F_n(x)$, $n = 0, 1, 2, \ldots$, in terms of the constant A. The constant A can then be determined (or closely approximated) using the normalizing condition $F(\infty) = 1$. Once $F(x)$ is obtained, we can determine $f(x)$ by substituting into formula (9.18).

9.3.6 Using "LCE" (Level Crossing Estimation) to Compute an Estimate for $f(x)$

As an alternative to the foregoing analytical solution, we can solve for $f(x)$ using "LCE" ("Level Crossing Estimation") via a single typical simulated sample path of $\{W(s)\}_{s\geq 0}$ over $(0,t)$ where time $t > 0$ is large. This alternative method gives an estimate of the pdf $f(x)$ for $0 < x < x_{M(t)}$, where $x_{M(t)}$ is the maximum state-space partition level reached by the single "long" sample path (see Sections 9.3 and 9.6 in Chapter 9 in [22] (2017)). (Details are outside the scope of LC techniques covered in this book.)

Remark 9.1 *"Theorem B" (see [22] (2017)) can be applied to compute the "time-dependent" (aka "transient") pdf and cdf of serum concentration, based on the three publications on LC transient techniques developed starting in 1983 and published in: [12] (1988), [13] (1988), and [14] (1088). Knowledge of the transient distributions may be useful in multiple dosing regimens where it is important to estimate the concentration after a short dosing period, or to* *manage dosing quantities between instillations. (Details are outside the scope of the LC techniques covered in this book.)*

Remark 9.2 *Some related stochastic models have characteristics in common with the pharmacokinetic model. One group of such models involves "consumer response" (abbreviated "CR") to non-uniform advertisements, which has been analyzed along similar lines using LC (see [18]) (1984).*

9.4 Observation of Simple Harmonic Motion at a Random Time Point

Wikipedia at (https://en.wikipedia.org/wiki/Simple_harmonic_motion) explains the mechanism of Simple harmonic motion. It's "a special type of periodic motion an object experiences due to a restoring force whose magnitude is directly proportional to the distance of the object from an equilibrium position and acts towards the equilibrium position". It "results in an oscillation that is described by a sinusoid which continues indefinitely (if uninhibited by friction or any other dissipation of energy)".

9.5 Simple Harmonic Motion ("SHM")

We analyze an elementary model of a "deterministic" simple harmonic motion, using LC.

Have you ever walked past a moving pendulum with a weight at the bottom. When you took a passing glance (or camera shot) of the moving weight, where was it?. Was it near the bottom (center) of motion, or near the outside edges of motion? This analysis shows, among other properties, where you are most likely to see the moving weight, and why. A similar property occurs in other instances of SHM.

Consider a particle moving according to simple harmonic motion ("SHM"). (See, e.g., p. 133 in [9] (1969); pp. 216–217 in [50] for discussions of SHM.) Let $X(t)$ denote the position of the SHM particle (weight in a pendulum) at instant $t \geq 0$, and let $X(0) = 0$. Let the particle move in the state space, which is the interval $\boldsymbol{S} = [-1, +1]$. In this version of the standard SHM model there is only "one" sample path, namely, where

$$X(t) = \sin(t), t \geq 0. \tag{9.22}$$

We fix attention on an arbitrary level x in the state space. We wish to determine the stationary pdf $f(x)$ and cdf $F(x)$ of $X(t)$ when we assume that the motion of the particle is observed over many "cycles", as $t \to \infty$.

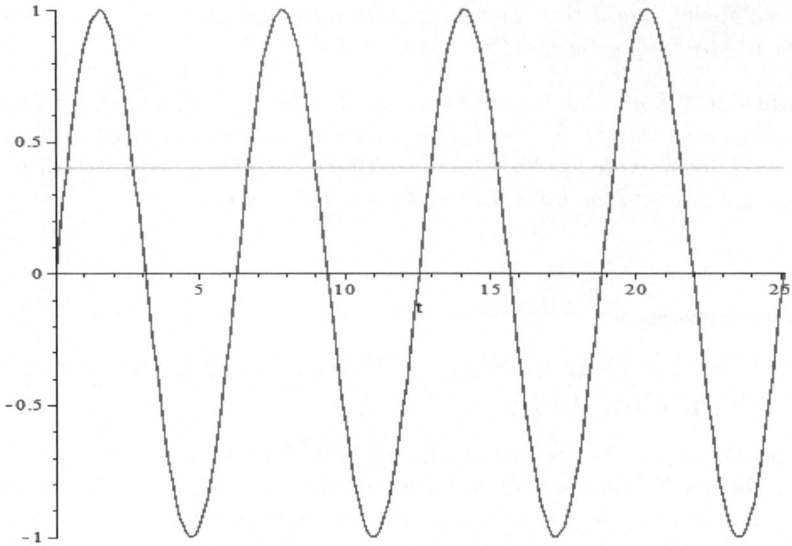

FIGURE 9.4
Sample path of simple harmonic motion (SHM) $X(t) = \sin t$. State space is $\boldsymbol{S} = [-1, +1]$. Shows "arbitrary level x" (at $x = 0.4$) in $\boldsymbol{S}$.

Consider the sample path $X(t), t \geq 0$ (Figure 9.4). The slope of the sample path at any Cartesian point $(t\,, x)$ on the sample path level x is given by

$$r(x) == \frac{d}{dt}\sin t|_{t=\sin^{-1}(x)} = \cos\left(\sin^{-1}(x)\right) = \sqrt{1-x^2}, x \in [-1, +1]\,. \tag{9.23}$$

To explain formula (9.23) $r(x) :=$ the cosine of the angle whose sine is x. (Recall from elementary calculus that $\frac{d\sin(x)}{dx} = \cos(x)$,) Consider levels x, $x + h \in \boldsymbol{S}$, where $h > 0$ is "small". The time required for the SP to ascend from level x to level $x + h$ is

$$\int_{y=x}^{x+h} \frac{1}{r(y)}dy = \int_{y=x}^{x+h} \frac{1}{\sqrt{1-y^2}}dy. \tag{9.24}$$

The symmetries of the sample path imply that the time required for the SP to descend from level $x + h$ to level x is also given by (9.24).

Applying (9.24), we see that the long-run "proportion of time" the SP spends in state-space interval $(x, x + h)$ in a "cycle of length 2π" time units is

$$\frac{2}{2\pi}\int_{y=x}^{x+h} \frac{1}{\sqrt{1-y^2}}dy = F(x+h) - F(x). \tag{9.25}$$

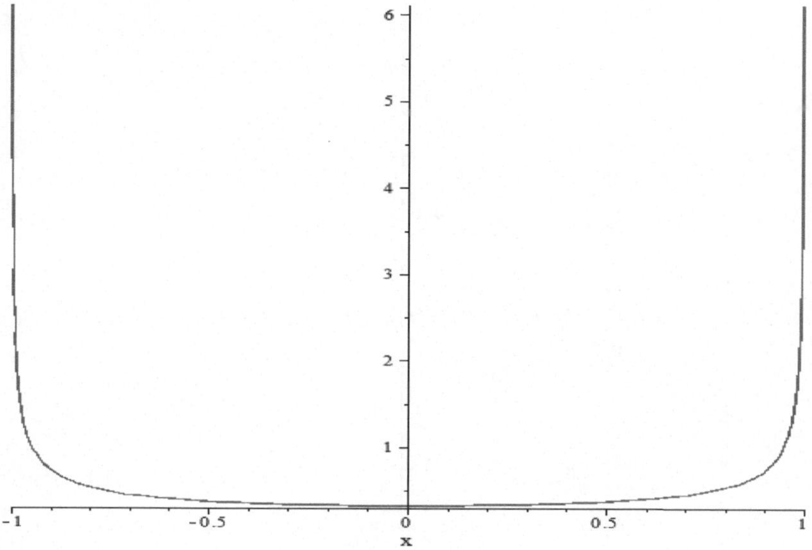

FIGURE 9.5
$f(x) = \frac{1}{\pi\sqrt{1-x^2}}, x \in [-1, +1]$, for particle moving in simple harmonic motion described by "$X(t) = \sin t, t \geq 0$".

Formula (9.25) leads to

$$\frac{1}{\pi} h \frac{1}{\sqrt{1-(x^*)^2}} = F(x+h) - F(x), \tag{9.26}$$

where $x < x^* <, x + h$, by the definition of $F(x)$ as the long-run proportion of time the process is in state-space interval $[-1, x]$. Dividing both sides of (9.26) by h and letting $h \downarrow 0$, yields

$$f(x) = \frac{1}{\pi\sqrt{1-x^2}}, x \in [-1, +1]. \tag{9.27}$$

The stationary pdf $f(x)$ in (9.27) is theoretically interesting and suggests intuitive insights (Figure 9.5). Note that $\lim_{x\downarrow(-1)} f(x) = \lim_{x\uparrow(+1)} f(x) = \infty$. Also, $\min_{x \in \boldsymbol{S}} f(x) = \frac{1}{\pi}$, when $x = 0$. The pdf $f(x)$ is "symmetric about $x = 0$", and is "convex" (see Figure 9.5).

From (9.27), the cdf is

$$F(x) = \int_{y=-1}^{x} f(y)dy, = \frac{1}{\pi}\left(\sin^{-1}(x) - \sin^{-1}(-1)\right)$$
$$= \frac{1}{\pi}\sin^{-1}(x) + \frac{1}{2}, x \in [-1, +1]. \tag{9.28}$$

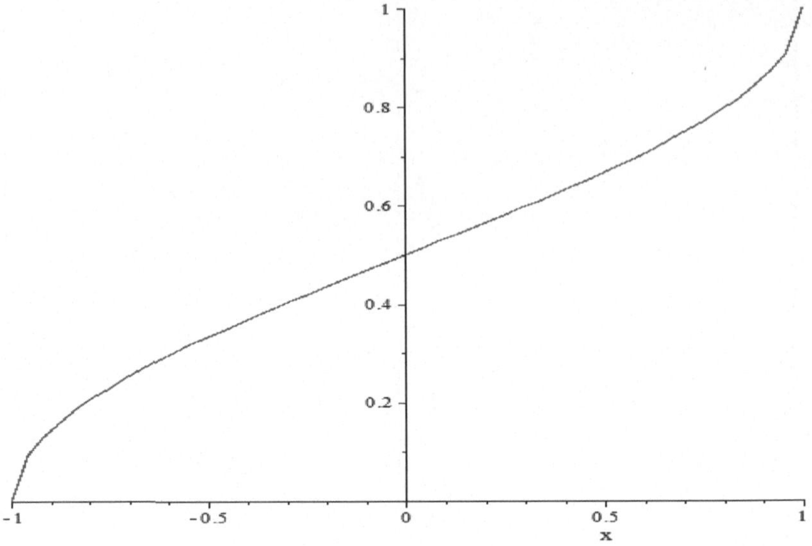

FIGURE 9.6
Stationary cdf $F(x) = \frac{1}{\pi}\sin^{-1}(x) + \frac{1}{2}, x \in [-1, +1]$, for particle moving in simple harmonic motion, described by $X(t) = \sin t, t \geq 0$.

9.5.1 Inferences Based on Pdf and Cdf in Simple Harmonic Motion

From (9.23), the speed of the particle is "$r(x) = \sqrt{1 - x^2}$", implying $r(\pm 1) = 0$ so that the particle comes to a complete stop at the outside edges $x = \pm 1$, and immediately changes direction of motion. Also, $r(0) = 1$, which is the "maximum speed in either direction". Hence, if one observes the particle at a random time point in the long run, it is much more likely to find the particle close to one of the outside edges (boundaries) of $\boldsymbol{S}$, i.e., $x = \pm 1$, rather than close to the center of $\boldsymbol{S}$ (the particle's range of motion). This fact suggests that the particle spends a much greater proportion of time near the edges $x = \pm 1$ than near the center $x = 0$. (That is, the path of the particle covers a small saste-space interval $\Delta \boldsymbol{S}$ about point $x = 0$ in a "very short time". The path covers the same small state-space interval $\Delta \boldsymbol{S}$ about point $x \approx \pm 1$ in a "much longer time". If the observer is passing by at a constant speed, there is a leger proportion of time in her/his pass-by to see the particle near the outer edges.

9.5.2 Numerical Example of Simple Harmonic Motion

From computations using formula (9.28), the proportion of time that the SP (particle) spends in the central interval $[-.5, +.5]$ is equal to $F(.5) - F(-.5) = 0.333$. The proportion of time the particle spends in the outer regions near the outer edges $[-1.0, -.5] \cup [.5, 1.0]$, is equal to $2 \cdot (F(1.0) - F(.5)) = 0.667$. The "median" symmetric outer intervals of $\boldsymbol{S}$ with respect to the time spent by the particle, is $\boldsymbol{A}_{0.5} \equiv [-1.0, -.707] \cup [.707, 1.0]$, i.e., $P(X(t) \in \boldsymbol{A}_{0.5}) = 0.5$. This indicates that it is equally likely to find the particle in two bands of equal width 0.293 touching the outside edges ± 1.0 (total width .586), as it is to find it in a central interval of width 1.414 about $x = 0$. Arbitrary observations on operating pendulum clocks, readily corroborate these theoretical computations.

Remark 9.3 *The type of LC analysis in this section, may be extended to analyze random trigonometric functions (e.g., like $A\sin(\theta t) + B\cos(\theta t), t \geq 0$, where A, B are random variables and θ is a constant). Extensions may also be applicable in some models of physics, and in the analysis of roots of equations.*

9.6 Exercises

Exercise 9.6.1 *Give three reasons why it is important to check the dimensions of the quantities in integral equations and algebraic equations.*

Exercise 9.6.2 *In formula (9.2) in Section 9.1, give the dimensions of the left-side and right-side terms.*

Exercise 9.6.3 *In Section 2.1.10 in Chapter 2, on the virtual wait in the $M_\lambda/M_\mu/1$ queue, compute the probability that a random arrival will wait $<$ 0.20000 time units before starting service, when the parameters are $\lambda = 2$, $\mu = 3$, (i.e., $1 - P_0 = \frac{\lambda}{\mu} = \frac{2}{3} > \frac{1}{2}$). Use a formula for the pdf $f(x), x > 0$, in formulas (2.14) and (2.15); of, for the cdf $F(x), x \geq 0$, in formula (2.16).*

Exercise 9.6.4 *In Section 9.2, determine the probability that an observer at a random time will see content of liquid beteen 4 and 5 content units.*

Exercise 9.6.5 *In Section 9.2, determine the probability that an observer at a random time will see content of liquid beteen 0 and 1 content units.*

Exercise 9.6.6 *In Section 9.3.1, In Equation (9.7), show why each side has dimension $\left[\frac{1}{Time}\right]$. Explain why the dimension of μ, i.e., $[\mu]$ in the analysis, is the same as the dimension of drug concentration [concentration]. (Distinguish between doses "prior" to dose instillation, and "posterior" to dose instillation of the drug, and why this distinction is important.)*

Exercise 9.6.7 *From the definition of Gamma(x), "$\Gamma(x) = \int_{t=0}^{\infty} e^{-t} t^{x-1} dt$", show that "$\Gamma(x+1) = x\Gamma(x)$". (Use integration by parts.) Show that $\Gamma(1) = 1$.*

Exercise 9.6.8 *In the immediately previous exercise, assume that "$x = 1$ (an integer)". Then:* $\Gamma(1) = 1$; $\Gamma(2) = 1\Gamma(1) = 1 \times 1 = 1$; $\Gamma(3) = 2\Gamma(2) = 2 \times 1 = 2$; $\Gamma(4) = 3\Gamma(3) = 3 \times 2 \times 1 = 6$, *etc. Prove by mathematical induction that* $\Gamma(x) = (x-1)!$, *i.e.,* $\Gamma(x) \equiv x$ *factorial, when* x *is a positive integer. (State the induction hypothesis, and continue to prove it is true.)*

Exercise 9.6.9 *Derive the values of* $E[W]$ *and* $E\left[W^2\right]$ *in formula (9.14) directly from the definition of the* n^{th} *moment* $E[x^n] \equiv \int_{x=0}^{\infty} x^n f(x)dx$, *where* $f(x), x > 0$, *is the pdf.*

Exercise 9.6.10 *Show that in formula (9.15) the variance is always a non-negative value.*

Exercise 9.6.11 *Check the validity of formula (9.12) in Section 9.3.4 using the formula for the stationary pdf* $f(x)$ *or stationary cdf* $F(x)$ *of drug concentration in multiple dosing, starting from formula (9.7). Suggested approach: (1) Take the derivative with respect* x *on both sides of (9.7), (2) Express the ratio* $\frac{f'(x)}{f(x)}$ *on the left side, in terms of other quantities in the expressions, on the right side, (3) Apply the definition of the Gamma function in formula (9.13), (4) Solve for* $f(x)$ *using the identity "*$e^{\ln(y)} = y$*", for* $y > 0$.

Exercise 9.6.12 *Show that the right side of Equation (9.12) has dimension* $\left[\frac{L^3}{M}\right] = \frac{1}{[concentration]}$.

Bibliography

[1] Abramowitz, M., Stegun, I.A. (1964). Handbook of Mathematical Functions, Dover Publications, Inc., New York.

[2] Ancker, C.J., Gafarian, A.V. (1961). Queueing with Multiple Inputs and Exponential Service Times, *Operations Research* **9**(3), 321–327.

[3] Azoury, K., Brill, P.H. (1986). An Application of the System-Point Method to Inventory Models Under Continuous Review, *Journal of Applied Probability* **23**(3), 778–789.

[4] Azoury, K., Brill, P.H. (1986). An Application of the System-Point Method to Inventory Models Under Continuous Review, Journal of Applied Probability 23(3), 778–789.

[5] Azoury, K., Brill, P.H. (1992). Analysis of Net Inventory in Continuous Review Models with Random Lead Time, European Journal of Operational Research, 39, 383–392.

[6] Bartle, R.G. (1976). The Elements of Real Analysis, 2nd ed, John Wiley & Sons, New York.

[7] (1999). Beard, R.E., Pentikämen, T., Pesonen, E. (1999). Risk Theory, Second Edition, The Stochastic Basis of Insurance, Halsted Press, Division of John Wilry & Sons, Inc., New York.

[8] Botta, F., Harris, C.M. (1986), Approximation with Generalized Hyperexponential distributions: Weak Convergence Results, Queueing Systems 1, 169–190.

[9] Boyce, W.E., DiPrima, R.C. (1969). Elementary Differential Equations and Boundary Value Problems, 2nd ed., John Wiley & Sons, Inc., New York.

[10] Brill, P.H. (1975). System Point Theory in Exponential Queues, Ph.D. Dissertation, University of Toronto. Available from University Microfilms International, Ann Arbor, Michigan, order number 8901129.

[11] Brill, P.H. (1975) . A New Methodology for Analyzing a Broad Class of Exponential Queues, 5th Conference on Stochastic Processes & their Applications, Univ. of Maryland, 1975, in Advances in Applied Probability, Volume 8, Issue 2 June 1976, p. 242 (invited by J. Keilson and R. Syski).

[12] Brill, P.H. (1988). A Technique for Transient Distributions in Stochastic Models, submitted to University of Windsor School of Business, WP #88–14 (20 pp.).

[13] Brill, P.H. (1988). The Time-dependent System Point Method in Queues, Dams and Inventories, submitted to University of Windsor School of Business, WP # 88–16 (25 pp.).

[14] Brill, P.H. (1988). TheTtime Dependent system point Level Crossing Method for Exponential Queues, Submitted to University of Windsor School of Business, WP #88–18 (33 pp).

[15] Brill, P.H. (1988) "Single-server Queues with Delay-dependent Arrival Streams", Probability in the Engineering and Informational Sciences, 2, 231–247.

[16] Brill, P.H. (2002). Properties of the Waiting Time in M/D/1 Queues and Related Models, submitted to University of Windsor, Dept of Mathematics and Statistics, WMSR # 02-01 (19 pp.).

[17] Brill, P.H., Harris, C.M. (1997). M/G/1 Queues with Markov-generated Server Vacations, Stochastic Models, 13(3), 491–521.

[18] Brill, P.H., Hornik, J. (1984). A System Point Approach to Nonuniform Adversising Insertions, *Operations Research*, **32** (1), 7–22.

[19] Brill, P.H., Huang, M.L., Hlynka, M. (2009). Note on an <s, S> Inventory System with Decay, IAENG International Journal of Applied Mathematics, 39(3), 171–174.

[20] Brill, P.H., Huang, M.L., Hlymka, M. (2019). On the Service Time in a Workload-barrier M/G/1 Queue with Accepted and Blocked Customers, European Journal of Operational Research, 283(1) 2020, 235–243.

[21] Brill, P.H. (2008). Level Crossing Methods in Stochastic Models, Springer, New York.

[22] Brill, P.H. (2017). Level Crossing Methods in Stochastic Models, Second Ed, Springer, New York.

[23] Brill, P.H., Moon, R.E. (1980). Application of Queueing Theory to Pharmacokinetics, Journal of Pharmaceutical Sciences, 89(5), 558–560.

[24] Brill, P.H. (2015). Note on the Service Time in an M/G/1 Queue with Bounded Workload, Statistics and Probability Letters, 96, 162–169.

[25] Brill, P.H., Posner M.J.M. (1981). The System Point Method in Exponential Queues: A Level Crossing Approach, Mathematics of Operations Research 6(1), 31–49.

[26] Brill, P.H., Posner, M.J.M. (1977). Level Crossings in Point Processes Applied to Queues: Single-server Case, Operations Research, vol. 26, No. 4, July–August.

[27] Brill, P.H., Yu, K. (2011). Analysis of Risk Models Using a Level Crossing Technique, Insurance: Mathematics and Economics, 49(3), 298–309.

[28] Britannica, T. Editors of Encyclopaedia (2021, August 17). Speed of light. Encyclopedia Britannica. https://www.britannica.com/science/speed-of-light.

[29] Courant, R., Robbins, H. (1956). What is Mathematics? An elementary Approach to Ideas and Methods, Oxford University Press, London.

[30] Cramer, H., Leadbetter, M.R. (1967). Stationary and Related Stochastic Processes – Sample Function Properties and their Applications, John Wiley & Sons, New York, (Dover Publications, Mineola, N.Y).

[31] Einstein, A. (1905). Does the Inertia of a Body Depend upon It's Energy Content? Annalen der Physik, 18(1905), 639–641.

[32] Embrechts, P., Klüppelberg, C., Mikosch T. (2003). Modelling Extremal Events for Insurance and Finance, Springer-Verlag, Berlin, New York.

[33] Erlang, A.K. (1909). The Theory of Probabilities and Telephone Conversations, in Brockmeyer, E., Halstrøm, H.L., Jensen, A. (1948). The Life and Works of A.K. Erlang, Transactions of the Danish Academy of Technical Sciences, 2, 131–137.

[34] Feller, W. (1968). An Introduction to Probability Theory and Its Applications, Volume I, 3rd ed., John Wiley, New York.

[35] Gavish, E., Schweitzer, P. (1977). The Markovian Queue with Bounded Waiting Time, Management Science, 23(12), 1349–1357.

[36] Gross, D., Shortle, J.F., Thompson, J.M., Harris, C.M. (2008). Fundamentals of Queueing Theory, 4th ed., Wiley.

[37] Hogg, R.V., Vraig, A.I. (1995). Introduction to Mathematical Statistics, 5th ed., Prentice Hall.

[38] Huang, M.L., Brill P.H. (1993). System Point Estimation of the Probability Distribution of the Waiting Time in Variations of $\mathrm{M/G}^B/1$ Queues, American Statistical Association, Proceedings of the Statistical Computing Section, San Francisco, August 8–12.

[39] Johnson, N.L., Kotz, S. (1970). Distributions in Statistics, Continuous Univariate Distributions – 1, John Wiley and Sons, New York.

[40] Karlin, S., Taylor, H.M. (1975). A First Course in Stochastic Processes, 2nd Edition, Academic Press, New York.

[41] Kleinrock, L. (1975). Queueing Systems, Volume 1: Theory, John Wiley & Sins, New York

[42] LeCorbeiller, P. (1966). Dimensional Analysis, Appleton- Century-Crofts Div. of Meredith Publishing Co., New York.

[43] Lindley, D.V. (1952). The Theory of Queues with a Single Server, Proceedings of the Cambridge Philosophical Society 48, 277–289.

[44] Little, J.D.C. (1961). A Proof for the Queuing Formula $\boldsymbol{L} = \lambda \boldsymbol{W}$, Operations Research 9(3): 383–387. doi:10.1287/opre.9.3.383. JSTOR 167570.

[45] Loney, S.L. (1947). The Elements of Statics and Dynamics, Part II Dynamics, Twenty-sixth Imprinting. Cambridge: at the University Press.

[46] Maple 17. Maplesoft, A Division of Waterloo Maple Inc., Waterloo, Ontario.

[47] Neuts, M.F. (1981). Matrix Geometric Solations in Stochastic Models, an Algorithmic Approach, The Johns Hopkins University Press, Baltimore and New York.

[48] Pinsky M.A., Karlin, S. (2011). An Introduction to Stochastic Modelling, 4th ed., Academic Press, Amsterdam.

[49] Posner, M. J., (1973). Single Server Queues with Service Time Depending on Waiting Time, Operations Research, 21(2), 610–616.

[50] Rainville, E. D., Bedient, P. E. (1969). A Short Course in Differential Equations, 4th ed., The Macmillan Co., New York.

[51] Resnick, S. (2005). Adventures in Stochastic Processes, 4th Printing. Birkhauser, Boston.

[52] Ross, S.M. (2010). Introduction to Probability Models, 10th ed., Elsevier, Amsterdam.

[53] Sclove, S.L. (2013). A Course on Statistics for Finance, CRC Press, Taylor and Francis Group, Boca Raton.

[54] Spivak, M. (1967). *Calculus*, W. A. Benjamin, Inc., New York.

[55] Sigman, K., Wolff, R. W. (1993). A Review of Regenerative Processes, SIAM Review, 35(2), 269–288.

[56] Takács, L. (1962). Introduction to the Theory of Queues, Oxford University Press, New York.

[57] Tijms, H. C. (2003). A First Course in Stochastic Models, Wiley, Amsterdam.

[58] Whitt,W., (1990). Queues with service times and interarrival times depending linearly and randomly upon waiting times, Queueing Systems, 6, 335–351. https://doi.org/10.1007/BF02411482.

[59] Wolff, R.W. (1982), Poisson Arrivals see Time Averages, Operations Research, 30(2), 223–231.

[60] Yu, K., Huang, M. L., Brill, P. H. (2012). An Algorithm for Fitting Heavy-tailed Distributions via Generalized Hyper-exponentials, INFORMS Journal on Computing, 24(1), 42–52.

Index